International Conference on Space Weather and Satellite Application (ICeSSAT 2018)

Journal of Physics: Conference Series Volume 1152

Shah Alam, Malaysia
7 – 8 August 2018

ISBN: 978-1-5108-8274-4
ISSN: 1742-6588

Printed from e-media with permission by:

Curran Associates, Inc.
57 Morehouse Lane
Red Hook, NY 12571

Some format issues inherent in the e-media version may also appear in this print version.

Printed by Curran Associates, Inc. (2019)

For permission requests, please contact the Institute of Physics
at the address below.

Institute of Physics
Dirac House, Temple Back
Bristol BS1 6BE UK

Phone: 44 1 17 929 7481
Fax: 44 1 17 920 0979

techtracking@iop.org

Additional copies of this publication are available from:

Curran Associates, Inc.
57 Morehouse Lane
Red Hook, NY 12571 USA
Phone: 845-758-0400
Fax: 845-758-2633
Email: curran@proceedings.com
Web: www.proceedings.com

Table of contents

Volume 1152

International Conference on Space Weather and Satellite Application (ICeSSAT 2018)
7–8 August 2018, Shah Alam, Selangor, Malaysia

Accepted papers received: 30 November 2018
Published online: 4 March 2019

Preface

Papers

IOP Conf. Series: Journal of Physics: Conf. Series **1152** (2019) 011001 doi:10.1088/1742-6596/1152/1/011001

1st International Conference on Space Weather and Satellite Application 2018 (ICeSSAT 2018)

[1]**Mohamad Huzaimy Jusoh** (Chair)

[1]**Norsuzila Yaacob** (Co-Chair)

[2]**Zahira Mohd Radzi** (Co-Chair)

[1]**Azrif Manut** (Publication)

[1]**Syed Abdul Mutalib Al Junid** (TPC)

[1]**Siti Amalina Enche Ab Rahim** (Secretary I)

[1]**Zatul Iffah Abd Latiff** (Secretary II)

[3]**Aznilinda Zainuddin** (Secretary III)

[1]**Adizul Ahmad** (Treasurer)

[1]**Mohd Fahmi Hussin@Mohamad** (Publicity)

[1]**Syed Farid Syed Adnan** (Local Arrangement)

[1]**Zulkifli Othman** (Local Arrangement)

[1]Faculty of Electrical Engineering, Universiti Teknologi MARA, Shah Alam, Selangor, Malaysia.

[2]National Space Agency (ANGKASA), Malaysia Space Centre, Banting, Selangor, Malaysia

[3]Faculty of Electrical Engineering, Universiti Teknologi MARA, Pasir Gudang, Johor, Malaysia

Abstract. On behalf of the Organizing Committee of ICeSSAT2018, and with great pleasure, we would like to welcome all delegates to Shah Alam, Selangor, Malaysia. Being held from 7th to 9th August 2018 at Faculty of Electrical Engineering, Universiti Teknologi MARA the event is organized by the Centre for Satellite Communication, Faculty of Electrical Engineering, Universiti Teknologi MARA (UiTM), Malaysia. The conference creates a forum for engineers, scientists and practitioners of varied disciplines and their associated applications. Throughout this event we hope to create an opportunity for researchers, academics and students to get together and, more importantly, to welcome new peers in diverse areas of expertise. On behalf of the organizing committee we would like to take the opportunity to express our gratitude to all reviewers who have been working hard to review the papers on time and share their expertise. Special mention goes to the distinguished Keynote Speakers for their willingness to spend their time and share their valuable knowledge. My deepest gratitude goes to the Organizing Committee who have been working hard behind the scenes for the success of this event. To all participants, we would like to encourage you to fully utilize this platform for knowledge sharing and networking while enjoying your stay in Shah Alam, Selangor, Malaysia.

IOP Conf. Series: Journal of Physics: Conf. Series **1152** (2019) 011002 doi:10.1088/1742-6596/1152/1/011002

Peer review statement

All papers published in this volume of *Journal of Physics: Conference Series* have been peer reviewed through processes administered by the proceedings Editors. Reviews were conducted by expert referees to the professional and scientific standards expected of a proceedings journal published by IOP Publishing.

IOP Conf. Series: Journal of Physics: Conf. Series **1152** (2019) 012001 doi:10.1088/1742-6596/1152/1/012001

Online Solar Observatory For Solar Learning Program Among Orang Asli's Community

[1]Lau Chen Chen and [2]Zahira Mohd Radzi

[1,2] National Space Agency (ANGKASA), Lot 2233, Jalan Turi, Kg. Sg. Lang, 42700 Banting, Selangor, Malaysia

lau@angkasa.gov.my, zahira@angkasa.gov.my

Abstract: Literatures indicate a few researches have been studied involving the Orang Asli community but a dearth is found focusing on ICT education and internet usage among the Orang Asli schoolchildren. These groups need the help of various parties, especially the government to ensure schools with Orang Asli children and community must be equipped with ICT facilities for them. Besides that, not all community or schools afford to have their own telescopes to do solar observation or participating in solar observation program at planetarium or science center. Therefore National Space Agency of Malaysia (ANGKASA) has taken initiative collaborate with National Planetarium (PLANET) and Department of Orang Asli Development (JAKOA) to conduct Solar Astronomy Learning Program among Orang Asli community through development of Online Solar Observatory (OSO). Orang Asli schoolchildren and community can learn solar astronomy virtually through the OSO. This pilot project will focus on Orang Asli community and schoolchildren in Selangor state and it can be expanded to all Malaysian's Orang Asli community and to all the public in the future. OSO can enhance the learning of solar astronomy subjects especially in the classroom. It does not more than just facilitate telescope use: a comprehensive range of support materials and educational resources is available. In this paper presentation, we would like to show the OSO system and portal for solar astronomy learning program.

1. INTRODUCTION

Internet has changed the learning processes, knowledge acquisition and the interaction between the learners and the learned. Using internet by the Orang Asli without education might lead to unethical utilization and misinformation. However, using the internet to acquire the education and information will lead to understand and knowing what the Orang Asli should know as the use of other technology such as television set and hand phone is a common within the Orang Asli community that are near to the urban area [4].

Literature indicates a few researches have been studied involving the Orang Asli community but a dearth is found focusing on Information Communication Technology (ICT) education and internet usage among the Orang Asli schoolchildren [1]. The survey findings indicate most of the Orang Asli are familiar with computer and internet but the level of usage varies [1]. From the study by [2], it was found that ICT played an important role in sustaining Orang Asli's cultural identity. These groups need the help of various parties, especially the government to ensure schools with Orang Asli children and community must be equipped with ICT facilities for them. Besides that, not all community or schools afford to have their own telescopes to do solar observation or participating in solar observation program at planetarium or science center [3].

Therefore, National Space Agency of Malaysia (ANGKASA) has taken initiative collaborate with National Planetarium (PLANET) and Department of Orang Asli Development (JAKOA) Selangor / Federal Territory to develop Online Solar Observatory (OSO) for solar astronomy learning program among Orang Asli schoolchildren and community through the provision of MOSTI Social Innovation Fund (MSI). OSO telescope system will use the existing facilities at the National Planetarium Observatory but with a little effort to upgrade the system. Via the OSO, Orang Asli schoolchildren and community can observe the Sun and study solar astronomy virtually where it can be accessed anywhere via the internet.

2. METHODOLOGY

The methodology of Online Solar Observatory's development as below:

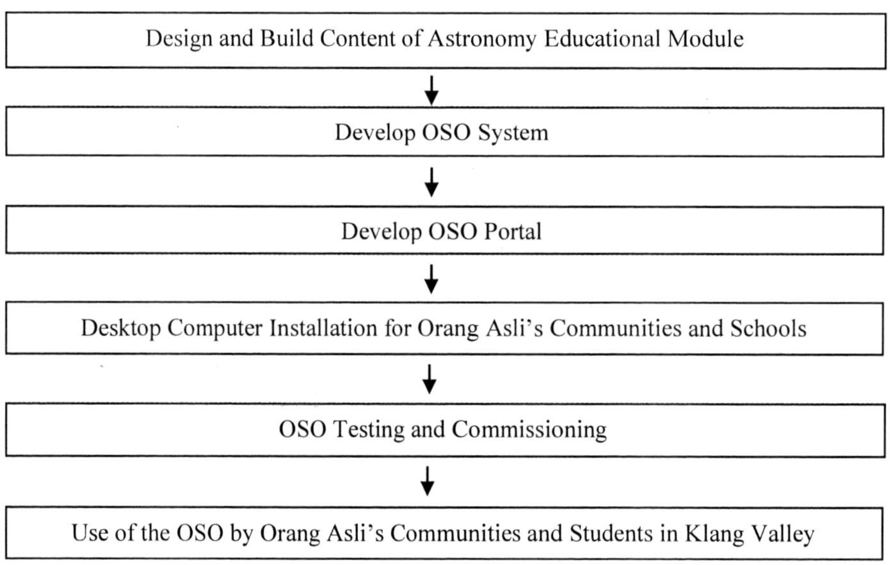

3. RESULTS AND DISCUSSIONS

a. ONLINE SOLAR OBSERVATORY (OSO) SYSTEM

Online Solar Observatory (OSO) is an integrated system for streaming live view of the Sun which is placed in the National Planetarium Observatory with a diameter of 5m dome and automatic control systems integrated between the dome, the telescope and the camera system. OSO is equipped by two units of telescopes with Hydrogen Alpha and White Light solar filters that enable the public to observe the Sun directly. Both telescopes are placed on the Software Bisque Paramount ME Mounting System which is controlled by The SkyX Professional software. Two units of camera have been installed on the telescopes; therefore the surface of the Sun's image will appear directly through the OSO portal.

OSO telescopes system comprising:
• Telescopes Takahashi FS-60CB FL355mm + RoboFocus + White Light Solar Filter + QHY-IMG132-E CCD Camera
• Telescopes Takahashi FS-102FSV + Robofocus + H-Alpha Solar Filter + QHY5P-II Color CCD Camera

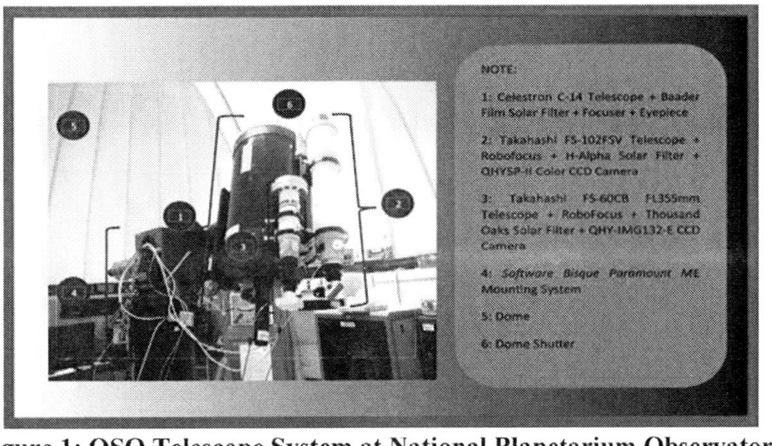

Figure 1: OSO Telescope System at National Planetarium Observatory

Besides that, OSO also equipped with all-sky camera to allow public to view the sky condition around the National Planetarium. Two webcams are installed in the Observatory to enable public to view the movement of the telescope system when in operation.

Figure 2: Weather Station and All-sky Camera at National Planetarium Observatory (outside)

b. ONLINE SOLAR OBSERVATORY (OSO) PORTAL

Online Solar Observatory Portal can be accessed through https://oso.planetariumnegara.gov.my. Via this portal, Orang Asli schoolchildren and community can live view the Sun's image, movement of the telescope system, sky condition and location of the Sun at National Planetarium. Introduction of the OSO system, the Sun's information, and video tutorial regarding how to register and capture the Sun's image through OSO portal also provided. They can view all the contents in the OSO portal without log in. They also can register as log in user if they want to capture and download the Sun's image when OSO in operation. Log in users will need to answer feedback survey after they log out from OSO portal. This feedback survey will help ANGKASA to improve OSO project in the future.

Figure 3: OSO Portal Homepage

1. The appearance of the Sun is highly dependent on the weather conditions. The Sun's image displays in the portal as shown in Figure 4 and Figure 5 below:
2.

• The display image is derived from alpha-hydrogen (H-alpha), where you will be able to see the splendor of the sun (prominence) moving from the surface of the Sun and flare sun (solar flares) that usually appears near patches of sun large and uneven and sun spots.

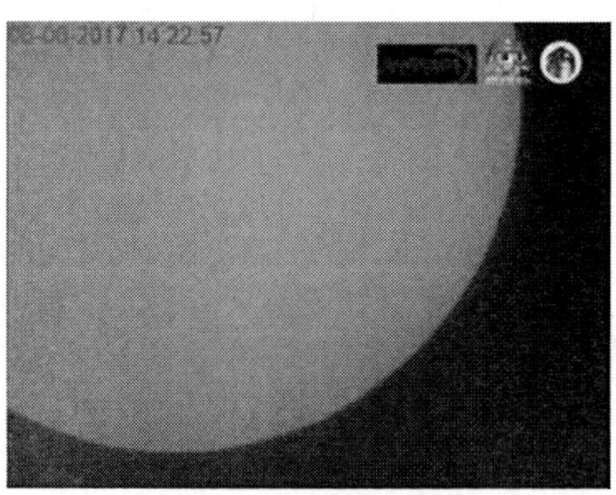

Figure 4: Prominence

• The display image is derived from white light beam (white light), where you will see Sunspots on the surface of the Sun.

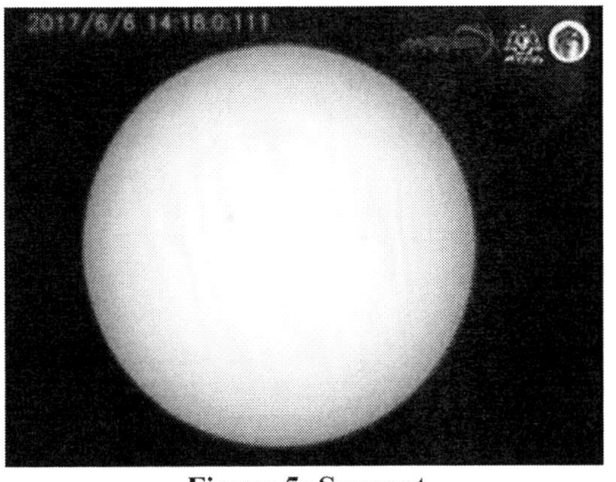

Figure 5: Sunspot

ANGKASA have installed six units of Desktop PC at SK Bukit Tampoi (Asli), Dengkil, SK Sg. Melut (Asli), Sepang, Kampung Orang Asli Sg. Melut, Sepang and Kampung Orang Asli Bukit Tadom, Banting to facilitate them in participating solar astronomy learning program. ANGKASA with the hope they acquiring sufficient knowledge and preparing them in the field of solar astronomy for enhancement of their teaching and learning activities in schools and community.

4. CONCLUSION

This OSO project is hoped to encourage and enhance Orang Asli schoolchildren and community in solar astronomy learning. It is hoped that OSO can be expanded to other Orang Asli community and schoolchildren in Selangor particularly and other states generally and finally to all the public in the future.

5. ACKNOWLEDGEMENT

We are grateful for the endowment MOSTI Social Innovation Fund (MSI) provided by Ministry of Science, Technology and Innovation under MSI16004. Token of appreciations also go to the National Planetarium (PLANET) and Department of Orang Asli Development (JAKOA), Orang Asli's schools and community in Selangor area.

6. REFERENCES

[1] Abdul Rahman, A., Yusri, A. (2011). The ICT Social Inclusion among Orang Asli Community in Gombak Area, IEEE Xplore, Electronic ISBN: 978-61284-294-3.

[2] Sarjit, S. (2015). Orang Asli's ICT Challenge, The Star 29 November 2015.

[3] Impey, C. and Danehy, A. (2017). A Website for Astronomy Education and Outreach, CAPjournal, No.22, September 2017.

[4] Kamini, R.S. (2013). Transforming Education for the Orang Asli. The New Straits Time 29 April 2013.

Development of Space-based IOT terminal for Nanosatellite UiTMSAT-1

S. N. Mohamad Rahim[1], N. A. Azahari[1], S.A. Enche Ab Rahim[1,3] and M. H. Jusoh[1,2,3]

[1]Faculty of Electrical Engineering, Universiti Teknologi Mara, Shah Alam, Selangor, Malaysia
[2]Applied Electromagnetic Research Group (AERG), Universiti Teknologi Mara, Shah Alam, Selangor, Malaysia
[3]Center for Satellite Communication (UiTMSAT), Universiti Teknologi Mara, Shah Alam, Selangor, Malaysia

E-mail: naddp1389@gmail.com; amirahazahari93@gmail.com

Abstract. This paper presents the development of space-based IOT terminal for Store and Forward (S&F) mission of nanosatellite UiTMSAT-1. The function of this ground terminal is to collect data from the sensor and send the data to the satellites over VHF amateur radio frequency band. In this case, the satellite is replaced by dummy S&F payload. To send the data from the terminal to satellite, a concept of communication in an amateur radio network is adapted. In an amateur radio network, the communication is done by a special microprocessor system called Terminal Node Controller (TNC) which implement a special protocol called AX.25 protocol. The hardware involves the development of the terminal are Control and Communication Unit (CCU) and Very High Frequency (VHF) Transceiver. CCU function as a device that collects the data from the sensor and sends the data to the VHF Transceiver, while VHF Transceiver function as a device that does the data processing before sending them to the satellites. The VHF Transceiver is built-in with TNC, AFSK Modulator, and FM transmitter. With the current setting, the total bytes allowed for sending the data packet to the satellite are 175 bytes with speed of 1200 bits per second. The data packet that was successfully received by the dummy S&F payload can be validated by using terminal emulator communication program such as real-term, here the data packet that is received by the dummy S&F payload will display on the real-term display terminal. The transmitting data from the proposed ground terminal that are successfully received by the dummy S&F payload can be seen at terminal emulator program (Real-Term).

Keywords— Store and Forward; AX.25 Protocol, nanosatellite

1. Introduction

The research on nanosatellite has moved to the fore, the construction of nanosatellite that made up of commercial off the shelf (COTS) device has cut the cost and reduce the development time due to this the research on high institute level can be done [1]. In satellite communication, there are three important elements which are the space segment, the ground segment and the mission control segment [2]. The proposed space-based IOT terminal is located in ground segment, where its roles are to collect the data from the sensor and be ready to communicate with satellite once the satellite approaches the location where the space-based IOT terminal lies. The function of space-based IOT terminal is to transmit or receive the information to/from the satellites in most reliable manner while maintaining the desired signal quality. The link for the process of data transmission from the space-based IOT terminal to the satellites is called an uplink, while the link for the process of data transmission from satellites to space-based IOT terminal is called downlink [3].

UiTMSAT-1 is a BIRDS-2 nanosatellite project involving Japan, Malaysia, Philippines, and Bhutan. There are six missions involved in this project, one of the mission is Store and Forward (S&F). The objective of this mission is to collect sensor data from space-based IOT terminal through the UiTMSAT-1, store them on-boards and download the sensor data to the Mission Control Centre (MCC) located at Kyushu Institute of Technology (Kyutech) in Japan or any BIRDS member ground station by using satellite's ultra-high frequency (UHF) antenna for the downlink [4]. Generally, S&F mission can be described as to receive and store message/data from one part of the earth and download them later over another part of the globe. Figure 1 shows the illustration of communication between S&F mission with space-based IOT terminal and MCC respectively.

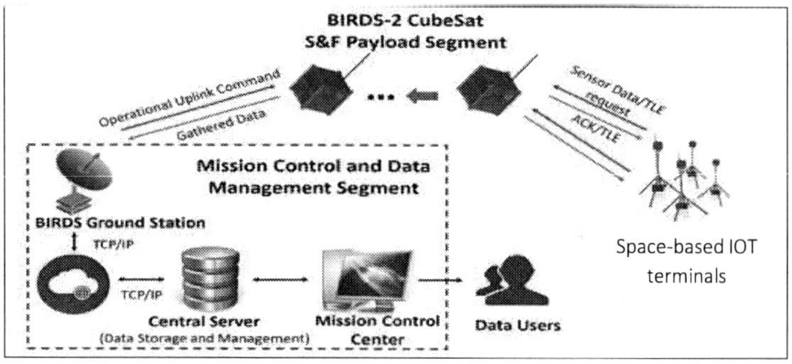

Figure 1. The System architecture of BIRDS-2 S&F nanosatellite-based remote data collection system [4].

To send the sensor data from the terminals to the BIRDS-2 CubeSat S&F Payload Segment, there are several communication layers involved, such as a physical layer, data link layer, and application layer. Figure 2 shows the communication layer for the space-based IOT terminal to communicate with S&F payload [3].

Application layer	Store & Forward data
Data link layer	AX.25 protocol
Physical layer	UART, NRZI, 1200bps AFSK Modem
	Radio & VHF transceiver

Figure 2. Communication Layer.

In application layer, CCU in the terminal will collect S&F data. The AX.25 protocol is used in data link layer, to accept and deliver the packet sensor data over a communication link. This protocol is a family of High Data Link Control (HDLC) protocol and modification result of radio amateur operation. The AX.25 protocol is arranged in a frame; each frame contains an individual datagram as given in Figure 3. The AX.25 protocol is a sophisticated protocol, that control and managing the data and build establish connection [5]. In the physical layer, the VHF antenna will be used to send the data to the satellites [6].

	START	DESTINATION CALL SIGN	SOURCE CALL SIGN	CONTROL	PID	INFORMATION FIELD	FCS	END
BYTES	1	2-8	9-15	16	17	18-172	173-174	175

Figure 3. AX.25 protocol format.

In this paper, we present the developed space-based IOT terminal for Store & Forward mission of UiTMSAT-1. This module was built by using COTS components. An experimental test was carried out in order to verify the end-to-end transmission, in which dummy S&F payload and dummy data were used. The hardware development consists of several devices, such as real-time clock (RTC), sensor, control, and a communication unit (CCU), SD card, VHF transceiver and an antenna. The real-time clock provides the facilities of real-time information to the microcontroller (MCU) and assists the MCU to manage the flow of data that involve a specific time. The sensor can be any sensor that is used as an input to the CCU. The SD card is used to keep the sensor data in the CCU and later the data will be sent to the VHF transceiver. From the VHF transceiver, the data packet will be transmitted to the satellite.

2. Methodology

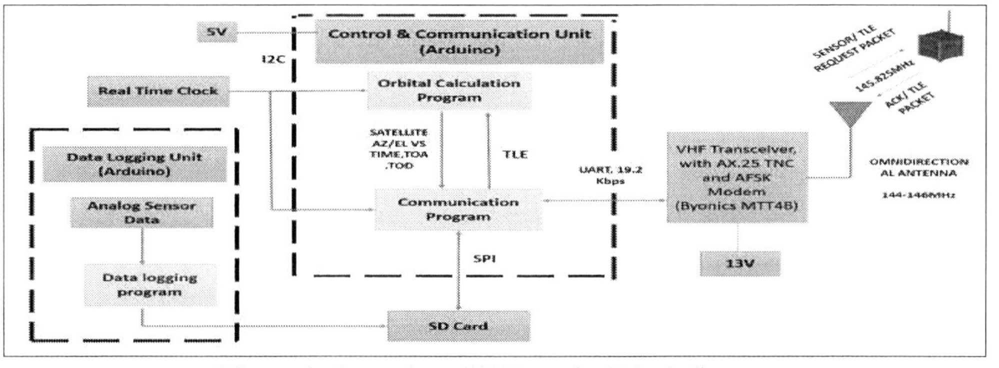

Figure 4. Space-based IOT terminal block diagram.

The block diagram of the developed space-based IOT terminal is illustrated in Figure 4. The terminal consists of 3 blocks which are Data Logging Unit with sensor integration, control, and communication unit and VHF transceiver [7]. However, in this work, we focus only on CCU and VHF transceiver. In this project, Arduino is used as CCU, where it contains the orbital calculation program and a communication program. The orbital calculation unit and communication will interact with RTC, for utility both unit to keep track with the real time. In the data logging unit, the analog sensor will transfer the output data to the SD card and later be used as data sensor for the CCU. The data sensor will be programmed according to the AX.25 protocol in the CCU. After the programming is done in the CCU, it will be sent to the VHF transceiver and transmitted to the payload.

2.1 Space-based IOT terminal operation.

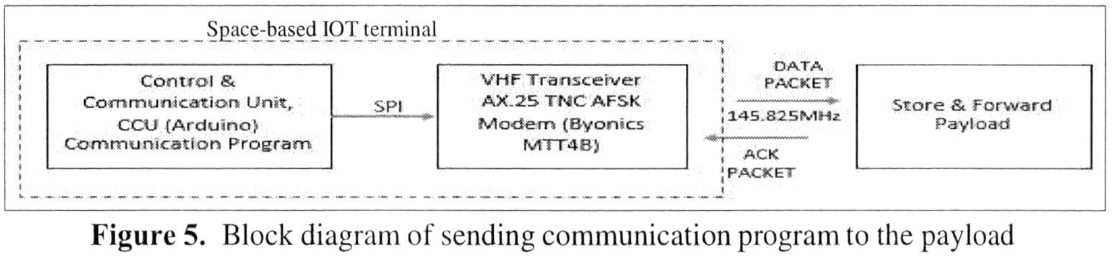

Figure 5. Block diagram of sending communication program to the payload

IOP Conf. Series: Journal of Physics: Conf. Series **1152** (2019) 012002 doi:10.1088/1742-6596/1152/1/012002

In Figure 5 the block diagram of communication between the space-based IOT terminal and S&F payload is given, which consists of CCU, VHF transceiver, and a dummy S&F payload. The transmission process from the terminal to the dummy payload is illustrated in the flowchart in Figure 6.

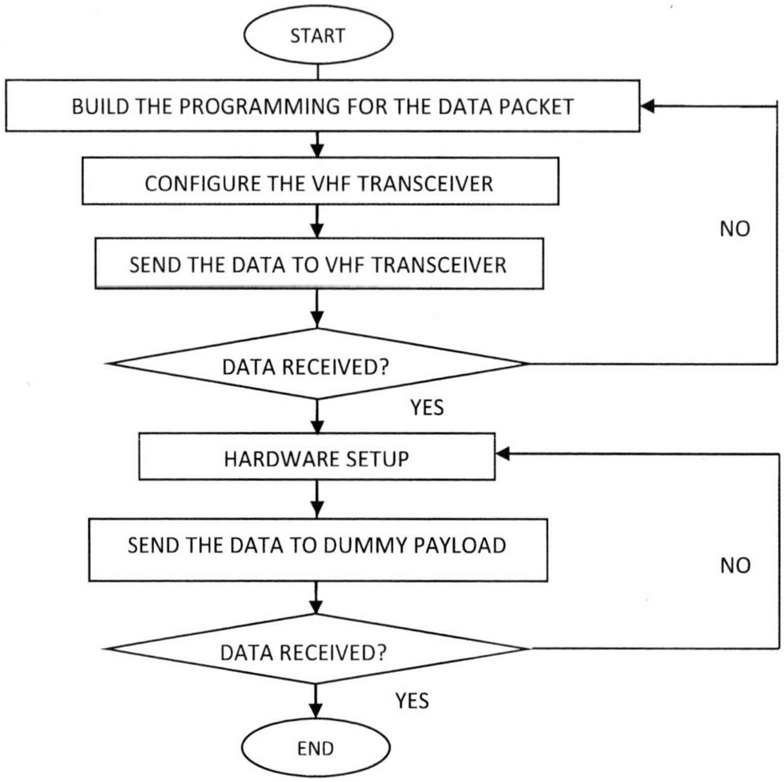

Figure 6. The flowchart for sending communication program to the payload.

2.2 AX.25 protocol

In transmitting the packet data to the satellite, several constraint factors need to be considered such as such as the size of bytes that allowed to be sent to the payload. By referring to the Table 1, the total bytes that are available for sending data packet for each time packet data is transferred to the payload is 175 bytes [8]. Although the available size of bytes is 175 in total, only 122 bytes are available for sensor data [3].

Table 1. Format of sensor data packet.

	Start	Destination Callsign	Source Callsign	Control	PID	Information Field				FCS	End
Byte #	1	2-8	9-15	16	17	18	19-20	21-50	51-172 (up to 122 bytes)	173-174	175
Content	Start flag	Satellite callsign	GST callsign or ID	Control (0x3E)	PID (0xF0)	Packet type (0x01)	Data Packet sequence #	Time packet transmitted, time packet generated, AZ and EL during packet transmission	Time data logged, sensor data, etc.	FCS	End flag
Format	HEX	ASCII string	ASCII string	HEX	HEX	HEX	HEX	ASCII string	ASCII string	HEX	HEX

The data packet was programmed in the CCU. The information in the data packet should include start, destination call sign, source call sign, control, PID, information field, FCS, and END.

2.3 CCU

In this work, Arduino is used as the CCU. For the purpose of testing, the data from the sensor is replaced by number 0 to 9. Data that was intended to send to the satellite is program comply with the AX.25 protocol by using C+ language. The programming is done in Arduino software ide and is called the data packet. It is called the data packet because it follows the AX.25 protocol which is in the form of the datagram, where each datagram contains a frame and each frame consists of a specific function. Fig. 7 shows the programming of the data packet which data sensor is replaced by number 0 to 9.

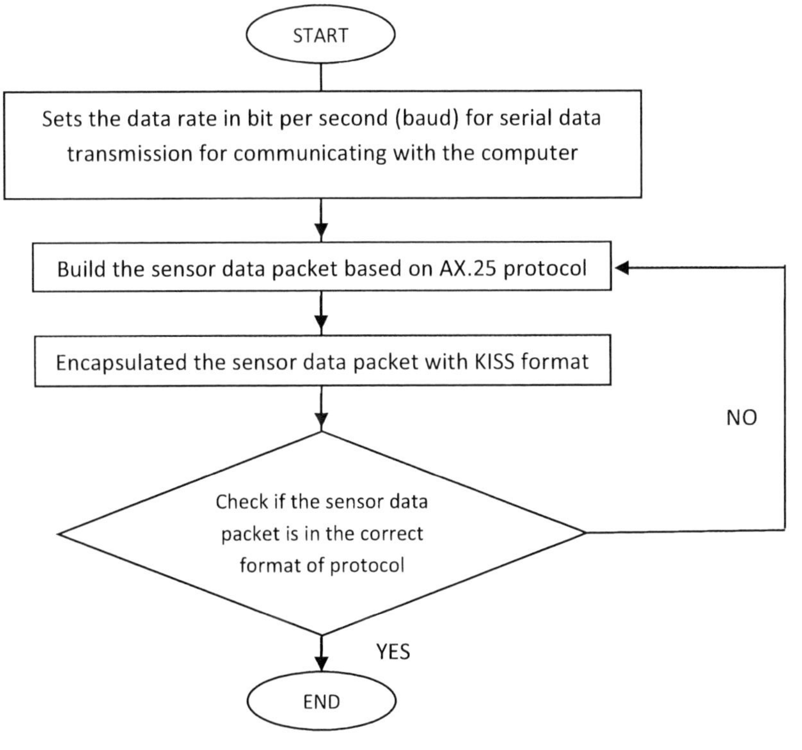

Figure 7. Flowchart for data packet programming

The data packet that was successfully constructed by can be seen on the serial monitor of Arduino ide software environment. The result will be shown in the result section. Next, data packet will be sent to the VHF transceiver

2.4 VHF transceiver

In this work, MTT4B is using as the VHF transceiver. The VHF transceiver must first be configured to the appropriate setting by using terminal emulator communication program such as Real Term. Figure

8 shows the terminal emulator communication program that is used for the configuration of VHF transceiver.

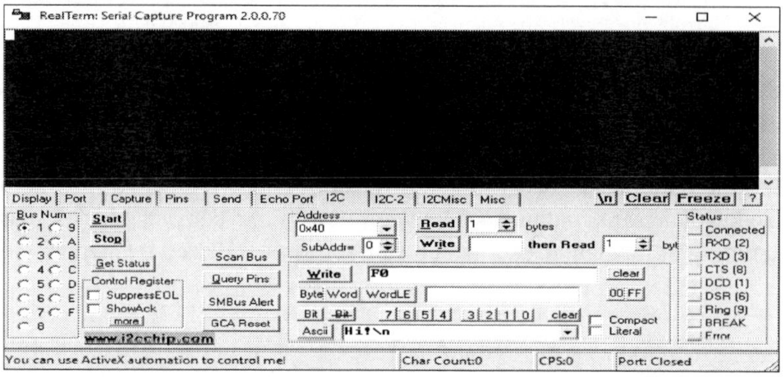

Figure 8. Terminal emulator communication program (Real-term.)

The VHF transceiver uses 13 volts power supply and 2 Amperes current and it is connected to the computer by using the DB9 connector. The communication between the transceiver and the computer is done through UART. To right port selection for configuring the VHF transceiver, can be check in the device manager.

Next, the baud rate is set to the suitable value, such in this case, the baud rate is set to 19200. In this project, the Byonics MTT4B is using as the VHF transceiver.This baud rate is set according to the Byonics MTT4B manual description. After setting the port and the baud rate. The configuration mode is entered by clicking the ESC button on the computer for three times and the programming for setting the VHF transceiver is made. Figure 9 shows the hardware required in configuring the VHF transceiver.

Figure 9. The hardware required for the configuration of VHF Transceiver.

In the VHF transceiver, the signal will be processed by the TNC, AFSK modem and later be transmitted by the FM transmitter to the payload. Figure 10 shows the block diagram of data processing in the VHF Transceiver.

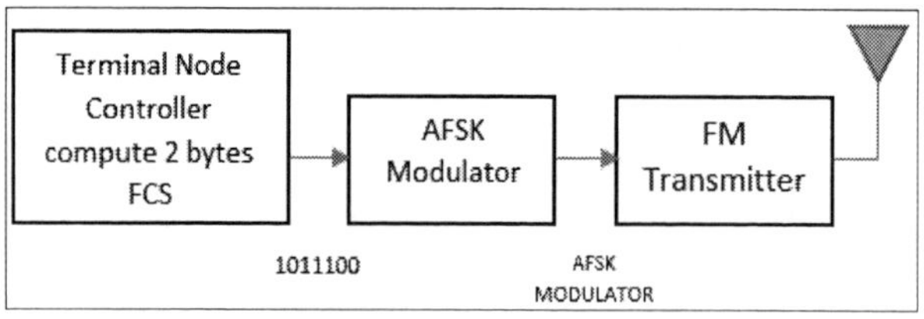

Figure 10. Block diagram of data processing in the VHF Transceiver

The data packet from the CCU is in the format according to the AX.25 protocol, but it is encapsulated in KISS frame, in order to ensure that the data packet can be transferred to the TNC over a serial link of the VHF transceiver. Besides, it allows the TNC to have a complete control and access to the data packet. Before sending the sensor data packet to the VHF transceiver, the connection between CCU and VHF transceiver is made. Figure 11 shows the connection between CCU and the VHF transceiver

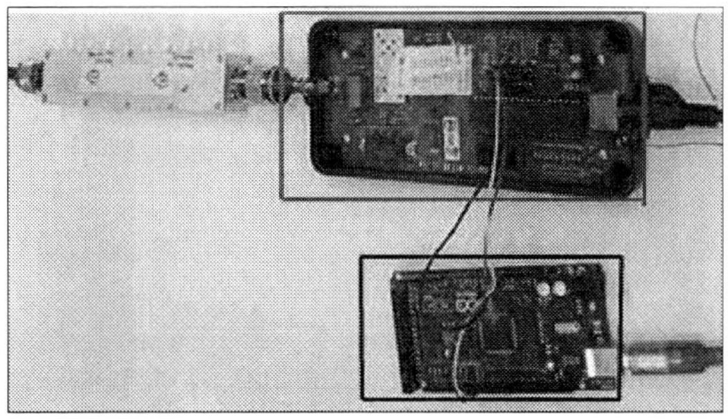

Figure 11. The connection between CCU and VHF transceiver.

After the data is received by the VHF transceiver, the TNC will perform frame check sequence calculation (FCS) to detect if there is an error in the data packet. Next, the AFSK modulator will modulate the data packet in which the digital data will be translated to the changes of frequency that make the encoded signal is suitable to transmit via the FM transmitter and later sent to the dummy payload via an internal antenna for short-range testing.

2.5 Sending the sensor data from terminal to the dummy payload

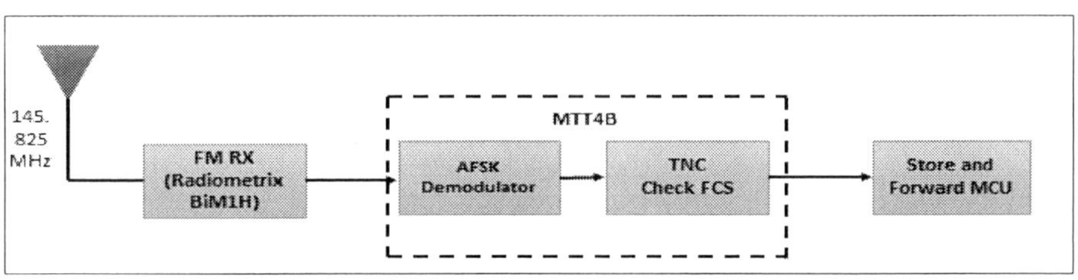

Figure 12. Block diagram of S&F payload

Figure 12 shows the block diagram of S&F payload. In fact, in S&F payload, the process of data received is the reverse process of what happens in the VHF transceiver in the terminal. The antenna receives the data packet, then the data packet goes through the FM receiver. Next, in the VHF transceiver of S&F payload, the data packet is demodulated by the Audio Frequency Shift Keying (AFSK) demodulator and sent to the Terminal Node Controller (TNC) for the process of Frame Check Sequence (FCS). When the process is completed, the data will be transferred to the S&F MCU.

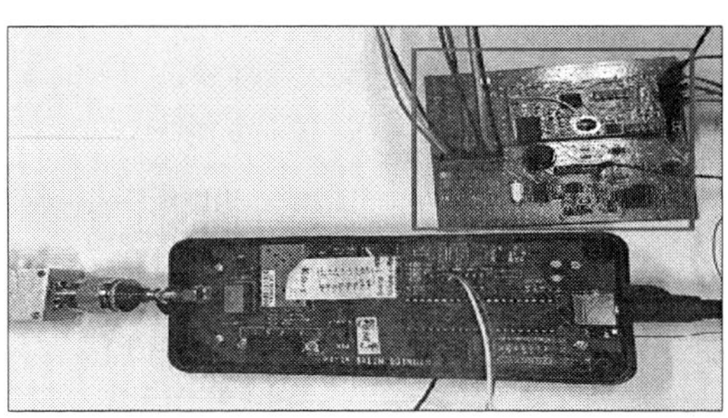

Figure 13. VHF transceiver is sending the data wirelessly to the dummy S&F payload.

Figure 13 shows the dummy S&F payload that represents the actual payload. The received data packet from the payload can be monitor through the Terminal emulator communication program by connecting the payload to the computer by UART. The VHF transceiver transmits the data wirelessly to the dummy S&F payload via the frequency 145.825 MHz.

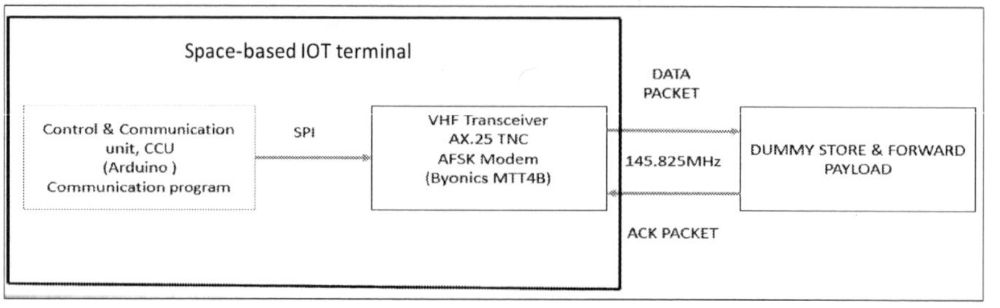

Figure 14. The Block diagram of transmission.

Before sending the data packet to the payload, the hardware setup required for this process must first be established based on block diagram on Figure 14.

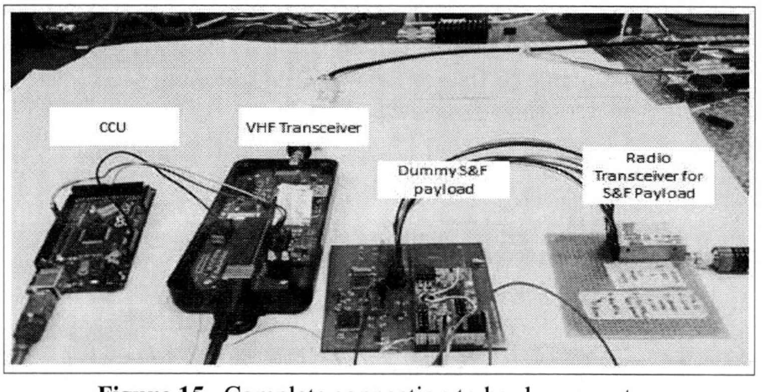

Figure 15. Complete connection to hardware system

Figure 15 shows the complete connection of the hardware system. In this work, the Arduino is supplying with 3v and 1ampere current, MTT4B is supplying with 13v and 5 amperes current while dummy payload is supplied with 5v and 2 amperes current.

In this testing, internal antenna from the MTT4B is used for the process of data packet transmission. The testing was done in short distance, and the main objective is to validate the data packet format that can be accepted by the payload.

For the installation of this space-based IOT terminal in the rural area, the data that will be collected is the Earth magnetic field. Power transmit for the MTT4B is set to 5 watts. To test either the MTT4B can have a variety of power transmission, the blue pin located on the board of MTT4B that is shown in Figure 16 is turned around, the MTT4B is connected to the spectrum analyzer to see the output power of the VHF transceiver.

IOP Conf. Series: Journal of Physics: Conf. Series **1152** (2019) 012002 doi:10.1088/1742-6596/1152/1/012002

Figure 16. VHF transceiver (MTT4B)

Figure 17 shows the testing done by check the power transmit by the MTT4B.

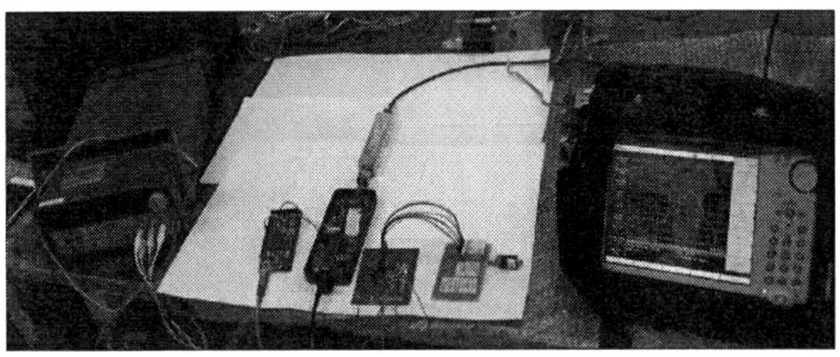

Figure 17. The hardware setup for testing the output power by the MTT4B.

3. Result and discussion

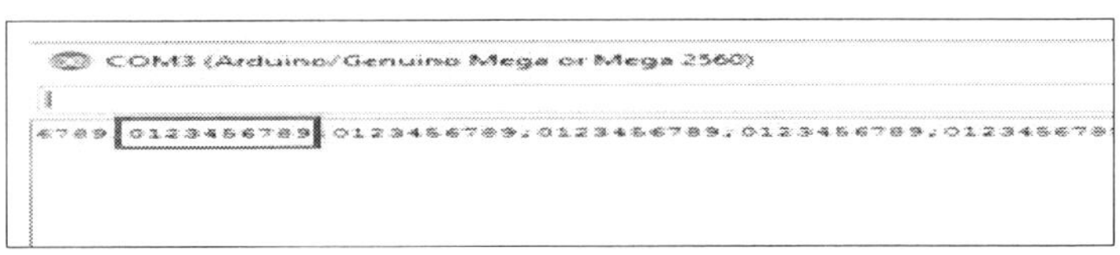

Figure 18. The serial monitor's output of the Arduino software ide

Figure 18 shows the result of the programming for a data packet in communication program that is built in Arduino software IDE, it shows that number 0 to 9 is displayed on the serial monitor.

Next, after programming was able to construct in Arduino software IDE, it was later sent to the VHF transceiver through UART by connected the CCU to the VHF transceiver, to validate if the data packet is according to the right format. The data packet will be displayed on the terminal emulator program.

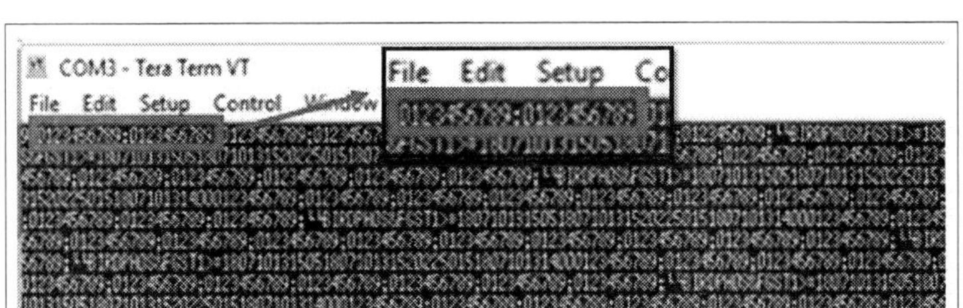

Figure 19. The data packet from MCU

In Figure 19 the number 0 to 9 was displayed at the terminal emulator program. When a data packet is transmitted to the VHF transceiver, the digital signal will be modulated by AFSK modulator. The spectrum of the modulated signal was verified by using a spectrum analyzer. Two signals with a different frequency are formed after modulation process is happening. The frequency of the signal will be 1200Hz and 2200Hz.

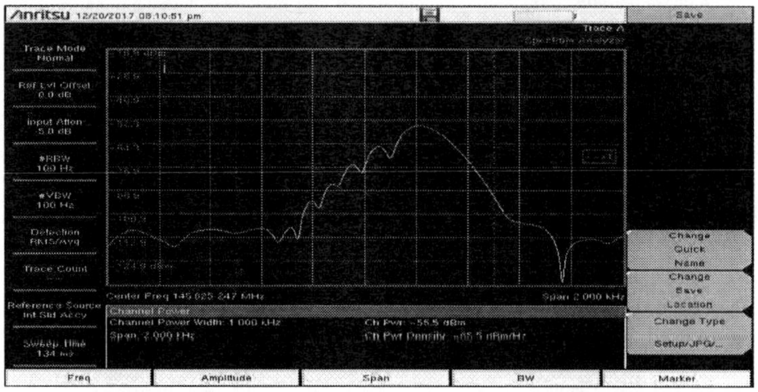

Figure 20. Signal spectrum of spectrum analyzer

In Figure 20 the amplitude of the spectrum is maximum at frequency 145.825MHz every time the data packet was sent to the VHF transceiver.

After sending the data packet to the VHF transceiver, the data was sent wirelessly to the dummy payload at frequency 145.825MHz. The data received by the dummy payload was observed at the terminal emulator program in order to verify if the transmitted data was successfully received.

Figure 21. The data packet received by the payload

Figure 21 shows the result at the payload where the numbers 0 to 9 was displayed at terminal emulator program. It shows that the data is successfully received by the payload.

Next, the transmission signal was tested with different power level supply by the VHF transceiver and signal wave of the data packet transmission to the payload is observed using a spectrum analyzer. A fixed attenuator of 40 dBm was connected to the VHF Transceiver. This testing was done to observe the signal power strength by using a variety of power level from the VHF transceiver. Channel power width value represents the signal power strength of the signal wave.

Table 2. Channel power width of the transmitted signal at different power level

Attenuator value (dBm)	Power level supply at VHF Transceiver(dBm) & (watts)	Channel power width at spectrum analyzer(dBm)
40 dBm	30.6 dBm / 1.14 watts	10.6 dBm
40 dBm	35.3 dBm/ 3.38 watts	4.7 dBm
40 dBm	38.4 dBm/ 6.91 watts	1.3 dBm

Table 2 shows the calculated channel power width value that should be achieved when testing with the different power level.

IOP Conf. Series: Journal of Physics: Conf. Series **1152** (2019) 012002 doi:10.1088/1742-6596/1152/1/012002

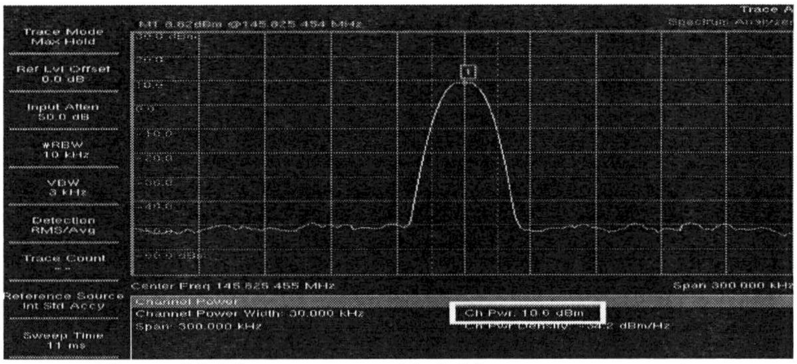

Figure 22. The spectrum with 1.14 watts power supply

Figure 22 the power supply is set to 1.14 watts, the channel power width of the signal shows the value of 10.6 dBm.

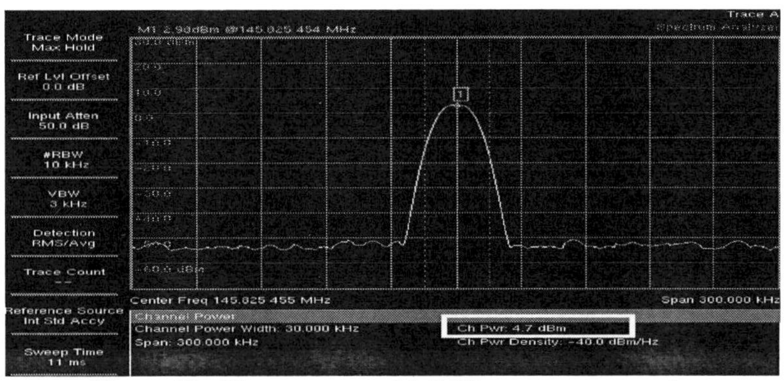

Figure 23. The spectrum with 3.38 watts power supply

Figure 23 shows when the power supply is set to 3.38 watts, the channel power width of the signal shows the value of 4.7dBm.

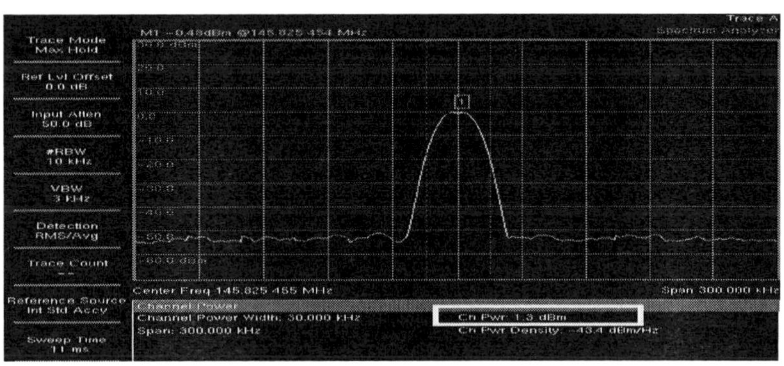

Figure 24. The spectrum with 6.91 watts power supply

Figure 24 shows when the power supply is set to 6.91 watts, the channel power width of the signal shows the value of 1.3 dBm.

From the result obtained in Figure 22 to 24, when VHF Transceiver is supplied with different power level, it shows that the spectrum analyzer can detect the signal wave produced by the packet data that was transmitted by the CCU.

4. Conclusion

This project is to shows the process of transmission of the data packet from the space-based IOT terminal to the dummy payload. By using COTS components, low-cost space-based IOT terminal can be built. The development of the proposed terminal at current stage has been tested and it shows a good agreement between the terminal as transmitter and dummy payload as a receiver. The data packet that was intended send to the payload, is successfully received. However, to show the overall project is successful several elements for the improvement of hardware and software should be added. The future recommendation such as orbital calculation and data logging should be added in software programming while instalment of VHF antenna for the hardware improvement can be added.

Acknowledgment

Authors would like to acknowledge the members of Electromagnetic Research Group (AERG) in the faculty of UiTM Shah Alam for mentoring in finishing this project within the limited time. A special recognition to Laboratory of Spacecraft Environment Interaction Engineering (LasSEINE), Kyushu Institute of Technology, Japan and the members of Joint Global Multi-Nation BIRDS-2 NanoSatellite project.

References

[1] U.Craig I, G.Richardson,J.Savignol, "In-orbit result from the SNAP-nanosatellites and its future", Philosophical transaction of the royal society of London, pp.199-203,2003
[2] S. Cakaj, "Practical horizon plane and communication for Low Earth Orbiting (LEO) satellites ground stations", WSEAS International Conference on Telecommunications and Informatics Transaction on communication, pp.63-67, 2016.
[3] Lorenzini. D & Violi D, "System design for comercial microsatellite missions", Utah state university Conference, pp.1-9, 1991.
[4] U.Craig I, G.Richardson,J.Savignol, "In-orbit result from the SNAP-nanosatellites and its future", Philosophical transaction of the royal society of London, pp.199-203,2003
[5] A.B.William, E.N. Douglas, T. Jack, "AX.25 Link Access Protocol for Amateur Packet Radio", Tucson Amateur Packet Radio Corporation, 1998.
[6] P.Karn, H. Price, R. Diersing , "Packet radio in amateur service", IEEE Journal in selected Areas in Communication , Vol 3, Issue:3, pp.431-439, 1985
[7] G. Maeda, T.Taiwo, M.Cho, "BIRDS Project Newsletter", Unpublished.

[8] A.Salces. "Communicaton protocol", Unpublished.

IOP Conf. Series: Journal of Physics: Conf. Series **1152** (2019) 012003 doi:10.1088/1742-6596/1152/1/012003

A Dual-Port Circular Polarized S-Band Antenna for Nanosatellite Communication System

Touhidul Alam[1], Mohammad Tariqul Islam[1], Mengu Cho[2]

[1]Centre of Advanced Electronic & Communication Engineering,
Faculty of Engineering and Built Environment, Universiti Kebangsaan Malaysia,
Bangi, Selangor D.E., 43600 Malaysia
[2]Laboratory of Spacecraft Environment Interaction Engineering (LaSEINE),
Kyushu Institute of Technology (KIT), Japan.

Email: touhid13@siswa.ukm.edu.my

Abstract. In recent years, the development of the nanosatellite is not for the just research purpose but commercial opportunities as well. The antenna is a very important part of the nanosatellite communication system. This paper presents an S-band circular polarized (CP) antenna for nanosatellite space mission communication system. The antenna is designed on Rogers 5880 substrate materials and the electrical dimension of the antenna is $0.39\lambda \times 0.39\lambda \times 0.03\lambda$. The result shows that the antenna has a-10dB reflection coefficient bandwidth of 15.625%, 3dB axial-ratio bandwidths of 70MHz and peak gain of around 6.68dB.

1. Introduction

Nowadays, Nanosatellite missions are equally popular in both academic and industrial research due to their remarkably low development cost and small development period compared to conventional satellite[1, 2]. A compact size antenna with the high gain antenna is very desirable for nanosatellite communication system[3].

In the last decades, several types of S-band antennas have been developed for small satellite communication systems. One of the earliest S-band antenna for the small satellite system was designed in [4], which was operated at 2.225GHz with 1.5% fractional bandwidth and linearly polarized. In [5], a high gain coaxial shorted annular patches antenna was developed for the European Student Earth Orbiter small satellite mission. The antenna size was $110\times130\times100$ mm^3 and the average required power was 3W, which is quite unsuitable for standard 1U and 2U nanosatellite system. A vertical patch CP antenna for satellite communication is presented in [6], which exhibited 3-dB axial ratio of 3.48% with an impedance bandwidth of 12.6%. Another circular polarized S-band antenna was reported in [7] for HORYU-IV nanosatellite. The antenna shows 10dB impedance bandwidth of 2.73% with 3-dB AR bandwidth of 13 MHz. In [8, 9], dual-port was demonstrated to achieve circular polarization.

In this paper, dual-coaxial fed patch antenna has been designed that combines several advantages making it a compatible for nanosatellite missions.

IOP Conf. Series: Journal of Physics: Conf. Series **1152** (2019) 012003 doi:10.1088/1742-6596/1152/1/012003

2. Antenna Design

The antenna schematic layout of the proposed antenna is illustrated in Figure 1. The antenna consists of three layers. The first layer is heptagon shaped radiator, the second layer is a rectangular shaped conductor and the third layer is the ground plane. The antenna is fed by two coaxial probes fed. The commercially available Rogers 5880 substrate material is used to design the antenna. The optimized design parameter is listed in Table 1.

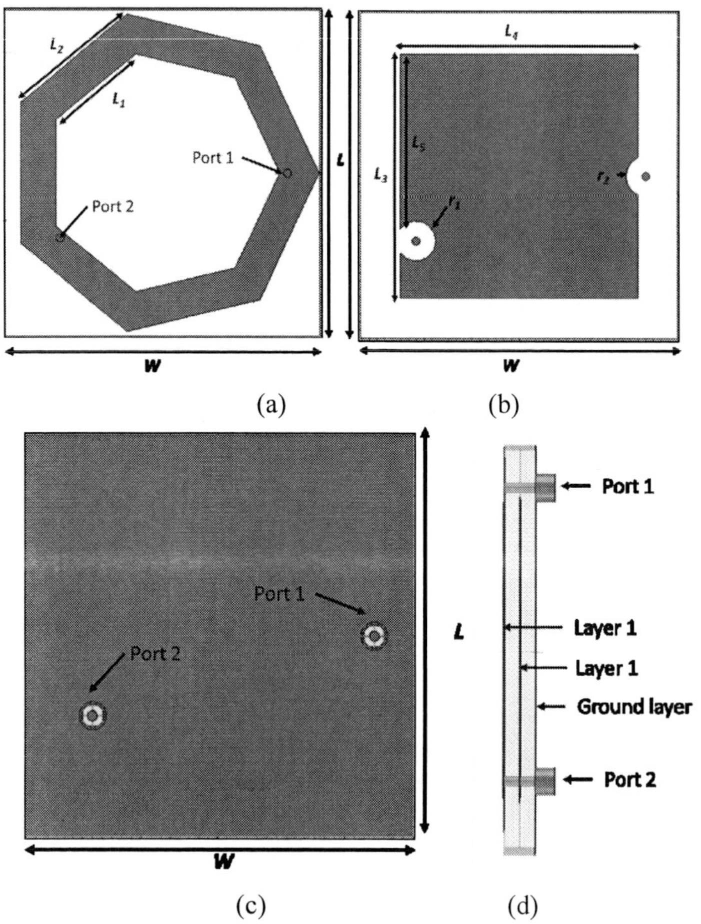

Figure 1. Antenna design layout (a) Antenna layer 1, (b) layer 2, (c) ground layer and, (d) side view

Table 1. Antenna design parameters.

Parameters	Value	Parameters	Value
L	40	L_4	30
W	40	L_5	21.5
L_1	13.02	R_1	1.25
L_2	17.36	R_2	1.2
L_3	30		

3. Antenna Results

The antenna is simulated using commercially available CST microwave studio. The reflection coefficient of the proposed antenna has been analyzed, shown in Figure 2. The antenna depicted -10dB impedance bandwidth of 500MHz (2.95 GHz to 3.45 GHz) for both ports. The axial ratio at Phi 0° and 90° have also been investigated, presented in Figure 1. The antenna achieved 3dB axial ratio of 70MHz (2.98-3.05 GHz) and 90 MHz (3.29-3.38 GHz0 at Phi 0°and Phi 90° plane, respectively. The axial ratio pattern at 3.05 GHz is presented in Figure 3 (a). In Figure 3(b), the 3D-radiation pattern at 3.05 GHz is presented, where it is showed that the main lobe magnitude is 6.68dB and the main lobe direction is at 0°(z-direction). The 3dB angular width is 86.6 dB.

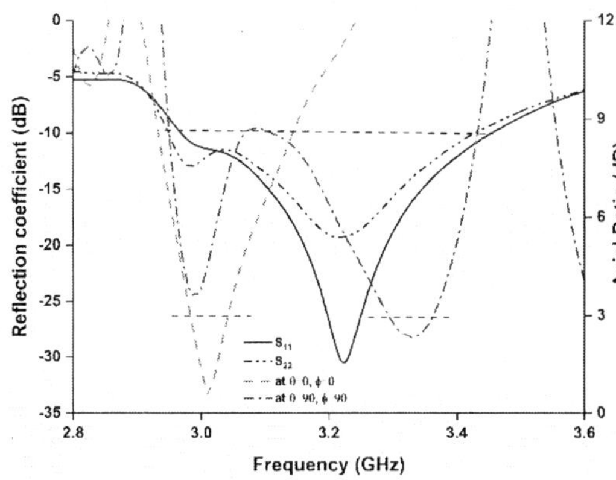

Figure 2. Reflection coefficient and axial ratio of the proposed antenna

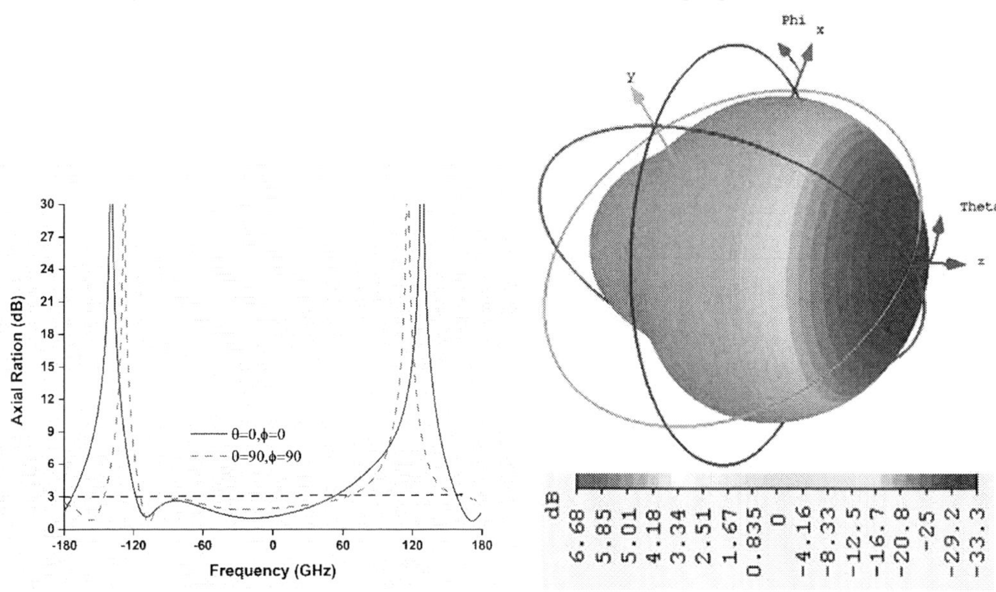

(a) (b)

Figure 3. (a) Axial ratio and (b) 3D radiation pattern of the proposed antenna at 3.05 GHz

4. Conclusion

In this paper, the design and performance analysis of a high gain circular polarized patch antenna is presented. The presented results show that -10dB impedance bandwidth of 500MHz, as well as 3 dB axial ratio of 70MHz and 90 MHz, was realized. The simple structured patch antenna with compact size, high gain, and high radiation efficiency is compatible with nanosatellite standard structure and easy fit with S-band nanosatellite payloads.

References

[1] S. P. Neeck, T. J. Magner, and G. E. Paules, "NASA's small satellite missions for Earth observation," *Acta Astronautica,* vol. 56, no. 1-2, pp. 187-192, 2005.
[2] R. Sandau, "Status and trends of small satellite missions for Earth observation," *Acta Astronautica,* vol. 66, no. 1-2, pp. 1-12, 2010.
[3] S. Gao *et al.*, "Antennas for modern small satellites," *IEEE Antennas and Propagation Magazine,* vol. 51, no. 4, 2009.
[4] M. Tanaka, Y. Suzuki, K. Araki, and R. Suzuki, "Microstrip antenna with solar cells for microsatellites," *Electronics Letters,* vol. 31, no. 1, pp. 5-6, 1995.
[5] E. Arnieri, L. Boccia, G. Amendola, and G. Di Massa, "A compact high gain antenna for small satellite applications," *IEEE Transactions on Antennas and Propagation,* vol. 55, no. 2, pp. 277-282, 2007.
[6] K.-B. Ng, C. H. Chan, and K.-M. Luk, "Low-cost vertical patch antenna with wide axial-ratio beamwidth for handheld satellite communications terminals," *IEEE Transactions on Antennas and Propagation,* vol. 63, no. 4, pp. 1417-1424, 2015.
[7] M. T. Islam, M. Cho, M. Samsuzzaman, and S. Kibria, "Compact antenna for small satellite applications [Antenna Applications Corner]," *IEEE Antennas and Propagation Magazine,* vol. 57, no. 2, pp. 30-36, 2015.
[8] K. Wei, Z. Zhang, Z. Feng, and R. Li, "A conformal S-band dual-port dual-polarized antenna for Satellite Digital Audio broadcast," in *Microwave and Millimeter Wave Technology (ICMMT), 2010 International Conference on*, 2010, pp. 11-14: IEEE.
[9] Y.-P. Hong *et al.*, "S-band dual-path dual-polarized antenna system for satellite digital audio radio service (SDARS) application," *IEEE transactions on microwave theory and techniques,* vol. 54, no. 4, pp. 1569-1575, 2006.

Investigation of angular distribution of solar energetic particles

V. E. Timofeev

Yu .G. Shafer Institute of Cosmophysical Research and Aeronomy, SB RAS, Yakutsk, 31 Lenin Ave.
vetimofeev@ikfia.ysn.ru

Abstract. Rotation is often used to stabilize the orientation of satellites in outer space. A method for recording of the angular distribution of energetic particles aboard a spacecraft oriented to the Sun is proposed. This is an example is shown of how to study the angular distribution of energetic solar particles using the own rotation of a spacecraft.

1. Introduction

To ensure that the orientation of a certain axis of the satellite is unchanged, an orientation system that uses the gyroscopic properties of rotating bodies is often used. Rotation orientation was used on the TIROS, ATS-B, ATS-C, some Explorer, Prognoz, and other satellites.

Currently small spacecraft are widely used more and more in both scientific and applied research Nervold (2016). For the last decade a problem of space weather develops intensively. In this problem the account of interactions between solar influences and particularly terrestrial processes is important. Studying of dynamics of the energetic charged particle fluxes in the nearby-Earth environmental space and interplanetary environment can give helpful information on this problem (Miroshnichenko 2008, Reames 2017). Investigations aboard small space vehicles can make a big contribution into solution of this problem. For these satellites, the devices of small sizes and weight and at the same time with good functional features are necessary.

Distribution of the charged particle fluxes in the solar wind is controlled by its magnetic field consisting of regular and irregular components.

In the range of small energies 1 - 15 MeV, particles of only solar origin, propagated in the interplanetary magnetic field or captured by it, are registered. The problem on free propagation of particles or their capture plays an important role in the physics of interplanetary environment.

For example, the captured particles at expansion of the interplanetary environment can "be cooled" quickly, whereas the "freely" propagated particles do not undergo such cooling.

There may also be quasi-stationary traps of particles in this energy range which rotate together with the Sun and do not lead to "cooling" of particles, and, probably, on the contrary, accelerate them. To classify cases of observation of the charged particle fluxes into specified types of their propagation, it is necessary to measure an anisotropy of particle fluxes. The main goal of our paper is to show the method of measuring the angular distribution of charged particles, which best suits the conditions of a scientific experiment, with a hard limit on weight, size, power consumption and also with high requirements on the flexibility of logic, reliability of operation and preliminary onboard processing of information.

IOP Conf. Series: Journal of Physics: Conf. Series **1152** (2019) 012004 doi:10.1088/1742-6596/1152/1/012004

2. Features of registration of low-energy component of charged particle fluxes

Such measurements almost in the same range were conducted aboard the spacecraft (SC) "Prognoz" and IMP-8 (Miroshnichenko, 2008, Reames 2017). As a result of these measurements the anisotropy components only in the plane of solar equator were determined.

To study the angular characteristics of investigated low-energetic flows aboard the IMP-8 "Prognoz" satellites a scientific equipment including sensors located at angles of 0^0, 45^0, 90^0, 135^0, to the axis of orientation was used.

Each detector registers the integral flow of particles only from a solid angle that "falls on the aperture of detector during the rotation of satellite around its axis. It is difficult to determine the anisotropy components in the plane perpendicular to the direction of the Sun using this technique. The measurement of these components provides the additional valuable information.

The mentioned anisotropy can be divided into the equilibrium and nonequilibrium components. The first one is caused by the movement of cosmic rays with the magnetic field and substance of the solar wind. This anisotropy is connected with a transport velocity of particles. It can be directed radially. In this case, the particles are transferred by the radial solar wind and the magnitude of anisotropy accounts for ≈ 20 %.

Let's estimate anisotropy amplitude. Anisotropy caused by the movement of cosmic ray gas in the magnetic field with the velocity U is equal to: $A = (\gamma+2)\frac{U}{C\beta}$. Here γ is an indicator of differential spectrum of pulses, C is the velocity of light, β is the relative velocity of the particles equal to $\sqrt{\frac{2T}{E_0}}$, _0 where E_0 is a rest energy, T is the kinetic energy. For particles 1-15 MeV $\gamma = 5$, $\beta = 0.05$. The solar wind velocity U = 400 km/s. The rotational velocity of the Sun related to the Earth's orbit is 400 km/s too. Hence A = 20 %.

The nonequilibrium anisotropy can be connected with its own movement of particles along the field lines and direction along the magnetic field and its value must be> 20 %.

In a case of sudden changes of the radial or azimuthal particle density gradients there should appear a nonequilibrium anisotropy in the south-north direction proportional to the value of gradient.

The study of anisotropy in the plane perpendicular to the direction of the Sun, gives an opportunity to investigate the nonequilibrium anisotropy connected with the deviation of direction of the interplanetary field from the plane of solar equator. The high-energy particles are most often observed in the form of the short-term increases following the fronts of shock waves, or after the solar flashes. The increases of flows of particles with energies of 0.5 - 15 MeV have a much larger time scale. In these cases the sequences of increases in 27 days like recurrent events can be observed.

In a number of works (Reames D.V. 2017, Gallagher, Simpson 1966, K.G. McCracken, U.R .Rao 1967) these sequences are interpreted as a repeated passage of spacecraft through the solar wind sector filled by energetic charged particles. The existence of such stationary structures in the solar wind, filled with low- energetic charged particles, constitute a new phenomenon, significantly different from the discrete flashes of cosmic rays. The nature of this phenomenon is not yet established.

One can assume two possible scenarios. One of them explains the existence of constant sectors filled with cosmic rays by means of continuous emission of particles by areas in the solar atmosphere. According to another one, the particles are considered as if captured in a "bag" that rotates together with the Sun. Such "bag", basically, can be formed by crowding of magnetic field lines at some distance from the Sun appearing as a result of collision of fast and slow streams of the solar wind. The area between this crowding and the Sun is a magnetic "bottle", which is stationary in a rotated system of reading.

Both of these development scenarios predict a quite different anisotropy of charged particle fluxes of the solar origin.

In the first scenario, the particles propagate mainly along the field lines outwards (**Figure.1**). And in the second one, the particles take part in the rotation of the Sun and their anisotropy has a direction of maximum from the east.

IOP Conf. Series: Journal of Physics: Conf. Series **1152** (2019) 012004 doi:10.1088/1742-6596/1152/1/012004

The separation of these scenarios by types of anisotropy maybe implemented using the following experiment.

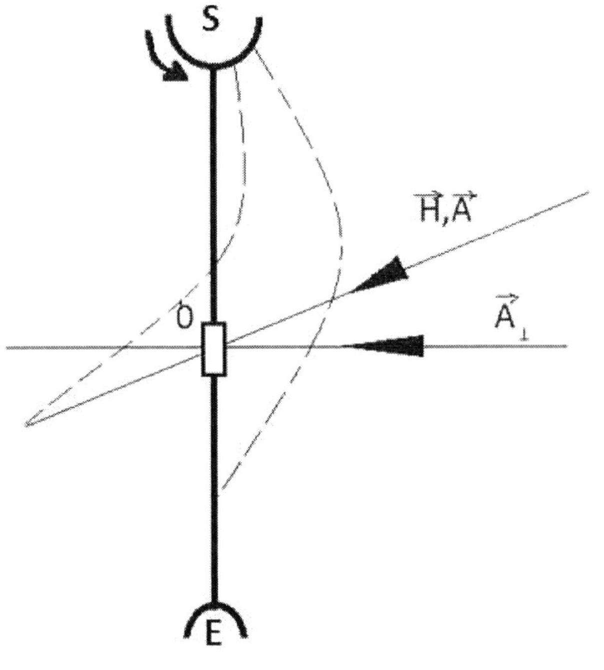

Figure. 1. Scheme explaining to anisotropy lying in a plane perpendicular to a direction towards the Sun. The dashed lines show the sector filled with low-energetic cosmic rays; 0 is the object, E - S is the Earth-Sun line, H, A is the direction of vector of the magnetic field and anisotropy, A is the cross-section component of anisotropy vector .

We propose to measure the angular distribution of cosmic ray intensity in the plane perpendicular to the Sun's direction. These measurements make it possible to determine a component of anisotropy of charged particle flows lying in this plane.

And the first scenario predicts anisotropies for this plane, one of them is directed to the east, and the second one – to the west.

Such measurements are conveniently carried out with a rotated spacecraft with a rotation axis towards the Sun using the technique of scanning.

When the scanning accuracy is not worse than ≈1%, the important information can be also obtained for solving the following scientific problems:

1. Detection and confident identification of small flows (a few tens % from the galactic flows) of charged particles of solar origin against the background of the galactic ones. The methodical side of this problem is based on the fact that the galactic cosmic rays form a current directed from east to west, whereas a solar particle flow has a component directed opposite, from west to east.

2. Estimation of the distribution of the "diffusion resistance" along the distribution channel, representing, roughly speaking, the number of scatterings of particles towards the observer.

The matter is that in the first case when scatterings occur, mainly, on the last section of the way, temporal characteristics of registered radiation differ a little. At the same time these two extreme cases are essentially various on the anisotropy amplitude. Thus, comparing the anisotropy with a temporal change of intensity, it is possible to judge a distribution of sources of particle scattering along a solar radius.

3. Anisotropy of charged particles of solar origin could give information about the fine structure of interplanetary medium. The anisotropy undergoes oscillations in directions with the orientation change of the field tubes of the interplanetary magnetic field. The frequency spectrum of anisotropy oscillations must be, in some degree, adequate to the spectrum of inhomogeneities.

The value of measurements of the low-energetic components increases, if aboard the same spacecraft, which conducts the scanning, the solar wind parameters such as the wind velocity and magnetic field intensity will be measured.

The experiment can be carried out aboard the spacecraft when it is outside of the magnetosphere in the undisturbed solar wind by it.

It is supposed to use a spectrometer consisting of semi-conductor detectors for registration of the radiation, whose detectors are placed in a protective collimator (Timofeev et. al. 2017, Brilkov 2014, Tsoulfanidis 2015). The effective space angle from an entrance window accounts for 0.5 ster.

The scanning facility of the device synchronized by impulses from the transducer of angular velocities of object divides each turn around its own axis, irrespective of its duration, into N of equal sectors and the special device connects the transducer sequentially to N memory channels. Thus, the information accumulated in these channels will reflect the anisotropy of radiation registered by the transducer.

The device should automatically monitor the time of turn of the object and be adjusted anew as it changes. To estimate the total number of registered particles necessary for definition of one point of anisotropy with the given accuracy we will make the following calculations.

The anisotropy can be defined as Fourier-components of the angular distribution of cosmic rays. In this case, its value is calculated by the formula:

$$A = \sum_{i=1}^{N} k_i I_i; \quad (1)$$

Where I_i are intensities in each of i - channels expressed in relative units, and k are harmonic coefficients.

The specified formula is the approximate expression of the following integral ratio:

$$A = \frac{\int_0^{2\pi} \sin\theta I(\theta) \partial\theta}{\int_0^{2\pi} \sin^2\theta \, \partial\theta}; \quad (2)$$

From a comparison of these two expressions at N>>I it is seen that

$$K_i = \left(\frac{2}{N}\right) \sin I_i \ \theta_i; \quad (3)$$

Let's assume that the relative error in each of channels is equal to σ. Then the error of anisotropy component A equals:

$$\sigma_A = \sqrt{\sum_{i=1}^{N} k_i^2 \sigma^2} = \sigma\sqrt{\sum_{i=1}^{N} k_i^2} = \sigma\sqrt{k^2}\sqrt{N} \quad (4)$$

Considering that, $\sin^2\theta = \frac{1}{2}$, we obtain from (3) and (4)

$$\sigma_A = \sigma \frac{2}{N}\sqrt{\frac{1}{2}}\sqrt{N} = \sigma\sqrt{\frac{2}{N}} \quad = (5)$$

If we designate the total number of particles registered by all channels together through n_0, then the number of particles registered by each channel, will be approximately equal to $\frac{n_0}{N}$, and a relative error of each channel is

$$\sigma = \sqrt{\frac{N}{n_0}} \quad (6)$$

Thus, from (5) and (6) we obtain: $\sigma_A = \sqrt{\frac{2}{n_0}}$ (7)

In order to reliably measure the anisotropy of 20 %value with a triple reserve of error, it is required:

$$\sigma_A \approx 0.07$$

Hence, it seen that, it is necessary $n_0 = \frac{2}{\sigma_A^2} \approx 400$, i.e. to achieve the required precision for each point of the anisotropy, the sensor must register ≈ 400 particles.

Based upon these numbers and assumed geometrical factors of sensors, we will estimate the required time of exposure for each point of anisotropy.

In the chosen energy ranges the geometrical factor of each detector is G = 0.5 cm^2 ster.

The intensity of such particles by measurements (Miroshnichenko, 2008, Reames 2017, Longair) during quiet time accounts for1 particle / cm^2 s sr.

Hence, the required time is: $t = \frac{n_0}{\Gamma * I} \approx 10$ - 15 min.

It is seen that measurements of anisotropy can be conducted with a rather good temporal resolution.

In order for this resolution to be achieved in practice, the rotation velocity of the object should be not less than 1 rotation/min. The specified method of measurement of the anisotropy will give additional errors in the case of fast increase or decrease in the radiation intensity. The greater is the rotation velocity of the object, the smaller are these errors. A rough estimation of this error can be obtained by the formula:

$$\Delta A = \frac{dI}{dt} * T$$

Where T is a period of rotation of the object.

This error can be reduced by a subsequent analysis in each particular event.

3. Conclusion

A specific example shows how to study the angular distribution of the charged particle fluxes. For this purpose the own rotation of small spacecraft is used. As an example it is shown how to measure the angular distribution of charged particle fluxes in the interplanetary space. This method can be used to study the angular distribution inside the Earth's magnetosphere and also in the investigation of the radiation belts.

References

[1] Nervold A.K., J. Berk, J. Straub, D. Whalen A Pathway to Small Satellite Market Growth Advances in Aerospace Science and Technology, 2016, V1, PP.14 - 20

[2] Kremnev R. S., A. I. Smirnov, and S. S. Gorkin Brief Description of the Prognoz-M2 Spacecraft in the INTERBALL Project //Cosmic research Vol. 34, No. 4, pp334-341 1996

[3] Reames D V 2017 Solar Energetic Particles: A Modern Primer on Understanding Sources, Acceleration and Propagation (New York: Springer)

[4] McCracken K.G, U.R.Rao A Digital System for Accurate Time Sector Division of a Spin-Stabilized Vehicle. IEEE Trensitions on Aerospace and Electronics Systems V. AES-3, N 2, March 1967.

[5]Timofeev V.E., A.G.Khristoforova, P.N.Nemchinov. Patent of the Russian Federation №2117129. Spectrometer of the charged particles.

[6] Gallagher J.J.O, J.A.Simpson Anizotropic Propagation of Solar Protons Deduced from Similtaneous Observations by Earth Satellites and the Mariner-4 Space Probe. //Phys. Rev. Letters, V.16, №26, 1212 1966.

[7] Miroshnichenko Leonty <http://istina.msu.ru/workers/1602655/> Solar Cosmic Rays: Fundamentals and Applications, Second Edition//2014. Springer Cham Heidelberg New York Dordrecht London P. 521

[8] Miroshnichenko L.I. Solar Cosmic Rays in the System of Solar-Terrestrial Relations (Review). J. of

Atmospheric and Solar-Terrestrial Physics (Special Issue of ISROSES Proceedings), 2008, v.70, P.450 - 466.
[9] Brilkov I., Vedenkin N., Panasyuk M., Amelyushkin A., V., Nechayev O., V. DEPRON dosimeter for "Lomonosov" satellite//40th COSPAR Scientific Assembly. Held 2-10 August 2014, in Moscow, Russia. Abstract F2.3-33-14.
[10] Longair, M.S. High energy astrophysic Vol.1 Particles, photons and their detections.-2 nd ed. 2004. P.418. Cambridge University Press
[11] Tsoulfanidis Nicholas. Measurement and Detection of Radiation,-4 th Edition. 2015. pp381-406

Acknowledgments
The work was financially supported by the Russian Fund for Basic Research (the project № 01- 02-17278, the project № 01-02-96206, the project №09-02-98507, the project №18-42-140006)

IOP Conf. Series: Journal of Physics: Conf. Series **1152** (2019) 012005 doi:10.1088/1742-6596/1152/1/012005

Solar radio burst in National Space Agency of Malaysia (ANGKASA)

[1]Asnor Nadirah Ishak, [2]Nur Zulaikha Mohd Afandi, [2]Roslan Umar, [3]Nor Hazmin Sabri, [1]Zahira Mohd Radzi, [4]Asnor Juraiza Ishak, [5]Radial Anwar

[1]National Space Agency (ANGKASA), Lot 2233, Jalan Turi, Kg. Sg. Lang, 42700 Banting, Selangor, Malaysia

[2]East Coast Environmental Research Institute, Universiti Sultan Zainal Abidin, Gong Badak Campus, 21300 Kuala Nerus, Terengganu, Malaysia

[3]School of Fundamental Science, Universiti Malaysia Terengganu, 21030, Kuala Terengganu, Terengganu, Malaysia

[4]Department of Electrical and Electronic Engineering, Faculty of Engineering, Universiti Putra Malaysia, 43400 UPM Serdang, Selangor, Malaysia

[5]Department of Telecommunication Engineering, School of Applied Science, Telkom University, Jl. Telekomunikasi Terusan Buah Batu, Bandung 40257, Indonesia

asnor@angkasa.gov.my, zulaikhafendy@yahoo.com.my, roslan@unisza.edu.my, norhazmin@umt.edu.my, zahira@angkasa.gov.my, asnorji@upm.edu.my, radialanwar@tass.telkomuniversity.ac.id

Abstract. The sun is an active star that produces large-scale energetic events, such as solar flares and coronal mass ejections (CMEs). These phenomena are observable across the electromagnetic spectrum, from gamma rays at hundreds of MeV to radio waves with wavelengths of tens of metres. Solar flares and CMEs can excite plasma oscillations which can emit radiation at metric and decametric wavelengths. These radio bursts are classified in five main types. This paper gives a solar radio burst type III (16 July 2017) in National Space Agency (ANGKASA), Banting, Selangor, Malaysia. Compact Astronomical Low-Cost Instrument for Spectroscopy in Transportable Observatories (CALLISTO) is used in this system to observe solar radio activities. The main applications for this CALLISTO is to observe a solar radio bursts and radio frequency interference (RFI) monitoring for astronomical science, education and outreach. The CALLISTO natively operates between 45MHz to 870MHz and its observed daily.

IOP Conf. Series: Journal of Physics: Conf. Series **1152** (2019) 012005 doi:10.1088/1742-6596/1152/1/012005

1. Introduction

The sun is the star at the heart of the solar system, where it is by far the largest object, which consists of hot plasma with magnetic field surrounding it, mostly made up of hydrogen and helium. Its gravity holds the solar system together, keeping everything from the biggest planets to the smallest particles of debris in its orbit. Electric currents in the Sun generate a magnetic field that is carried out through the solar system by the solar wind, a stream of electrically charged gas blowing outward from the Sun in all directions. The connection and interactions between the Sun and Earth drive the seasons, ocean currents, weather, climate, radiation belts and aurora.

Monitoring the solar activities are now being the most active research for solar radio in Malaysia. Commonly, solar flare and coronal mass ejections (CMEs) exists as couple with the emission of solar radio burst during active sun and can be categorized as rapidly drift varies over period of second or minutes. Emissions of burst occurred when high velocity of charged particles from the flare activity ejected away from sun and moved toward outer atmosphere (corona) of the sun. Based on the activity, charged particles from moving solar flare will excite the radio wave in corona, there are some findings found that unsavory radiation may affect human daily life on earth. Due to the disturbance caused by solar flare, the investigation of solar radio burst is important for future research on the importance of space weather.

In addition, with continuous meteorological and space weather observation can give the early awareness of solar flare and coronal mass ejections (CMEs) to space weather. Today the prediction of space weather become one of important to human as well as increasing the role of technology in daily life. Therefore, there are number of radio receivers and magnetometers that monitor the activity of the solar flare from the sun. The receiver must be able to detect within radio frequency

In this paper, the main objective is to study the solar radio burst using Compact Astronomical Low-Cost Instrument for Spectroscopy in Transportable Observatories (CALLISTO) in National Space Agency of Malaysia (ANGKASA), Banting, Selangor, Malaysia. ANGKASA is one of the government agency under Ministry Science, Technology and Innovation (MOSTI) is responsible in leading and observing the development of space science in Malaysia.

CALLISTO is a radio spectrometer used extensively for solar radio astronomy and radio frequency interference monitoring. CALLISTO spectrometer is used extensively as heterodyne receiver designing and leading by Christian Monstein and Radio and Plasma Physics Group from ETH Zurich, Switzerland in order to study the magnetic activity of a wide range of the Sun. Many CALLISTO spectrometer have been deployed more than 80 locations with users from more than 144 countries. The e-CALLISTO system has had its first success already in December 2006, when the last large flares of the present cycle occurred.

2. Methodology

The CALLISTO system is an instrument that was operated by ETH Zurich back in several years ago to observe solar radio activities by solar radio spectrometers. This system are used to monitor solar radio activities between 45MHz until 870MHz. This instrument can be operated by using a basic system that consisting of the receiver, linear polarized antenna system and a logging software as in Figure 1. Based on a new instrument developed in China, the new device comes with different and improved specifications and characteristics that functions to observe the solar radio observation using different frequency range of CALLISTO which was called as Solar Broadband Radio Spectrometer. The specification are the most important parameter for performance (Fu et al. 2004).

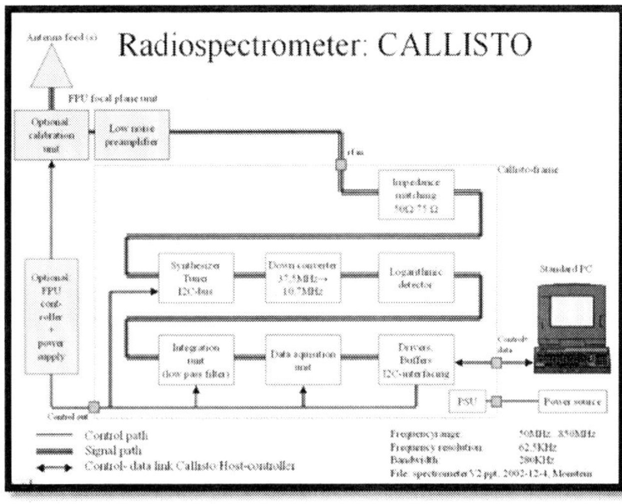

Figure 1: Configuration CALLISTO system

The data was collected from official CALLISTO station through downloading the data. The Log Periodic Dipole Antenna (LPDA) antenna is used for this system. The conversion of the fits file are necessary in order to use the data. Data was collected for 16 July 2017 and its show the type III solar radio burst.

The data was extracted using zip file software. Figure 2 and Figure 3 shows the data is analyzed using MATLAB by converting the fits files into intensity data in decibel (dB). The average intensity level was obtained through coding. The spectrum was recorded the color code intensity in frequency (MHz) vs time in second. The time for each spectrum was standardize in 15 minutes (900sec).

3. Result and Discussion

Both spectrum in Figure 2 and Figure 3 have been recorded the burst at National Space Agency of Malaysia referring to Malaysia Banting station with 2°46'58.8"N 101°30'25.2"E. In order to enhance the appearances of burst on the spectrum, the pre-processing process was applied by using MATLAB software. The solar radio burst type III (storm) is clearly seen through red circle in Figure 2 and Figure 3. This burst was confirmed with Solar Weather Prediction Center by NOAA as noted in the green rectangle of Figure 4. The burst was start 015940 UTC until 020150 UTC with frequency range from 650 MHz until 870 MHz. Solar flare C3.1 from 0158UTC until 0257UTC recorded in the purple rectangle of in Figure 4. The image in Figure 5 was taken from website https://www.solarmonitor.org is similar with the Reg#2665 in Figure 4. The solar radio burst type III (storm) is associated with solar flare C3.1.

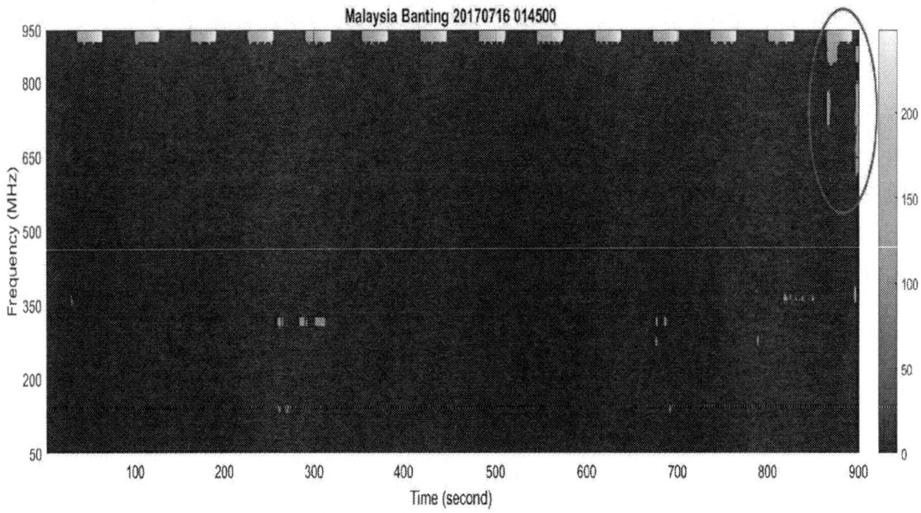

Figure 2: The spectrum of Malaysia Banting in 20170716 at 014500 UTC after enhancement using MATLAB software

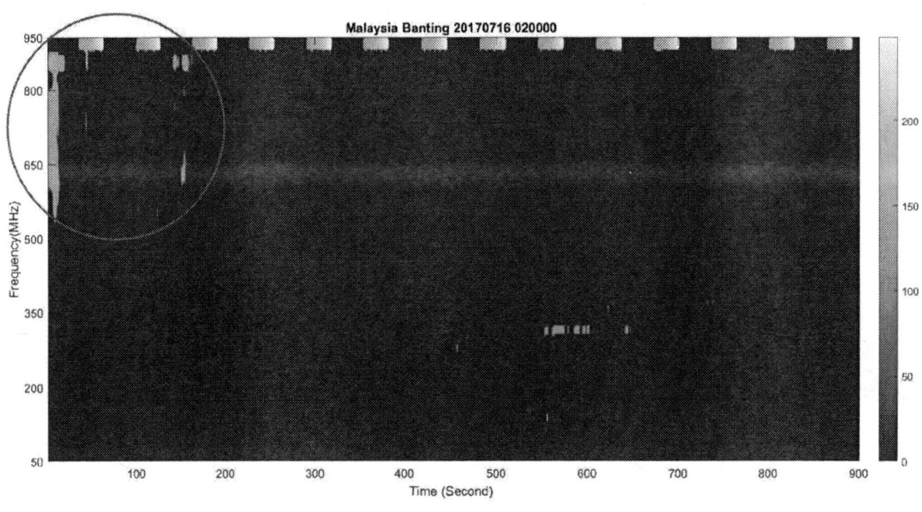

Figure 3: The spectrum of Malaysia Banting in 20170716 at 0200 UTC after enhancement using MATLAB software

```
:Product: 20170716events.txt
:Created: 2017 Jul 16 2327 UT
:Date: 2017 07 16
# Prepared by the U.S. Dept. of Commerce, NOAA, Space Weather Prediction Center
# Please send comments and suggestions to SWPC.Webmaster@noaa.gov
#
# Missing data: ////
# Updated every 5 minutes.
#                         Edited Events for 2017 Jul 16
#
#Event    Begin    Max       End  Obs  Q  Type  Loc/Frq   Particulars       Reg#
#-------------------------------------------------------------------------------

 480 +     0043    0043      0044  LEA  G   RBR  245       190

 490       0118    ////      0118  PAL  C   RSP  025-072   III/1
 490 +     0118    0118      0118  LEA  G   RBR  245       280

 510 +     0158    0235      0257  G15  5   XRA  1-8A      C3.1    9.2E-03  2665
 510 +     0159    0200      0204  LEA  G   RBR  245       17000            2665
 510       0159    ////      0202  PAL  C   RSP  025-180   III/3            2665
 510 +     0159    0200      0202  LEA  G   RBR  410       4200             2665
 510 +     0200    0200      0200  LEA  G   RBR  610       220              2665
 510       0202    0202      0250  LEA  3   FLA  S06W61    SF      DSD      2665

 520       0229    0229      0229  LEA  G   RBR  245       110

 580 +     0251    0510      0555  LEA  G   RNS  245       350

 530 +     0314    0314      0315  LEA  G   RBR  410       120

 550       0322    0322      0322  PAL  G   RBR  245       120

 560       0431    0431      0432  SVI  G   RBR  245       270

 570       0544    0547      0549  G13  5   XRA  1-8A      B7.6    1.2E-04  2665

 590       0641    0641      0641  LEA  G   RBR  245       210

 600 +     0723    0726      0728  G15  5   XRA  1-8A      C1.1    1.6E-04  2665
 600       0725    0726      0728  LEA  3   FLA  S06W64    SF               2665
 600       0725    ////      0725  SVI  C   RSP  025-180   III/2            2665
 600 +     0725    0725      0725  LEA  G   RBR  245       12000            2665
 600 +     0725    0725      0725  LEA  G   RBR  410       1500             2665
```

Figure 4: The data from Space Weather Prediction Center for 16 July 2017 event

Flare Locator Image

Difference: AIA20170716_023600_0094.fits - AIA20170716_015800_0094.fits

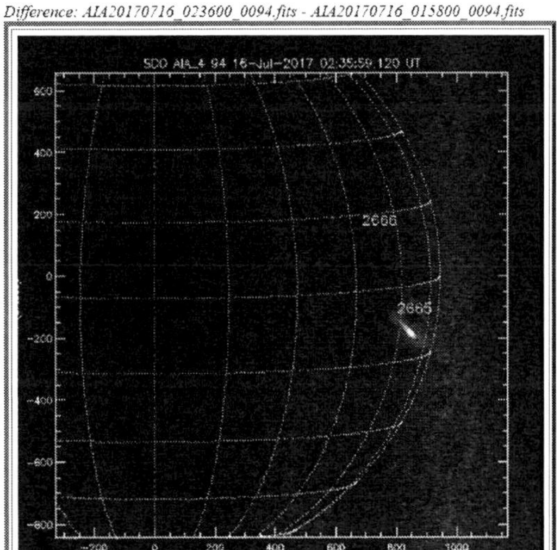

Figure 5: The image from https://www.solarmonitor.org

4. Conclusion

As a conclusion, the solar radio burst type III has been collected at National Space Agency of Malaysia (ANGKASA) using CALLISTO system with LPDA antenna. The antenna can be operated with frequency 45MHz until 870MHz. The solar radio burst type III (storm) is associated with solar flare C3.1 and this burst was confirmed with Solar Weather Prediction Center by NOAA. Malaysia has a very good potential to monitor solar activities due to the consistency of period observation throughout a year.

References

[1] N.Z.M. Afandi, S.N. Hazmin, R. Umar, A.N. Ishak, 2017. Automated Solar Radio Burst Detection on Radio Spectrum: A Review of Techniques in Image Processing. Journal of Fundamental and Applied Sciences. ISSN 1112-9867.

[2] Z.S. Hamidi, N.N.M. Shariff, Z.Z. Abidin, Z.A.Ibrahim and C. Moinstein, 2012. Coverage of Solar Radio Spectrum in Malaysia and Spectral Overview of Radio Frequency Interference (RFI) by Using CALLISTO Spectrometer form 1MHz to 900MHz. Middle-East Journal of Scienctific Research 12(6): 893-898. doi: 10.5829/idosi.mejsr.2012.12.6.1733*

[3] Abidin, Z. Z., Anim, N. M, Hamidi, Z. S., Monstein, C., Ibrahim, Z. A., Umar, R., Shariff, N. N. M., Ramli, N.,Aziz, N. A. I. & Sukma, I., 2015. Radio frequency interference in solar monitoring using CALLISTO. New Astronomy Reviews. 67, 18-33.

[4] Benz, A.O., Monstein, C.,Meyer, H.,Manoharan, P. K., Ramesh, R., Altyhynster, A.,Lara, A., Paez, J. & Cho, K. S., 2009. A World Wide Net of Solar Radio Spectrometer: e-CALLISTO. Earth Moon Planet. 104, 277-285.

[5] Hamidi, Z. S., Abidin,Z. Z., Ibrahim, Z. A., N. N. M., Shariff & C. Monstein., 2012. Signal detection performed by Log Periodic Dipole Antenna (LDPA) in Solar Monitoring. International Journal of Fundamental Physical Sciences. 2, 2, 24-26.

IOP Conf. Series: Journal of Physics: Conf. Series **1152** (2019) 012006 doi:10.1088/1742-6596/1152/1/012006

Thermal Validation Testing of an Automatic Identification System (AIS) Receiver for Low Earth Orbit (LEO) CubeSat

Emir Husni, Nazmi Febrian

School of Electrical Engineering and Informatics, Institut Teknologi Bandung, Bandung 40132, Indonesia

Corresponding author's e-mail address: ehusni@lskk.ee.itb.ac.id

Abstract. The advancements in electronics technology make nanosatellite missions more varied. One of its missions is to retrieve Automatic Identification System (AIS) data from vessels on the ground. The type of nanosatellite made in this study is a CubeSat with a size of 1.5U. The AIS receiver is created using RFIC Si4362 and MSP430G2553 Microcontrollers to control the device. The main mission of AIS Receivers is to receive and decode AIS signals at the frequency of 161.975 MHz and 162.025 MHz into the AIVDM NMEA0183 standard format. The decoded AIS data is sent to a Raspberry Pi Compute Module as On-Board Data Handling (OBDH) CubeSat before it is sent to the ground station. All of the CubeSat's subsystems must be able to pass thermal testing before being launched into orbit. Thermal testing is performed to observe the performance of subsystems in extreme conditions similar to CubeSat conditions in space. Thermal testing is carried out at the Satellite Technology Center (Pusteksat) of the National Aeronautics and Space Agency (LAPAN) facility. The thermal test is done by running the device at temperatures of -20°C - 60°C for 3 hours and 56 minutes. The thermal testing indicates that the AIS Receiver can work in the orbital temperature range. After testing, the data reception success rate is 36.76% with a minimum RSSI value of -96dBm.

1. Introduction

An AIS Receiver is a device that is designed to receive AIS signals from vessels and decode the signals into the AIS AIVDM data format. AIS Receivers achieve better reception results located at 650km altitude. Many satellites in various sizes have been launched with an AIS receiver as their payload [1].

CubeSat was chosen because it is one of the satellite platforms developed at lower cost and shorter development time than other large satellites. Various universities worldwide and in ASEAN, in particular, have developed CubeSat for a variety of useful missions [2].

CubeSat has several subsystems such as an Electrical Power System (EPS); Communications (COM); an Attitude Determination and Control System (ADCS); Sensors and GPS; On-Board Data Handling (OBDH); and payload [3]. This study focuses on the development of a space-based AIS receiver as the CubeSat payload.

The main challenge in designing a CubeSat is the use of COTS components that can work in satellite orbit conditions [4]. In order to be classified as a space-based AIS receiver, it is required to pass some environmental tests of the robustness of the device in conditions approaching the satellite conditions when orbiting in space.

Common environmental tests are vibration tests, vacuum tests, thermal tests, electromagnetic interference/compatibility (EMI/EMC) tests, total ionizing dose (TID) tests, etc. [5]. This paper focuses only on the test results of the thermal test.

2. AIS receiver design and implementation

In 2016, an AIS receiver has been designed using RFIC Si4362 from Silicon Labs. The AIS receiver prototype was built from COTS components. The components of the AIS receiver prototype are not thermal and vacuum qualified because the receiver was merely built to verify its functional specifications. The functional test is completed by running the device at Tanjung Priok, North Jakarta [6]. The functional test shows that the AIS receiver can receive AIS signals from vessels.

To complete the thermal test, the components of the AIS receiver prototype must be replaced by industrial-grade components that have working temperatures between -40°C and 85°C. Electrolytic capacitors need to be removed and replaced by ceramic capacitors. Every through-hole component, except the SMA socket antenna, was replaced by Surface Mount Device (SMD) components. Figure 1 shows the AIS receiver for thermal testing in this paper.

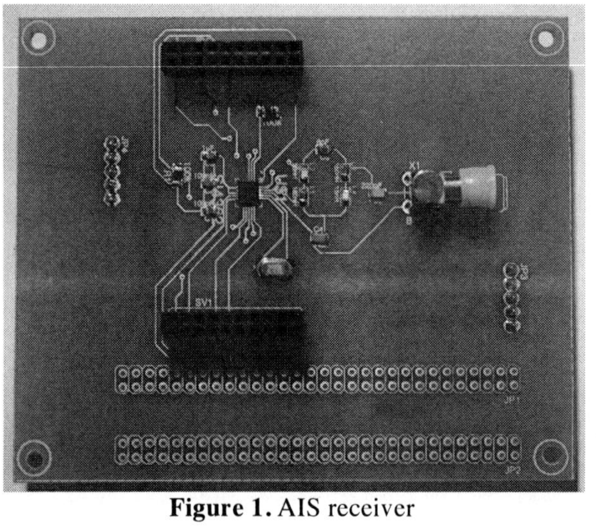

Figure 1. AIS receiver

AIS receivers require a microcontroller to function. MSP430G2552 is selected as the microcontroller and it communicates using I2C protocol. MSP430G2552 is set in a low power state if no data is received by the AIS receiver. The process of receiving and translating data into the AIVDM format is done during the interrupt process by the AIS receiver.

The AIVDM formatted AIS data translated by MSP430G2552 is delivered to CubeSat OBDH, a Raspberry Pi Compute Module (RPI CM) with working capability at -40°C - 85°C. MSP430G2552 is connected to RPI-CM via a USB connection to send AIS RAW data. RPI CM stores AIS data on a file in the internal memory to be further accessible via a Linux-based Raspbian OS.

3. Thermal testing design and implementation AIS

Thermal testing for satellite components must be completed in order to meet the space requirements. Orbit temperatures can reach -40°C at minimum and +85°C at maximum temperatures. For the thermal testing of this AIS receiver, the temperature was set at -20°C until 60°C for one cycle.

The AIS receiver was placed in the thermal chamber with only the antenna cable extending out of the chamber. The antenna cable needs to extend outside the chamber because the VHF antenna used in this test exceeds the size of the chamber. During the test, several data are sent simultaneously by the AIS receiver to the master PC via Bluetooth communication. The HC-05 Bluetooth module was attached to the device using serial communication. Two HC-05 modules were used in this test. The first module was installed for the RPI CM and the second one for the power board.

The thermal chamber that was used in this test is a Votsch VC^2 4018. The chamber is located at the Satellite Technology Center (Pusteksat) of Indonesia's National Institute of Aeronautics and Space (LAPAN). The minimum temperature is set at -20°C and the maximum temperature at 60°C. The minimum and maximum temperatures are maintained for one hour.

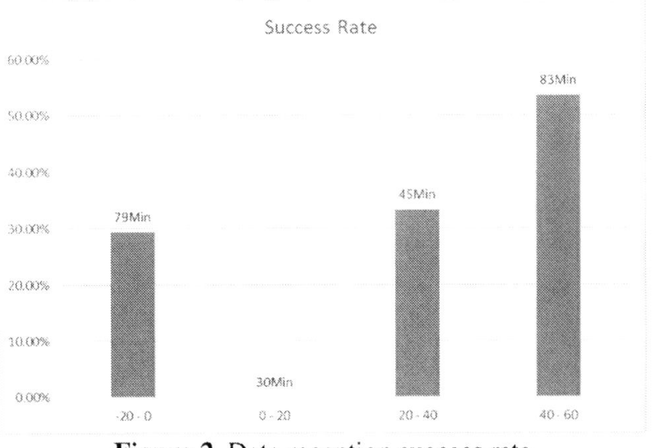

Figure 2. Data reception success rate

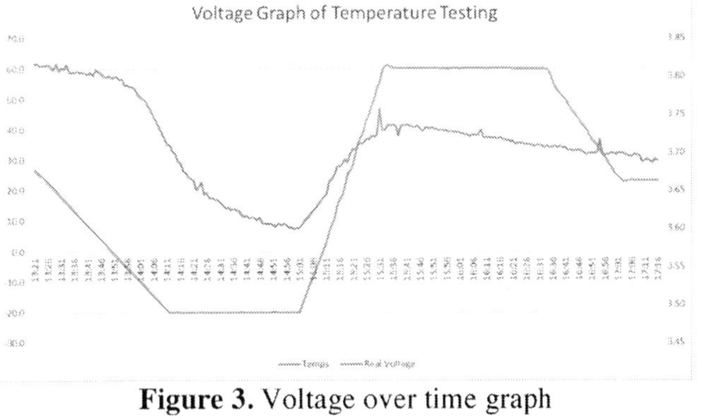

Figure 3. Voltage over time graph

4. Experimental results

The thermal test for the AIS receiver takes 3 hours and 56 minutes for a maximum temperature of 1 hour and a minimum temperature of 1 hour. The RPI-CM that acts as OBDH receives AIS data in AIVDM format and stores it in a file. The RPI-CM is controlled using the HC-05 Bluetooth module and can be monitored when receiving AIS data. The AIS receiver is installed into the chamber with RPI-CM, Power board, and CubeSat structure. Figure 6 shows the device placed into the chamber during the test. After testing, it was observed that no part of the DUT sustained any physical damage.

The success rate of data transmission at a certain temperature interval can be seen in Figure 2. Based on Figure 2, the success rate of the data transmission is influenced by the duration of the test. This happens because the process of sending AIS data static is only done every 3 minutes for AIS class B. AIS class B also has the disadvantage of the absence of a booking time slot after one transmitting signal is finished.

Through figure 2, it can be concluded that at maximum and minimum extreme temperatures, the AIS receiver can still work with data reception rates of 53.37% and 29.41% respectively. These numbers are quite good considering that the number of vessels within the orbiting CubeSat coverage could amount to thousands of ships in a single pass through certain ocean territories [1]. The success of the data reception is seen after the AIS RAW data format has been received by the PC and it is decoded using online an AIS decoder. The correct data shows the data message ID, user ID and coordinate position of the correct AIS transponder.

Figure 3 shows that battery voltage decreases 0.13V in 3 hours and 56 minutes testing. At low temperature, battery voltage dropped below the final battery voltage. The lowest voltage detected was 3.60V at the low temperature. The battery voltage drop occurs due to the chemical characteristics of the battery. At low temperatures, the battery's ohmic resistance increases, this reduces the battery capacity. The utilizing of a battery heater could solve the battery voltage drop at low temperature. The current consumption shows almost stable values and did not have a temperature.

5. Conclusion

This test proves that the AIS receiver can work in extreme temperatures between -20°C and 60°C. The thermal test design in this research could collect various data such as AIS signal, RSSI, humidity, temperature, voltage and current consumption by the device. The AIS receiver has an overall 37.5% success rate in receiving AIS data. At the highest temperature test, the AIS receiver can receive 53.57% and 29.41% at the lowest temperature. The receiver success rate could be enhanced if several transponders are used during the test to increase the AIS signal over time. Creating a customized AIS transmitter could be a perfect solution. It could be designed to transmit AIS signal every second and transmit in two channels of AIS simultaneously. During the test, the current consumption of the device was almost constant at 200mA.

References

[1] Hoye G K Eriksen T Meland B J and Narheim B T 2008 Space-based AIS for global maritime traffic monitoring *Acta Astronautica* **62** 240-245.

[2] Zafrane M A Boudjemai A and Boughanmi N ALCUBESAT-1 2015 *Proceeding of 7th International Conference on Recent Advances in Space Technologies (RAST)* (Istanbul, IEEE) 807-812.

[3] Johl S Lightsey E G Horton S M and Anandayuvaraj G R 2014 *Proceeding of IEEE Aerospace Conference* (Big Sky, IEEE) 1-13.

[4] Sethi A Gajanur N Sadasivan K S Thakurta V Cheela B S and Hosangadi R 2017 *Proceeding of IEEE Aerospace Conference* (Big Sky, IEEE) 1-11.

[5] Kingsbury R Schmidt E Cahoy K Sklair D Blackwell W Osarentin I and Legge R 2013 *Proceeding of Radiation Effects Data Workshop (REDW)* (San Fransisco, IEEE).

[6] Husni E 2016 Design of automatic identification system (AIS) receiver for low earth orbit (LEO) satellite *International Review on Modelling and Simulations (IREMOS)* **9** 435-441.

Design, Analysis and Testing of Monopole Antenna Deployment Mechanism for BIRDS-2 CubeSat Applications

S B M Zaki[1, 2], M H Azami[1,2], BIRDS-2 Project Members[1], T. Yamauchi[1], S Kim[1], H Masui[1], M Cho[1]

[1]Laboratory of Spacecraft Environment Interaction Engineering, Kyushu Institute of Technology, Kitakyushu, Fukuoka, Japan
[2]Faculty of Electrical Engineering, Universiti Teknologi Mara, Shah Alam, Selangor, Malaysia

Email: q595907b@mail.kyutech.jp

Abstract This paper presents an approach to design a feasible and reliable monopole antennae deployment mechanism for CubeSat applications. Nano-satellite has faced problems in designing antennae deployment system due to space and deployment mechanism constraints. In BIRDS-2 CubeSat, the nichrome wire burning release mechanism was used when activated, it will be heated up and thermally cut through fishing line, allowing the two monopole antennas to be deployed on +Y board and at +Z axis direction of the satellite. These antennae involved amateur radio frequencies in UHF and VHF bands due to its low cost and high accessibility by the end users. The UHF frequency band was used for command uplink, mission, telemetry, and CW beacon downlink whilst the VHF frequency band was used specifically for APRS-DP/S&F-RDC mission for both uplink and downlink. In order to achieve the effectiveness of the BIRDS-2 CubeSat communication system agreed with the link budget estimation, the dimensions of antennae on the CubeSat structure depend on the required gain and operating frequency. The paper discusses detailed results of mechanical and electrical interfaces of the two monopole antennas deployment mechanism with the satellite body and the nichrome wire burning release mechanism analysis. The tests results of the mechanism were analyzed particularly on the deployment time and the nichrome wire temperature differences.

1. Introduction

The antenna is an essential part of a nano-satellite. It facilitates communication between the nano-satellite and the ground station. A majority of the nano-satellites launched so far are intended to operate in the amateur VHF (144-146 MHz) and UHF (435-440 MHz) frequency band [1], so as in BIRDS-2 CubeSat. Regardless all the challenges in the 1U dimension limitation, both antennae designs should be implement miniaturization and envelope techniques of deployment mechanism to comply with the constraint. The development of BIRDS-2 antennae deployment mechanism is categorized into Antenna Securing (Stowage) and Antenna Deployment. Antenna deployment subsystem mechanism elements for BIRDS-2 CubeSat are discussed in more detail in the following sections.

2. Mechanism Elements

The mechanism depends on the main part called the secure part, which is a 3D printing material (white plastic) where antennae will be attached to. These are the elements needed for the mechanism's operation which is located at the external configuration of +Y board which includes the secure parts (3D printing material), antennae retention fishing line (dedicated mounting screws) and an antennae deployment detection switch. Figure 1 shows the deployment mechanism's location at the external configuration of +Y board. The burner circuit will be placed at the internal configuration of +Y board as in Figure 2 where the nichrome wire will be penetrated through a dedicated hole and appears at the external configuration of +Y board. The operation of the burner circuit mechanism is very simple.

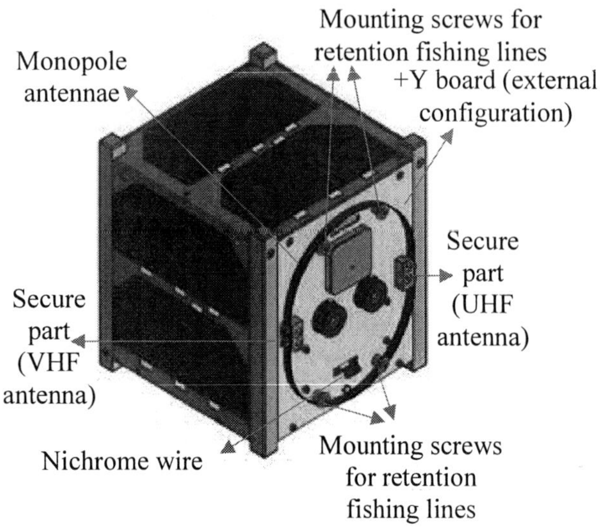

Figure 1 Deployment Mechanism in Stowed Position at External Configuration of BIRDS-2 CubeSat

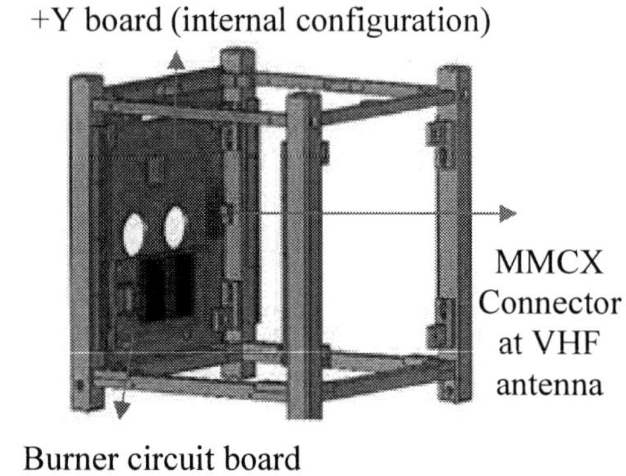

Figure 2 Burner Circuit Mechanism at Internal Configuration of BIRDS-2 CubeSat

The retention fishing lines allow antennae to be secured, retained and keep antennae folded until the time for deployment. At that moment BIRDS-2 CubeSat unregulated power supply is turned on the burner circuit be activated and finally heat up the nichrome wire, cut the fishing line, and therefore deploying antennae, as shown in Figure 3.

2.1. Monopole Antennae Characteristics

The monopole antenna for UHF and VHF frequency band were chosen due to its ease of deployment, omnidirectional radiation pattern and higher gain to establish communication according to link budget analysis. Both UHF and VHF monopole antennae will be folded during the journey to the space and deployed once the satellite is in orbit. The mounting screw has enough force to keep both antennae folded and tied (triple knots) by the retention of a fishing line (Polyethylene (PE)), which will keep it secure. Both monopole antennae have the same width of 4mm and thickness of 0.3mm; with different length according to quarter wavelength calculations. UHF monopole antenna length is 175mm whilst VHF monopole antenna have a length of 501mm. Monopole antennae

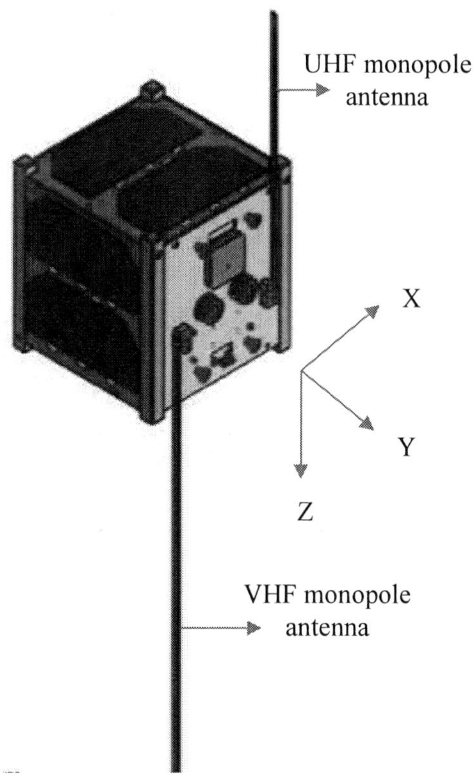

Figure 3 BIRDS-2 CubeSat Structure with Deployed Antennae

are driven using a discrete source with the conducting BIRDS-2 CubeSat body, which acting as the ground plane for both antennae. The proposed material for both monopole antennae is carbon tool steels, SK85 (SK5) [2].

2.2. Retention System

The element used for the retention of the antennae for securing purpose is polyethylene (PE) material, known as fishing line. This type of fishing line can endure a maximum weight of 37kg and unlikely to have a failure of the retention system due to yielding of fishing line. Antennae were initially in the secured position and held at the dedicated mounting screws by using fishing line. In this mechanism, the free end of each antenna was tied with a single fishing line and after securing them in proper tensile [3], it was passed over the nichrome wire coils and fixed to the two dedicated mounting screws as a final point. These mounting screws were provided for the nichrome wire in order to hold the retention of the fishing line while the nichrome wire as in tension in the secured position of antennae. This support not only provides the means to fix the nichrome wire too but also provides a path for the fishing line. The retention of fishing line passes over the nichrome wire in a straight formation so that there will be maximum contact area between the nichrome wire and the fishing line and effective heat can be transferred between them. The current was passed through the nichrome wire and as a result, the nichrome wire heats up which in turn leads to the melting of the fishing line.

2.3. Burning Mechanism

The nichrome burn wire release mechanism uses a nichrome burn wire which when activated, it will be heated up and thermally cut through fishing line, allowing both deployable antennae on the satellite to actuate. The entire burner circuit mechanism will remain inside of except for the nichrome wire which will be outside the satellite. The burner circuit is powered with unregulated supply from EPS and 3.3V signal from (Attitude Determination and Control System) ADCS microcontroller. After 30 minutes releasing from ISS, the EPS turn on the unregulated power. ADCS Microcontroller will detect power of unregulated power line and start send the 3.3V signal to the burner circuit, thereafter heat up the nichrome wire, release the fishing line and deploy both monopole antennae. Both monopole antennae will be deployed on +Y panel and at +Z axis direction of the satellite.

2.3.1. Nichrome wire. The nichrome wire resistance value is depending on the length of the nichrome wire itself. In order to select the correct nichrome wire from those available commercially, many room temperature tests were conducted by using different types of nichrome wire until the suitable length are determined. After rigorous tests, for an input voltage of 3.7V which is the voltage commonly provided by the batteries used in CubeSats, it was decided that the length of the nichrome wire is 10.5cm with a resistance value of 8.2Ω was the best and suitable for the deployment mechanism. The deciding parameters comprised of maximum temperature reached by the nichrome wire, burning time of the fishing line and compatibility with the Overcurrent Protection circuit (OCPC) located at Electrical Power System (EPS) board which limits the current of the maximum of 2.77A.

2.3.2. Burner Circuit Board. The burner circuit consists of a high side Metal Oxide Semiconductor Field Effect Transistor (MOSFET) switch which is designed to protect loads from damage due to a latch up. It can also be used as a controllable switch (signal 3.3V) by the Attitude Control and Determination System (ADCS) Micro Controller Unit (MCU) of the CubeSat to switch off loads in case of an emergency or in the case of the Power Management Algorithm of the CubeSat demands it. The algorithm for the antennae deployment will be programmed in ADCS MCU, depending on parameters of nichrome wire burning time and maximum temperature required to burn up the nichrome wire.

2.3.3. Power Supply from Batteries. After the length of the nichrome wire and its resistance value were determined and the position of antennae was fixed, more than 30 tests were conducted to estimate the required time for deployment. All tests were conducted by using BIRDS-2 batteries which at the worst condition (less than 10% charging capacity). As room temperature tests continued, batteries capacity and supply voltage were monitored. From the results of these tests we detected and observed that the nichrome wire, when subjected to lower voltages than the nominal ones (3.7V), underwent deterioration and a slight variation in their properties when the current was passed through them. This was helpful when determining the point at which the nichrome wire was capable of burning the fishing line in a much colder environment in space and not destroy itself at room temperature. For this, the deterioration was related to the variation in the current flow by maintaining the voltage constant over time. This fact can only be due to the variation in the nichrome wire resistance value over time due to the deterioration through overload. A test is conducted to observe the time taken for antenna deployment by increasing the current supply gradually in order to determine the suitable current supply value. The average value was found to be 4.56 seconds for deployment using 2.75A of current supply. To ensure deployment, the burning time was set to the maximum of 10 seconds and maximum temperature was set to 213°C in the ADCS MCU algorithm with the function to control and activate antennae deployment within its time. This is important in order to prevent batteries from draining out if the nichrome wire continuously keeps burning.

2.3.4. Temperature Sensor. The nichrome wire coils around the retention wire and functions with the help of a burner circuit that is designed to be switched on for a calculated period of time so as to allow current to pass through it and the temperature sensor would correspond to the temperature reached by the nichrome and when the temperature is slightly above the melting point of fishing line, its resistance would be a definite value and the current would provide a definite voltage across it. This circuit has a feedback system that uses temperature sensor which would let the ADCS MCU knows when the temperature reached the optimum level.

3. Monopole Antennae Measurement

The electrical characteristics of an antenna that are of interest to obtain by direct measurement can be categorized into 3 measurements:

3.1. Reflection Coefficient Measurement (Frequency Tuning)

Both UHF/VHF antennae will be connect to Port 1 of calibrated Vector Network Analyzer to be tuned at the desired resonance frequency and S_{11} parameter will be measured. This test was done iteratively by trimming the antenna length till a resonance at the frequency operation for UHF Downlink/Uplink and VHF were obtained [4]. The measurement setup will consider the horizontal and vertical orientation for both UHF/VHF and also for the transmit antenna, in the condition of both antennae is not deploy and deploy. The optimum S_{11} parameter measurement results for both UHF/VHF antennae are as shown in in Table 1. As per the reflection coefficient measurement, for the intended resonance frequency operation for both UHF (downlink and uplink) and VHF, the antenna length of 175mm (for UHF) and 501mm (for VHF) provided the maximum efficiency. The efficiency of the UHF Downlink antenna is 0.08962dB, or 91.04% while for UHF Uplink antenna is 0.0836dB, or 91.64%. Both -10dB bandwidth is 28MHz. The S_{11} characteristics of the VHF monopole antenna shows the efficiency of 0.08974dB, or 91.03%. By referring to above-mentioned efficiency of UHF/VHF antennae, about 9% of the incident power are reflected by antennae and roughly, about 91% of the available power is transmitted into antennae.

Table 1. S_{11} parameter of BIRDS-2 UHF and VHF antennae.

Monopole Antennae	Resonance Frequency, f_r (MHz)	S_{11} (dB)
UHF Downlink	437	-11.158
UHF Uplink	435	-11.962
VHF	145	-11.143

3.2. Gain measurement

The technique used to conduct the monopole antennae measurement is by using gain comparison method; using two antennae with reference using 3 antennae. According to the link budget calculation, the estimation gain of the UHF/VHF monopole antennae should be in the margin of 0 – 2.2 dBi and the desired radiation pattern should be omni-directional pattern. The result of gain of both UHF/VHF antennae calculated is shown as in Table 2. By referring to the gain result in Table 2, all the calculated gain meet the link budget estimation, thus can establish good communication and perform missions successfully from ground station to BIRDS-2 CubeSat and vice versa.

Table 2. Gain of Antennae Under Test (AUT) BIRDS-2 CubeSat.

Type of Antenna	Condition of Antenna	Frequency (MHz)	AUT Gain (dBi)
UHF Downlink	Deploy/Not Deploy	437	4.45/5.15
UHF Uplink	Deploy/Not Deploy	435	6.59/5.39
VHF	Deploy/Not Deploy	145	7.57/4.27

3.3. Antennae Radiation Pattern Measurement Results

The optimum results of radiation pattern is when antennae are both in deployed condition, transmit antenna is in horizontal position and satellite in a vertical position (+Y panel is facing the transmit antenna). From the results of radiation pattern with these conditions, it shows that the monopole antennae perform omni-directional patterns. The 2-D polar plot results are as presented as in Figure 4 (a), (b) and (c). The peak directivity in the azimuthal plane (constant θ = 90°) for UHF Downlink antenna is 4.45dBi while for UHF Uplink antenna is 6.59dBi. Both peaks directivity are highest at the angle of 240° position. From the 2-D polar plot for VHF antenna in Figure 4(c), it can be seen that the VHF antenna has a distorted radiation pattern within the angle of 120° to 180°, and 300° to 330°, with slightly higher directivity which is 7.57dBi at 240° angle position. Besides of the satellite orientation during the radiation pattern measurement, the presence of the metal structure has a significant effect on antennae performance. Therefore, they were neglected as most of the orientation position of the satellite still can give higher gain and omni-directional radiation pattern results.

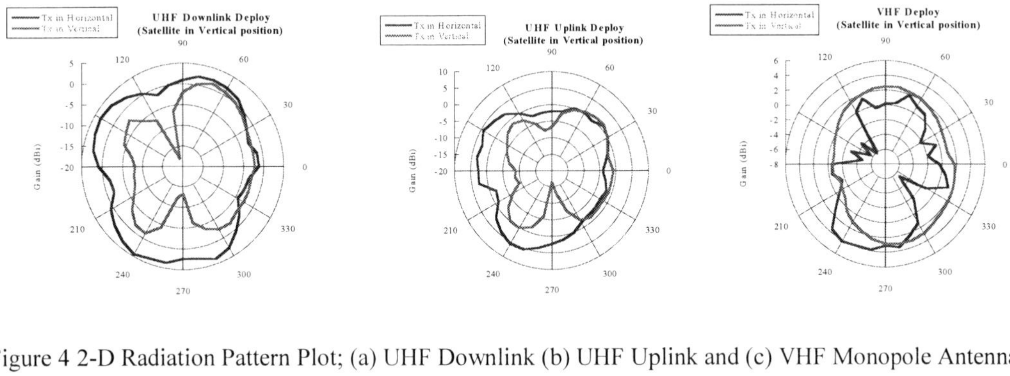

Figure 4 2-D Radiation Pattern Plot; (a) UHF Downlink (b) UHF Uplink and (c) VHF Monopole Antenna

4. Space Environment Test

4.1. Vibration Test

BIRDS-2 CubeSats underwent vibration test utilizing HTV and SpaceX launch vehicle profiles to ensure that it will survive the launch environment and still function afterwards. The satellite will endure vibrations and loads very similar to the ones it will experience on the launch vehicle to ensure that the design is solid and nothing will break. It also verifies that the lowest resonance frequency of the satellite is higher than the exciting frequency during the launch. Specifically, vibration test for antenna configuration on a shaker is mainly to verify the Flight Model (FM) +Y panel structural design and mechanical requirements. The BIRDS-2 CubeSats are tested in Random QT level of 6.526G and Sine burst AT of 18.1G. All BIRDS-2 FM structural vibration test result revealed that both antennae; UHF and VHF did not prematurely deployed as the single fishing line could hold the rolled antennas in a fixed position. Vibrations did not caused the antenna to slip from the mounted screw and the fishing line survive the constant rubbing against the sharp edges of the antenna.

4.2. Cold Environment Test

The antenna deployment test in cold environment in a thermostatic chamber for the BIRDS-2 FM CubeSat demonstrates the ability of the antenna deployment mechanism to meet qualification requirements under low temperature conditions (extreme temperature) which simulates thermal stressing (extreme cold environment) of the BIRDS-2 CubeSat. It was tested in the chamber within the temperature range of -20 to -40°C concerning five units of burner circuits. By referring the Table 3 which showing the result of antenna deployment in cold environment, the nichrome wire is capable to burn the fishing line and antenna can be deployed within 13 to 18 seconds (average) under the extreme low temperature condition with 80% battery charging capacity. ADCS algorithm is verified to be able to turn 'ON' and 'OFF' the 3.3V signal to/from the burner circuit within its time. The antenna deployment time, current consumption (during the nichrome wire is burned), initial and final battery voltage (before and after nichrome wire is burned) are analyzed.

Table 3. Result of Antenna Deployment in Cold Environment.

Burner Circuit Label	Temperature (°C)	Current Consumption (A)	Deployment Time (s)	Battery Voltage (V_i/V_f)
1	-30	2.068	15.77	4.018/3.830
2	-25	2.154	15.93	3.99/3.90
3	-35	2.034	82.22	4.00/3.811
4	-40	2.045	17.75	4.00/3.793
5	-20	2.007	13.21	4.018/3.832

Conclusion

This paper describes the approach on designing the antenna deployment mechanism, analysis and testing of monopole antenna on BIRDS-2 CubeSats. Although it is so attractive due to its simple design, low cost and fast development time, it offers many technical challenges especially on the size, power and communication link budget. Due to this, deployable antennae are becoming more of a necessity because the electrical length at UHF/VHF bands and enhancing gain according to the link budget estimation. The chosen material for the deployable antennae also must be flexible and robust for operation in space. Issues addressed in this paper also proved the rigidity of the mechanical support structures [5], accuracy of the deployment time, stowage of antennae during harsh launch environment, and the ability for the antennae to achieve high gain and omni-directional radiation pattern for successful communication.

Acknowledgement

We would like to thank all the BIRDS partners, especially Center for Satellite Communication UiTM for supporting this development. We also would like to extend our appreciation to BIRDS-1 team members who helped to complete this project.

References

[1] Bouwmeester, Jasper, and J. Guo. "Survey of worldwide pico-and nanosatellite missions, distributions and subsystem technology." Acta Astronautica 67.7-8 (2010): 854-862.
[2] http://www.nagai-giken.com/snonfm18.html
[3] Bharadwaj, B. Kapil, and Yash Vardhan Gupta. "Design, analysis and testing of Antenna Deployment mechanism for CubeSat Applications."
[4] Sadasivan, Kshitij Sandeep, et al. "Design and analysis of antennas for a nano-satellite." Aerospace Conference, 2017 IEEE. IEEE, 2017.
[5] Keny, Pramath, et al. "Development of antenna deployment circuit for nano-satellites." Circuit Theory and Design (ECCTD), 2013 European Conference on. IEEE, 2013.

IOP Conf. Series: Journal of Physics: Conf. Series **1152** (2019) 012008 doi:10.1088/1742-6596/1152/1/012008

BIRDS-2: A Constellation of Joint Global Multi-Nation 1U CubeSats

M H Azami[1, 2], BIRDS Partners, BIRDS-2 Project Members[1], G. Maeda[1], P. Faure[1], T. Yamauchi[1], S. Kim[1], H. Masui[1], and Mengu Cho[1]

[1]Laboratory of Spacecraft Environment Interaction Engineering, Kyushu Institute of Technology, Japan
[2]Faculty of Electrical Engineering, Universiti Teknologi Mara, Shah Alam, Selangor, Malaysia

E-mail: q350939h@mail.kyutech.jp

Abstract BIRDS-2, the second generation of the Joint Global Multi-Nation Birds (JGMNB) project is a constellation of 1U CubeSats from Laboratory of Spacecraft Environment Interaction Engineering (LaSEINE), Kyushu Institute of Technology, Japan. The BIRDS-2 project consists of three identical 1U built by a group of students from four nations; Japan, Malaysia, Philippines, and Bhutan. It consists of 6 missions; Camera, Automatic Radio Packet Service-Digipeater (APRS-DP), Store and Forward (S&F), COTS GPS, Anisotropic Magneto Resistance Magnetometer (AMR-MM), and detection of single-event latch-up (SEL). The objective of BIRDS-2 is to provide an opportunity to learn the entire satellite system cycle, to lay down the foundation of the sustainable space program, and to create international networks of the ground station to assist the infant space program of the participating country. A total of 11 members learns the satellite development in a lean concept of 30m radius of the work area. The paper will describe the bus system of the BIRDS-2 CubeSats and the development process throughout the one year and three months of the project timeline. Completion of BIRDS-2 flight models remarks the successful design of the CubeSats bus system and waiting for the launch and deploy from International Space Station (ISS) around 3rd quarter of this year.

1. Introduction

BIRDS project started in 2015 led by Professor Mengu Cho from Kyushu Institute of Technology, Japan. It begins with a successful constellation of five 1U CubeSats built by registered students from Japan, Ghana, Nigeria, Bangladesh, and Thailand. The prime mission is to "by successfully building and operating the first satellite of the country, make the first step toward indigenous space program" [1]. The BIRDS-1 CubeSats have been deployed from International Space Station (ISS) in July 2017 and still operating ever since then. The legacy of the BIRDS project keeps moving forward to BIRDS-2, BIRDS-3, and so on. By decreasing the development process about one year and three months, it is a very good platform for students while studying at the university.

CubeSat, by its name, means a cubic-like nano-satellite size originated by California Polytechnic State University and Stanford University in 1999. It has a standardized platform of 1U, which is weighing not more than 1.33kg and volume of 10cm cubic. The main objective of CubeSats platform is to help the researchers in universities to perform space exploration in a lean system. Even though CubeSat is considered as miniaturized traditional satellite, but it has been very similar subsystem such as Electric Power System (EPS), Onboard Computer (OBC), Attitude Determination Control (ADCS), Communication subsystem, and also payload subsystem. The payload will define the objective and mission of the CubeSat. Since CubeSats' trend is remarkable growth, the most demand mission payload is by using Commercial off-the-shelf (COTS) components, which are ready –made and available for sale to the general public. The BIRDS-2 CubeSats consist of 6 missions, which suggested by the participating countries at the early stage and have been tested and verified during the development process. We begin the paper with the description of BIRDS-2 missions, the bus system and other subsystems, the development process, and lastly conclusion and future works.

2. Satellite Missions

The BIRDS-2 CubeSats sustain the same ultimate objective of the BIRDS project which is to build and operate the first satellite of the nation [2]. Students who are in the BIRDS-2 members have been given the golden opportunity to learn the complete development process from mission planning to satellite disposal. The constellation of three CubeSats has identical six missions, which are (1) Camera (CAM), (2) APRS Digipeater Demonstration on CubeSat (APRS-DP), (3) Demonstration of a CubeSat's Store-and-Forward (S&F), (4) COTS GPS technology demonstration, (5) Anisotropic Magneto Resistance Magnetometer (AMR-MM), and (6) Single Event latch-up detection (SEL).

Camera (CAM): The mission is divided into three successive levels, which are the minimum success is to capture images of Earth's surfaces, the full success is to take pictures of the four countries in desired high resolution, and the extra success is to take a short video. There are 2 COTS camera modules installed onboard capable of capturing 5 Megapixel (MP) of images, store the image data in flash memory and downlink the data to the ground station.

APRS Digipeater (APRS-DP): The objective of the mission is to demonstrate the functionality of low-cost COTS APRS digipeater payload onboard in 1U CubeSat in space. By using handheld radio of VHF communication, the mission can provide real-time digital communication service to the amateur community and also provide alternative communication services during emergencies's situation.

Store-and-Forward (S&F): The mission is combined with APRS-DP mission, which using the same payload and antenna. The objective is to demonstrate the use of a CubeSat constellation-based store-and-forward system for remote data collection. Ground sensor terminal (GST) is developed by each participating country to support the functionality of the mission (as shown in Figure 1).

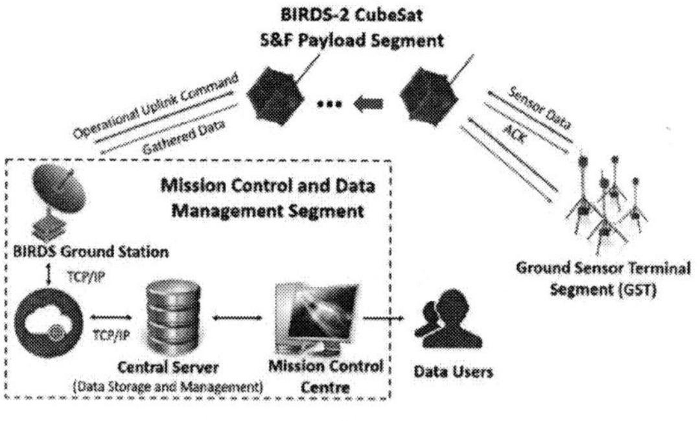

Figure 1. S&F Mission Block Diagram.

COTS GPS: Due to power-hungry of the GPS module, it is very rare to see a 1U CubeSat installed with a GPS onboard. This COTS GPS uses the low power 80mW for the technology demonstration for BIRDS-2 CubeSats. GPS satellite will continuously beam data to CubeSats where COTS GPS onboard is allowing to locate its position wherever it is in orbit.

Anisotropic Magneto Resistance Magnetometer (AMR-MM): A scientific mission to conduct measurement of magnetic field in space. It can provide an alternative method of magnetic-field measurement in space, which works as a support data for geomagnetic mapping of the South Asian region.

Single Event latch-up detection (SEL): A continuous mission from BIRDS-1 for the database on SEL occurrence rate of the microprocessor. The mission objective is to identify any correlation of SEL occurrence with satellite locations, time, and space weather. In the future, we can predict SEL rate in orbit using the database as scientific merit of developing less radiation risk of CubeSat.

3. Satellite Configuration

Three identical BIRDS-2 CubeSats has a 1U design, volume of 10cm cubic and mass of 1kg. The BIRDS-2 architecture is inherited from the BIRDS-1 design which originated from the design of UWE-3 of University Wurtzburg, Germany. It consists of an aluminium alloy skeleton structural, a base bottom board called the backplane board which integrates a battery box and six boards of front access board (FAB), onboard computer and electrical power system board (OBC/EPS), communication 1 & 2 board (COM1 & COM2), mission board, and rear access board (RAB) as shown in Figure 2. Five solar panels are installed at the external side of the CubeSat where another +Y board is attached the antenna burner circuit to hold two folded antennas of UHF and VHF using string wire. By using this configuration, we can minimize the harness inside the CubeSat. This will make a simple design and optimize the interface between all the subsystems onboard.

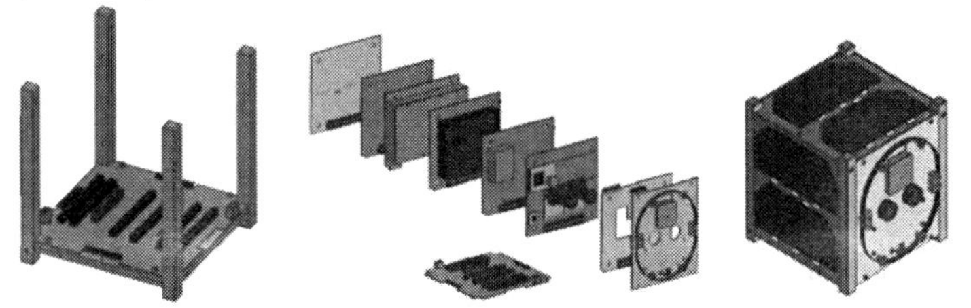

Figure 2. BIRDS-2 Configuration.

The following part will describe more details regarding the BIRDS-2 subsystems, explaining an overview of the block diagram of the satellite. We divided the explanation into one mission payload and other bus systems.

3.1. Mission Payload

A total of six mission payload is installed on mission board except for SEL mission where the micro-controller unit (MCU) is implemented in OBC/EPS board. The schematic and board are designed such that it can fit the mission modules within 90 x 86 mm PCB board. The mission modules include two units of COTS camera modules, one unit of COTS GPS module and one unit of a terminal node controller (TNC) adapter for APRS-DP and S&F mission as can be seen in Figure 3. Further explanation is discussed as follows;

3.1.1 *COTS Camera.* Two identical cameras named *OVCAM* modules are installed onboard, which have the narrow-angle lens and wide-angle lens. Both camera modules are capable of taking an image of 5 MP using the OV5642 sensor. By having the focal length lens of 25mm, the camera can have a spatial resolution of 25m. The camera module is connected to ATmega2560 by using interface of SPI and I2C. It also has a size of 34 x 24 x 30mm, the mass of 20g, and power consumption of 0.9W.

3.1.2 *APRS-DP and S&F.* These two missions are sharing single MCU connects to AX.25 TNC adaptor by UART interface. The BIM1H radio of VHF-TRX is installed on

COM2 board, which connects to the VHF monopole antenna. For the S&F mission, it needs the ground sensor terminal (GST) to communicate with the onboard payload, save the sensor data to the onboard flash memory too and downlink to the BIRDS ground station. On the other hand, the APRS-DP mission needs a minimum of handheld radio, which has APRS functionality to use this application which beneficial to the amateur radio community.

3.1.3 *COTS GPS.* A low-power Venus838FLPx-L using interface of UART is connected to the ADCS MCU. Even though the power consumption is low, the GPS module is turned ON only by an uplink command from the ground station. It will not fully turn ON during satellite operation due to saving the power budget for the other mission. This kind of GPS is the first time used in space, which will demonstrate the functionality of the GPS.

3.1.4 *AMR-MM.* The magnetometer used is HMC5883L commercially sells in the electronics store. This sensor is also connected to the ADCS MCU using interface of I2C. Due to very low power consumption, the sensor will turn ON all the time. The sensor data will be saved into the flash memory of ADCS subsystem.

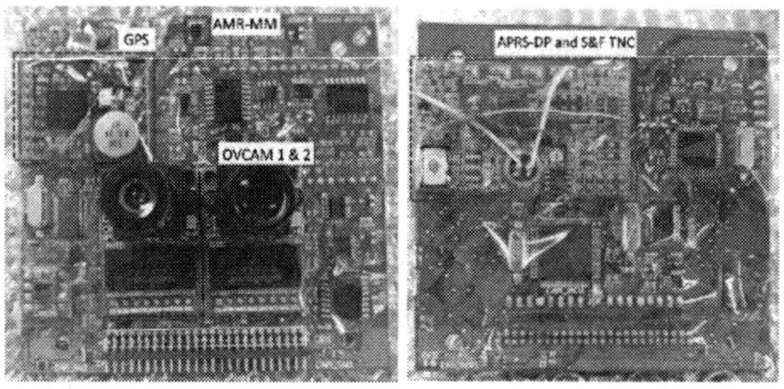

Figure 3. BIRDS-2 Mission Board Front-View (left) and Rear-View (right).

3.2. Onboard Computer and Electrical Power System

The brain and power system of the satellite are integrated into one board. OBC consists of two H8 MCUs for housekeeping data and to control the other MCUs installed onboard. The reason for having two OBC MCUs is because to support the burden of H8-Main functions and for redundancy.

The electrical power system is divided into two parts: solar arrays and battery. Main power generation comes from five units of solar panels which each consists of two series connected with 30% efficiency of Triple junction solar cells during sunlight. Meanwhile, during an eclipse, the power generated from 3S2P Ni-MH Eneloop Plus batteries. The EPS supplies unregulated line, +5V, and +3.3V line to OBC and other subsystems through ON/OFF controlled and overcurrent protected lines. It has three inhibit in the system, two from deployment switches installed below of two rails, and another one is integrated into the battery.

3.3. Attitude Determination Control System

BIRDS-2 CubeSats are applying passive attitude control. Four permanent magnets (ALNICO-5) are glued to each rail, and two hysteresis damper (HYMU80) are glued to the back side of –Z solar panel. These permanent magnets attached to the structure of the satellite are to align the satellite with the Earth's magnetic-field direction, and the dampers will reduce the oscillation and rotational motion of the satellite. By applying this kind of method, we can control the spin rate about the local magnetic direction.

3.4. Communication Subsystem

To communicate between the ground station and satellite, the uplink and downlink communication must be successful work by the transceiver onboard. Using within 435-438 MHz command uplink and a 437 MHz command downlink, the amateur radio bands are used to control and receive the data from the satellite. To receive an uplink command from the ground station, the satellite uses UHF (9600 bps GMSK) with AX.25 protocol. To send telemetry, image, and other mission data to the ground station, the frequency is 437.375 MHz (9600 bps GMSK). The BIRDS-2 have 10 operation network ground stations in different countries (Japan, Malaysia, Philippines, Bhutan, Nigeria, Ghana, Mongolia, Bangladesh, Thailand, and Taiwan) to communicate with the satellite.

4. Development process and testing result

The BIRDS-2 project went through three important reviews; Mission Definition Review (MDR), Preliminary Design Review (PDR), and Critical Design Review (CDR). Started from project kick-off meeting in November 2016 up to finish the flight model (FM) in February 2018, it took over one year and three months of the development process. It can be divided into three major parts where the beginning part of MDR was the mission idea and development stage, the middle part of PDR and CDR were the Engineering Model (EM) development stage, and the last part of FM was the flight model stage and satellite delivery to JAXA. The first stage was to discuss and develop the mission idea based on stakeholder needs and also based on system requirement and modes. The second stage was to build the breadboard model (BBM) for each subsystem and create the engineering model by fabricating the PCB boards. Lastly, the flight model stage was the very important phase to integrate all three identical CubeSats and launch to space.

The satellite was mainly tested with two environment testing, which were thermal vacuum testing (TVT) and vibration testing (VT).

4.1 Thermal Vacuum Testing

There are two main objectives of TVT, firstly, is to meet (function within) functional requirements under vacuum condition and extreme temperature and secondly is to withstand the thermal stressing environment under vacuum condition. The test will simulate the extreme hot and cold temperatures in vacuum condition of the space. In EM phase, the CubeSat was tested from the worst hot of +55 degrees Celsius to the worst cold of -25 degrees Celsius as shown from the result of six thermocouples attached to the external panels (+X, -X, +Y, -Y, +Z, -Z), two thermocouples at the battery box, and a thermocouple at the communication board (COM-1) in Figure 4 below. The graph verified the satellite had been tested on the range of temperature mentioned.

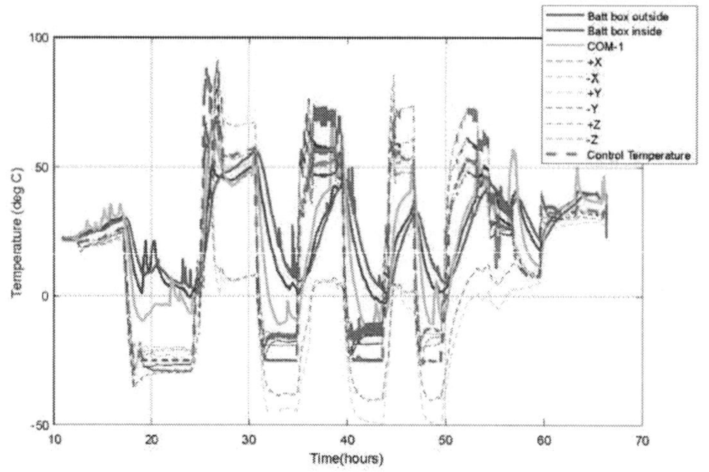

Figure 4. Temperature Difference on External Panels, Battery Box, and COM-1 board.

4.2 Vibration Testing

The objective of this testing is to imitate the environment in the rocket launcher. For BIRDS-2 CubeSats, the vibration testing is using HTV and SpaceX launch vehicle profiles. The CubeSats is tested in Random QT level of 6.526 G, and Sine burst AT of 18.1 G. For instance, the camera subsystem, the critical part to be observed after this test is the camera's lens, whether it is moving or remains fix. We used a lock ring of S- mount lens and attached some glue to strengthen the lens. Torque mark on the screws was remained fixed from the initial position before the testing. We can conclude the result of VT from the Table 1. It was shown that every axis had greater natural frequency than the required 100 Hz.

Table 1. Vibration Testing Result.

Axis	Natural Frequency [Hz]
X	278
Y	231
Z	552

Thus, the results were passed and moreover, the satellite survived the environment in the launch vehicle, and in space based on JAXA Safety Review documents. Table 2 summarizes the environment carried out at each phase of the satellite project. In FM, the environment testing had different parameters due to minimal exposure to the harsh space environment which applied in EM satellite. Random vibration and two cycles were sets as sufficient for the JAXA Safety Review documents.

Table 2. Summary of Environment Test.

	Vibration	Thermal vacuum
EM ver.1	Random (6.526Grms) and Quasi-static (sine-burst) 18.1G	Not tested
EM ver.2	Random (6.526Grms) and Quasi-static (sine-burst) 18.1G	4 cycles -25 ~ +55 degC
FM	Random (4.832Grms)	2 cycles -25 ~ +65 degC

Figure 5 shows the photograph of flight models. The satellites will be delivered to JAXA after going through regulation inspection and will be launched to ISS.

Figure 5. Flight models; Bhutan-1 (left), Maya-1 (middle), and UiTMSAT-1 (right).

5. Conclusion

A BIRDS-2 CubeSats approaches the final development process of the flight model in one year and three months. The satellites have been going through all the testing and verified with the space environment testing. The bus system and the mission payload are tested in integrated of 1U CubeSats. The evaluation of all mission's data downlink after BIRDS-2 CubeSats deployed from the ISS is very useful to complete the verification of the whole project so that it will be also beneficial for the continuity of the BIRDS project in the future.

Acknowledgment

We would like to thank all the BIRDS partners, especially Center for Satellite Communication UiTM for supporting this development. We also would like to extend our appreciation to BIRDS-1 team members who helped to complete this project. A part of the S&F mission and the constellation operation was supported by JSPS Core-to-Core Program, B. Asia-Africa Science Platform.

References

[1] Cho, M., JGMNB project members, N. Kuruhara, "International Network Operations of Five CubeSats Constellation," Conference on Small Satellites, Utah, 2016.

[2] Arifur R. Khan *et al.*, "Five-nations CubeSat Constellation; An inexpensive test case for learning and capacity building," IAA-CU-15-01-16.

[3] Kiril A. Dontchev *et al.*, "M-Cubed: University of Michigan Multipurpose MiniSatellite with Optical Imager Payload", AIAA Paper.

IOP Conf. Series: Journal of Physics: Conf. Series **1152** (2019) 012009 doi:10.1088/1742-6596/1152/1/012009

Northeast Monsoon Effect on Ultra High Frequency (UHF) Signal Attenuation at Kusza Observatory

Nor Hazmin Sabri[1], Sharifah Nurul Aisyah Syed Zafar[2], Roslan Umar[2*] and Roshidah Mat[2] ,

[1]School of Fundamental Science, Universiti Malaysia Terengganu, 21030, Kuala Terengganu, Terengganu

[2]East Coast Environmental Research Institute (ESERI), Universiti Sultan Zainal Abidin, Gong Badak Campus, 21300, Gong Badak, Kuala Terengganu, Terengganu

E-mail: roslan@unisza.edu.my

Abstract. Northeast (NE) monsoon season in Malaysia was developed due to conjunction with cold air outbreaks from Siberia produce heavy rains, strong wind, and high humidity. This weather condition can affect the radio signal transmission which is crucial to spectrum user. This work investigates the effect of the wet season due to Northeast season on radio signal power within Ultra High Frequency (UHF) band at KUSZA Observatory (KO). Three of weather parameters (wind speed, humidity and rain rate) was considered with the radio signal variation. Measurement of these weather parameters and radio signal power level (dBm) were collected using weather station and spectrum analyzer respectively. From the result obtained, we found that wind speed played a role in radio signal attenuation indirectly. The combination of wind speed (km/h) along with high humidity (%) and rain rate(mm/s) during NE monsoon season have given a significant impact to a radio signal. These findings is useful to radio astronomer to determine the best period for radio astronomical observation with minimum interference considering weather condition. This study will also beneficial to spectrum user such as mobile telecommunication, wireless internet and radio broadcasting in managing and improving their services.

1. Introduction

Electromagnetic (EM) waves comprised of radio waves, infrared, visible light, ultraviolet, x-ray and gamma-ray [1]. As noted by John Seybold, 2005, radio waves have wavelengths ranging from 1 millimeter (0.039 in) to 100 kilometers (62 mi) and frequencies from 300GHz to as low as 3kHz[2]. Generally, the radio signal was referred as radio waves that propagate from transmitting antenna to receiving antenna in three identical methods namely as following: line of sight, reflection and refraction [3]. In radio band, International Telecommunication Union, (ITU) have divided the radio spectrum into 12 bands as stated in Table 1 based on their frequencies and wavelength [4]. This frequency band has a different way of radio propagation [5]. UHF is mainly propagated in direct waves or called as Line of Sight (LOS) and ground reflection. Figure 1 illustrates the propagation mechanism of UHF band. Television broadcasting, cell phones, personal radio service (Wi-Fi) and walkie-talkies is the example of UHF band users. The main issue is the radio signal interference that interrupt the transmission of the radio signal. It is crucial to identify factors affecting the radio signal propagation, so the radio signal transmission

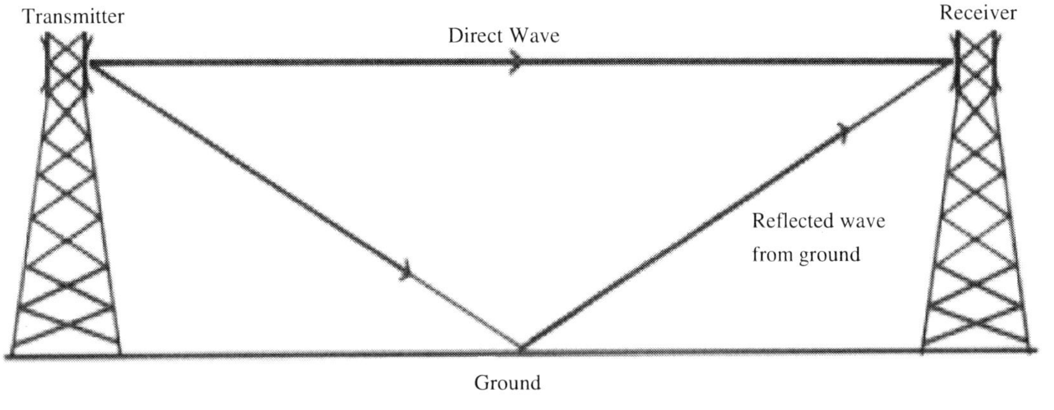

Figure 1. UHF propagation.

can be managed to reduce attenuations and path loss of radio signal. The interference usually comes from man-man interference such as radar, electronic equipment, AM/FM radio and service provider[6]. Apart from man-made interferences, atmospheric condition as temperature, humidity, and wind also can lead to the changes of radio signal propagation [7]. The propagation of radio waves through atmospheres is affected by the atmospheric condition which is come from different weather component. Several studies have been conducted on the effect of weather condition on radio signal by Umar et al. 2015[8]. They studied the effect of rain on radio signal attenuation and found the radio frequency interference (RFI) difference was increased with decreases of rain rate. They have stated that rain gave the significant effect to the attenuation of the radio signal. Apart from rain, the humidity also contributes to the attenuation of the signal. Hamza et al. 2017 has investigated the troposphere parameter (temperature, relative humidity and atmospheric pressure) effect on microwaves link[9]. They found the humidity is the major factor affects the refractivity index of the atmosphere. These high in refractivity causes to the greater attenuation of microwaves. Likewise, Amajama et al. 2016 found that humidity indicates the high correlation with radio signal with a value of r = -0.93ref[10]. Besides, wind speed also plays an important role in losing signal strength. Wind speed is found indirectly affect the radio signal by Meng et al. 2009, wind speed caused the movement of leaves and branches within foliage channel and obstructed the propagation [11]. Thus, attenuated the signal strength. Also, Chua et al. 2010 studies found that wind speed and direction was positively correlated with received signal strength for wireless link research . In this paper, we investigated the effect of NE monsoon on UHF signal attenuation at KO which is located on the east coast of Peninsular Malaysia. Three weather component has been considered namely wind speed, humidity, and rain rate, measured in km/h, % and mm/s respectively. They are believed to contribute to the radio signal power level attenuation. NE monsoon in Malaysia received high precipitation with high rain rate, high humidity and strong wind. The combination of these three weather parameter causes the attenuation of the strength of radio signal. This research will benefit radio astronomer for selecting a site with very minimum RFI by considering weather condition. Besides, this will benefit active spectrum user such as satellite broadcasting, AM, FM & TV transmission, internet service provider to manage and to improve their system performances.

2. Methodology

The 24-hour measurement was done at 21th January 2016 and 3thFebruary 2016 which both were in NE monsoon season. NE Monsoon received high precipitation with heavy rain and strong

Table 1. Designation of radio band by ITU

Radio Band	Frequency range	Wavelength range
Very Low Frequency (VLF)	3-30kHz	10 to 100km
Low Frequency (LF)	30-300kHz	1 to 10km
Medium Frequency (MF)	300-3000kHz	100 to 1000m
High Frequency (HF)	3-30MHz	10 to 100m
Very High Frequency (VHF)	30-300MHz	1 to 10cm
Ultra High Frequency (UHF)	300-3000MHz	100 to 1000cm
Super High Frequency (SHF)	3-30GHz	10 to 100cm
Extremely High Frequency (EHF)	30-300GHz	1 to 10mm
Tremendously High Frequency (THF)	300-3000GHz	0.1 to 1mm

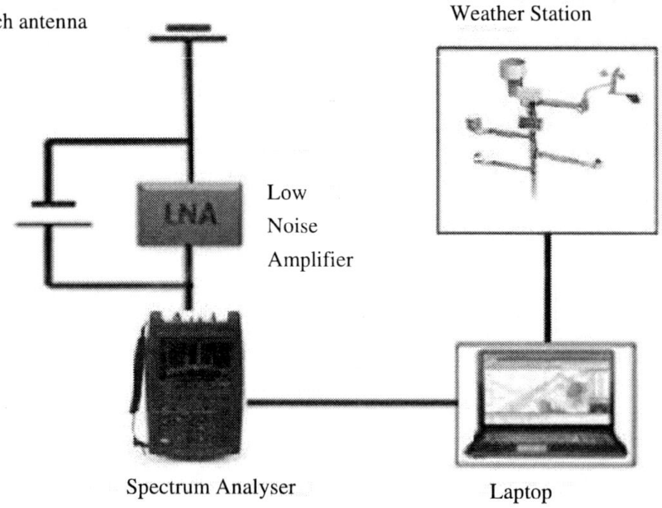

Figure 2. Receiver system of measurement [13]

wind which happened from November to March. These measurements were carried out at KO which located on top of the hill, surrounded with forest, rural area and the nearby South China Sea. This feature of KO allows receiving strong wind, high humidity and far from human-made interferences. Weather parameter was collected using Davis Vantage Pro 2 weather station. At the same time, radio signal power level (dBm) data had also has been collected by a spectrum analyzer. The data was collected for each minute, so the total weather data (wind, humidity and rain rate) were 2880 each. Similarly, the radio signal strength data were 2880 so the plot of wind speed and radio signal power level against time could be determined. The measurement receiver system was comprised of an antenna, low noise amplifier (LNA) and 9 GHz spectrum analyzer (Keysight N99150) connected to the computer. The antenna was kept in vertical polarization with height taller than trees. Figure 2 illustrate the receiver system of this study.

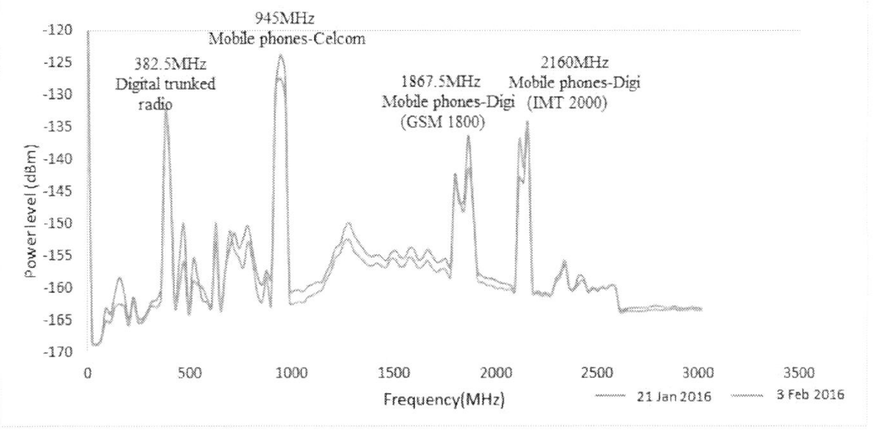

Figure 3. Graph of RFI at frequency up to 3000MHz at KO with 4 interferences peaks and sources were labelled

3. Results and Finding

The radio signal power level detected at KO for all frequencies up to 3GHz is shown in Figure 3. Based on the graph, four (4) prominent peaks were detected. They were peaks at frequencies of 382MHz, 945MHz, 1867.5MHz and 2160MHz. These frequencies are in UHF band and mostly come from mobile phone and digital trunked radio sources. In this study, the effect of wind speed on a radio signal is discussed and also relate to other meteorological factors which are humidity and rain rate. Figure 4 and Figure 5 show the change in signal power level as the wind speed varies with time.

From Figure 4, we can see that the change of power level is not obviously seen by the change of wind speed, while in Figure 5, the drop of signal power level is noticeable around 1.24pm until 5.03pm at 3th February 2016. However, the average of the radio signal power level at each frequency had been plotted to see the overall performance of the signal power level (Figure 6). We can see the average signal power level is lower at 3th February 2016 compared to 21th January 2016. It was corresponds to the value of wind speed, humidity and rain rate for each day. All of the meteorological factors considered were higher at 3th February 2016 as illustrated in Figure 7.

From the plot, the wind speed, humidity and rain rate were 10.74484665 km/h, 2.238948171 % and 75.52820122 mm/s at 3th February 2016 respectively. These values are higher than in 21th January 2016 which are 9.335827897 km/h, 0.390631506 % and 74.44136017 mm/s respectively. These shows that, collectively, the wind speed, humidity and rain rate level give a significant effect on the radio signal strength. However, as discussed by Meng et al. 2009, the radio signal strength was affected by the wind speed indirectly. The analysis suggests that stronger wind speed attenuate the radio signal strength in the presence of high precipitation and high humidity level. It may because of the strong wind has brought voluminous rainwater due to high rain rate and humidity. The water content causes the absorption and scattering on a radio signal. Hence, the radio signal strength interferes. It were concluded that the combination of wind speed, humidity and rain rate affected radio signal in UHF radio propagation.

IOP Conf. Series: Journal of Physics: Conf. Series **1152** (2019) 012009 doi:10.1088/1742-6596/1152/1/012009

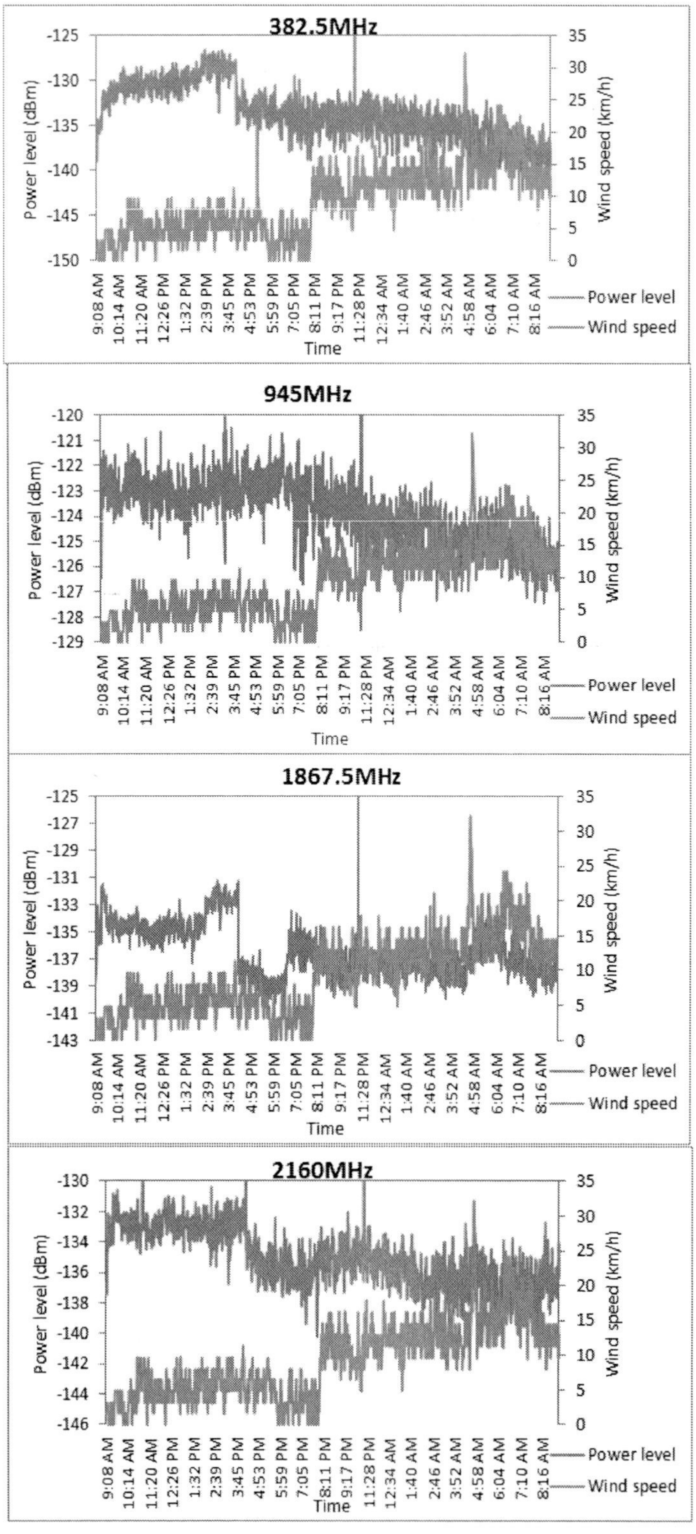

Figure 4. Graph of Wind speed, Power level against time for frequency 382.5MHz, 945MHz, 1867.5MHz and 2160MHz at 21th January 2016

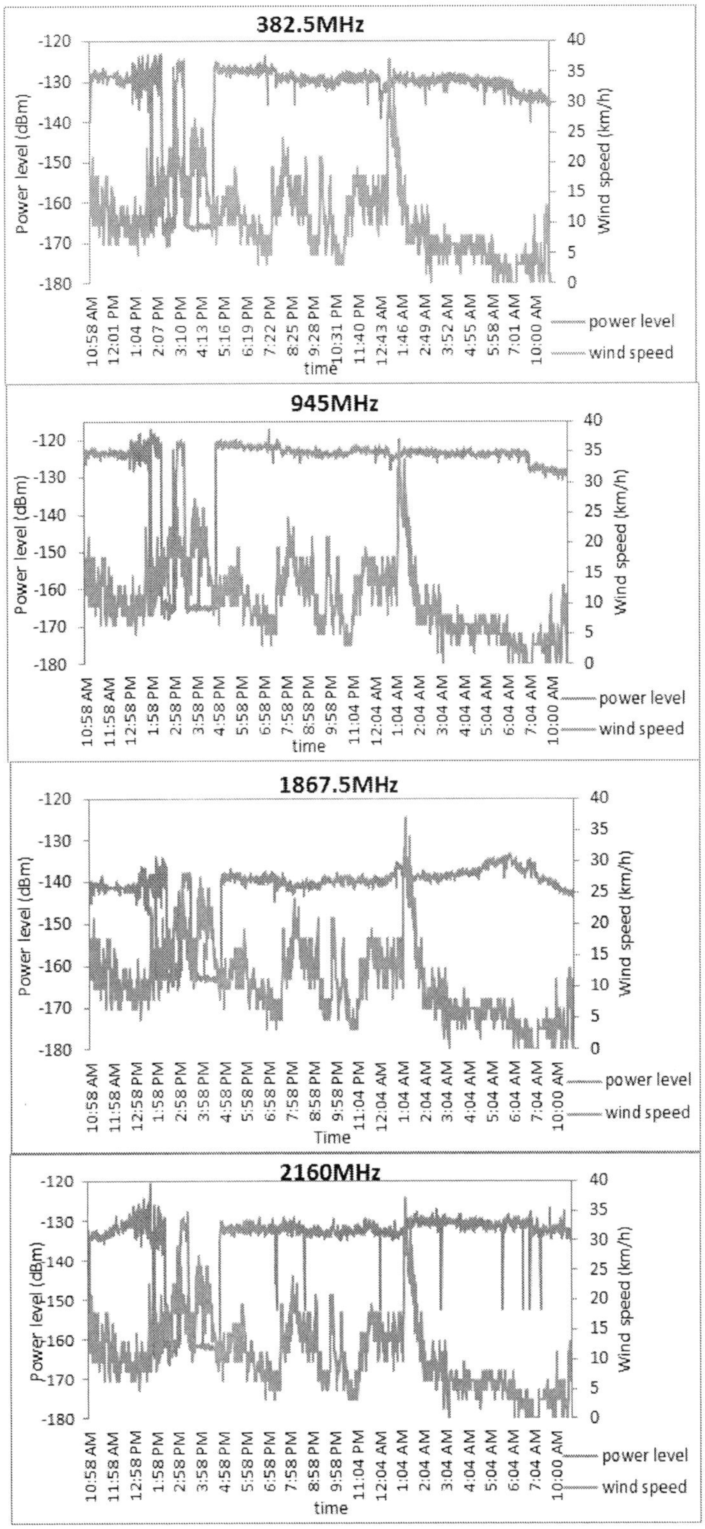

Figure 5. Graph of Wind speed, Power level against time for frequency 382.5MHz, 945MHz, 1867.5MHz and 2160MHz at 3th February 2016

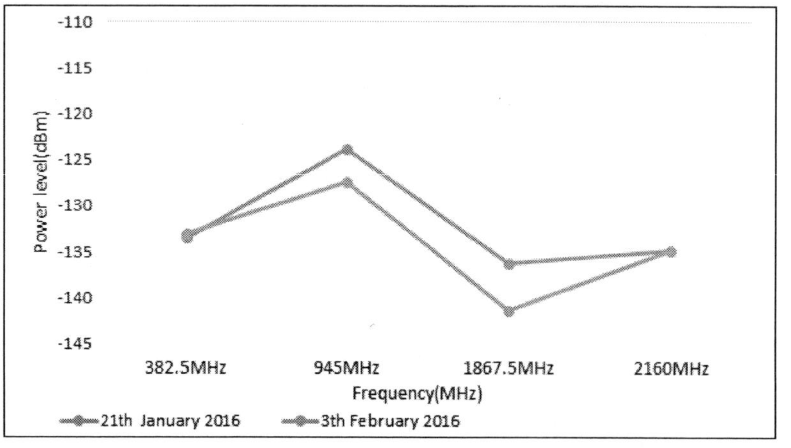

Figure 6. The average of radio signal power level in 21th January 2016 and 3th February 2016 for frequency of 382.5MHz, 945MHz, 1867.5MHz and 2160MHz.

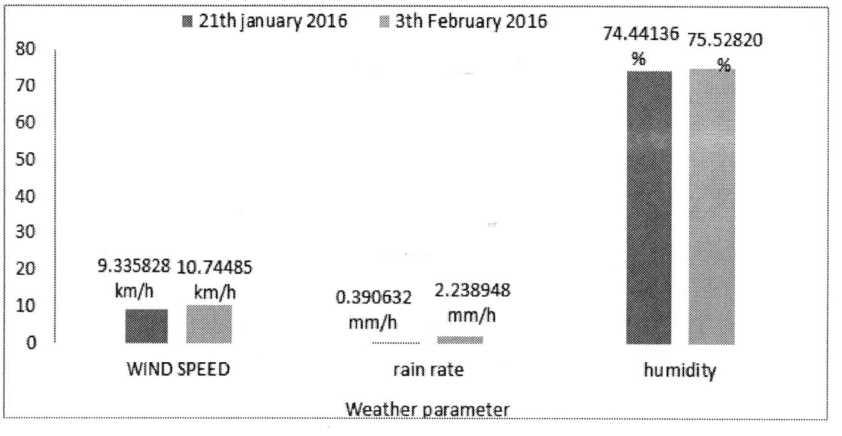

Figure 7. The average values of wind speed, humidity and rain rate in 21th January 2016 and 3th February 2016.

4. Conclusion

In this study, we have presented the effect wet season due to Northeast monsoon in Malaysia on the radio signal power level in UHF band. It was determined that the combination of weather parameters which are wind speed, humidity and rain rate affected radio signal in UHF band. These will benefit radio communication services and another related spectrum user. The data collected for 24 hours each day. Clouds data could be included to investigate the effect on radio signal strength in future work.

References

[1] Sasao, Tetsuo and Fletcher, André B.*Basic knowledge of radio astronomy*.Suwon: Ajou University, 2005
[2] Seybold, John S.*Introduction to RF propagation*.John Wiley & Sons, 2005

[3] Meeks, R and Beafore, FJ. *The science and study of radio wave reflection refraction difraction absorption polarization and scattering,* WS8B, 2011.

[4] Bakshi, Uday A and Bakshi, Ajay V. *Antenna and wave propagation.* Technical Publications, 2009.

[5] Haslett, Christopher. *Essentials of radio wave propagation.* Cambridge University Press, 2008.

[6] Umar et al. 2013. The Study of Radio Frequency Interference (RFI) in Altitude Effect on Radio Astronomy In Malaysia And Thailand. *Middle East Journal of Scientific Research, 14* (6), 861–866.

[7] Oluwole, Famoriji J and Olayinka, Oyeleye M. 2013. A test of the relationship between refractivity and radio signal propagation for particulates. *Research desk, 2*(4), 334-338.

[8] Umar et al.2015. Radio frequency interference: The study of rain effect on radio signal attenuation. *Malaysian Journal of Analytical Sciences, 19*(5), 1093-1098.

[9] Hamza, Mohd Ghazali and Latiff, Liza A and Mustafa, Eltayeb M and Gasim, Waleed M.2017. *Radio propagation and troposphere parameters effect for Microwave links in Sudan. Communication, Control, Computing and Electronics Engineering (ICCCCEE), 2017 International Conference on,* on (pp. 1-5). IEEE.

[10] Joseph, Amajama.2016. Force of Atmospheric Humidity on (UHF) Radio Signal. *International Journal of Scientific Research Engineering & Technology, 2,* 56-59.

[11] Meng, Yu Song and Lee, Yee Hui and Ng, Boon Chong.2009.The effects of tropical weather on radio-wave propagation over foliage channel. *IEEE Transactions on Vehicular Technology, 58*(8), 4023-4030.

[12] Chua, TH and Wassell, IJ and Rahman, T Abd.2010. Combined effects of wind speed and wind direction on received signal strength in foliated broadband fixed wireless links. *Antennas and Propagation (EuCAP), 2010 Proceedings of the Fourth European Conference on* (pp. 1-5). IEEE.

[13] Sabri NH, Syed Zafar SNA, Umar R, Mokhtar WZAW. 2015 .Radio Frequency Interference: The Effect of Ambient Carbon Dioxide (CO2) Concentration on Radio Signal for Radio Astronomy Purposes. *Malaysian Journal of Analytical Sciences, 19*(5), 1065-1071.

[14] Sabri NH, Atiq WA, Umar R, Shahirah SS, Ibrahim ZA, Mokhtar WZAW. 2015. The Effect of Solar Radiation on Radio Signal for Radio Astronomy Purposes. *Malaysian Journal of Analytical Sciences, 19*(6), 1374-1381.

The First Solar Flare Effect Observation at UiTM-SID Station during Solar Cycle 23-24

N A Zakaria[1], N N Anuar[1], S A E Abdul Rahim[1], W Z A Wan Mokhtar[2], M H Jusoh[1]

[1]Applied Electromagnetic Research Group (AERG), Faculty of Electrical Engineering, Universiti Teknologi MARA, 40450, Shah Alam, Selangor, Malaysia
[2]Department of Physics, Faculty of Science and Mathematics, Universiti Pendidikan Sultan Idris, 35900, Tanjung, Malim, Perak, Malaysia.

E-mail: huzaimy@salam.uitm.edu.my

Abstract. International Heliophysical Year (IHY; 2007-2009) has a remarkable impact on space and earth electromagnetism research and society. For Malaysian perspectives, it acts as a spiral point to the expanding of space weather-related research. The research activities involve installation of real-time Magnetic Data Acquisition System of Circum-pan Pacific Magnetometer Network, i.e. MAGDAS/CPMN for space weather study and application, which was deployed for the IHY; 2007-2009. In addition to MAGDAS system, to understand the characteristics of the lower ionospheric layer (60-150km) during space weather events, a Very Low Frequency (VLF) monitoring system has been developed. The MAGDAS and VLF monitoring system called UiTM-SID are located at Universiti Pendidikan Sultan Idris, UPSI (3.71', 101.53'), Malaysia. The first observation effects of solar flare during November 2016 to March 2017, for the period of a minimum solar cycle 23-24 have been examined on sub-ionospheric VLF signals from VTX4 (19.2 kHz) and NWC (19.8 kHz) transmitter monitored at UPSI. In general, the level of amplitude reading for VTX4 signals is higher than NWC signals for quite diurnal variations. The effects of the solar flare class C7.5 on 29th November 2016 shows an enhancement of amplitude due to increase in the lower region electron density by the radiation of solar flare. The average daily amplitude reading was compared to the average value of solar flare event, the result shows the percentage difference is 12% increased in VTX4 and NWC signals.

1. Introduction

Solar Flare is a sudden intense brightness seen at the Sun. The flares occur when speed charged particles collide with plasma in the corona and chromospheres. It converts magnetic energy to kinetic energy in the form of solar energetic particles. The unconnected magnetic field can expand to all direction including to the Earth. Thus, it shows that a solar flare erupts from the active regions on the Sun when the magnetic fields are much stronger than the average [1].

Solar flares cause perturbation in the daytime ionosphere and D region is perturbed the most. During daytime, the lower ionosphere D-region ionization is maintained by Lyman-α (121.6 nm) radiation from the Sun. It partially ionizes the nitric oxide, N_2O at ~70km altitude, a minor neutral constituent in the D-region. Under normal conditions, while there is no space weather event, the solar X-ray flux is too small to be an important source of ionization. However, during solar flares event, the increased X-ray flux from the Sun in the wavelength range 0.2 – 0.8 nm ionizes neutral constituents along with the major O_2 and N_2 species. This modifies the electron density in the D-region ionosphere and changes the transmission conditions of Very Low Frequency (VLF; 3 - 30 kHz) waves in the Earth-Ionosphere waveguide formed between the D-region ionosphere and the Earth's surface [2].

There are studies [3,4,5,6,] focusing the various changes in lower D region caused by solar flare using VLF waves, transmitted by the navigational transmitter station. Apart from that, there have been studies mainly focused on the comparative investigations of the changes in the VLF signal amplitude/ phase and time delay with respect to Solar X-ray flux. Other than that, there are studies on the D-region solar flare effects using Long Wave Propagation Capability (LWPC) code to estimate D-region reflection height *(H')* in km and D-region electron density gradient or sharpness factor (β) in km^{-1} established by Wait's parameters [7].

The present study deals with the D-region ionospheric perturbations caused by solar flares on 29th November 2016 on VTX4 and NWC transmitter signal 19.2 kHz and 19.8 kHz respectively recorded at Universiti Pendidikan Sultan Idris (UPSI), Malaysia (3.71', 101.53'). Solar flare effects on the horizontal component (*H*) of the Earth's magnetic field at the same location have also been studied from the same period at the solar cycle 23-24.

2. Experimental setup and data

A VLF receiver system is known as UiTM Sudden Ionospheric Disturbance (UiTM-SID) and Magnetic Data Acquisition System (MAGDAS) were installed at UPSI, Malaysia (3.71', 101.53'). The UiTM-SID system is used to record 19.2 kHz and 19.8 kHz VLF signals transmitted by South Vijayanarayanam, India and North West Cape, Australia (NWC) transmitter respectively. Figure 1 shows a map of transmitter station (indicates as a star) and receiver station (indicates as a circle) over a Great Circle Path approximately ~3000 km.

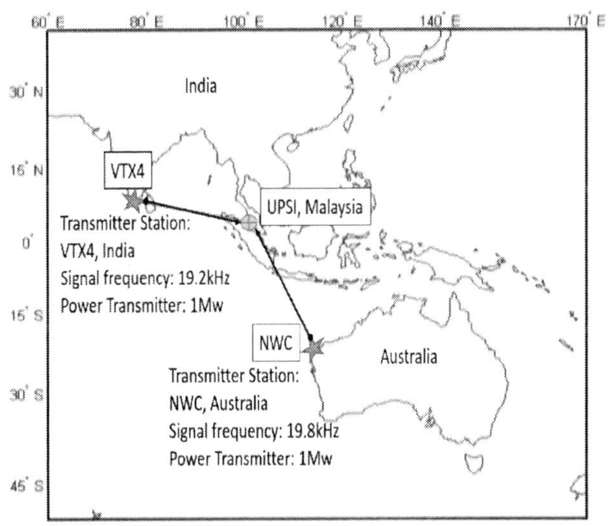

Figure 1. A map of VLF receiver system at Universiti Pendidikan Sultan Idris (UPSI). The system receiving signals from VTX4, India and NWC, Australia correspondinglyover a Great Circle Path.

The data of space weather were collected from November 2016 to March 2017. The daily average values of the amplitudes gained from UiTM-SID system have been utilized to display the typical changes in amplitudes VLF due to the flares. The daily quiet days were chosen based on Kp Index <5 and DST Index > −30 nT. Kp Index is known as geomagnetic index which indicates the geomagnetic disturbance in horizontal component of Earth's geomagnetic field that has a value between 0 to 9. The value from 0 to 5 is a quiet conditions while the value above 5 is represent geomagnetic storm [8]. Next parameter to indicate quiet condition is by observing DST Index. DST index is a measure of the strength of a magnetic storm that reflects the variations in the intensity of the symmetric part of the ring currents at the altitude ranging from 3 to 8 radii. The DST index is divided into four classes: DST $\leq$ -100 nT is intense, -100 nT < DST $\leq$ -50 nT is moderate, -50 nT < DST $\leq$ -30 nT is small storm and DST > -30 nT is no storm type correspondingly.

From the average quiet day amplitude value, the data were compared to the average amplitude value during solar flare event.The data that considered in this study are, a) DST Index, b) Kp Index and c) Amplitude enhancement during Solar Flare.

3. Observations

3.1. Quiet Day

Figure 2 (a) shows DST Index on 6 quiet days. The selected days are 27th November 2016, 28th November 2016, 3rd December 2016, 4th December 2016, 24th March 2017 and 25th March 2017. The DST Index on six quiet days is in a range of -23<DST>18. Figure 2 (b) presents Kp Index with range 0 <Kp Index> 2. The quietest are on 3rd and 4th December 2016 recorded a 0 Kp Index value.

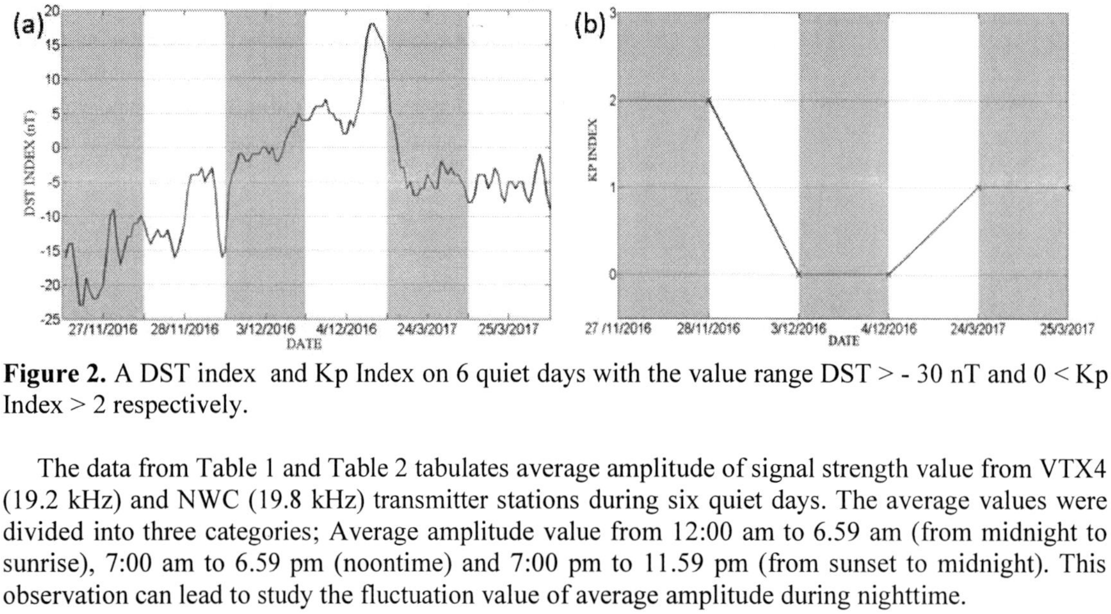

Figure 2. A DST index and Kp Index on 6 quiet days with the value range DST > - 30 nT and 0 < Kp Index > 2 respectively.

The data from Table 1 and Table 2 tabulates average amplitude of signal strength value from VTX4 (19.2 kHz) and NWC (19.8 kHz) transmitter stations during six quiet days. The average values were divided into three categories; Average amplitude value from 12:00 am to 6.59 am (from midnight to sunrise), 7:00 am to 6.59 pm (noontime) and 7:00 pm to 11.59 pm (from sunset to midnight). This observation can lead to study the fluctuation value of average amplitude during nighttime.

Table 1: Data of average amplitude value of the signal frequency 19.2 kHz transmitted from VTX4, India Station.

Date	Ave Amplitude 12:00 am – 6.59 am	Ave Amplitude 7:00 am – 6.59 pm	Ave Amplitude 7:00 pm – 11.59 pm
27 NOV 2016	174.8	170.21	176.15
28 NOV 2016	182.98	172.41	182.94
03 DEC 2016	182.12	147.53	145.40
04 DEC 2016	151.69	140	146.87
24 MAR 2017	77.14	85.34	68.76
25 MAR 2017	138.22	149.20	125.65

Table 2: Data of average amplitude value of the signal frequency 19.8 kHz from NWC, Australia Station.

Date	Ave Amplitude 12:00 am – 6.59 am	Ave Amplitude 7:00 am – 6.59 pm	Ave Amplitude 7:00 pm – 11.59 pm
27 NOV 2016	89.24	166.84	164.76
28 NOV 2016	182.98	172.4	182.94
03 DEC 2016	164.86	145.93	134.20
04 DEC 2016	137.46	138.37	128.49
24 MAR 2017	103.94	105.77	102.81
25 MAR 2017	101.47	107.9	106.10

3.2. Solar Flare Events

On 29th November 2016, the Sun unleashed C-class of solar flare at 15.03 Local Time. The event of the solar flare was detected by VLF receiver system at UPSI station. The data also was a good conformity with data reported by U.S. Department of Commerce, NOAA, Space Weather Prediction Center. Table 3 displays the average amplitude value during solar flare events from the transmitted signals station VTX4, India and NWC, Australia.

Table 3: Data of average amplitude value during solar flare event detected at UPSI, Malaysia receiver station on 29th November 2016.

Receiver Stations	Ave Amplitude 12:00 am – 6.59 am	Ave Amplitude 7:00 am – 6.59 pm	Ave Amplitude 7:00 pm – 11.59 pm
VTX4	200.84	181.94	176.94
NWC	200.84	179.18	165.75

4. Results and Discussions

During solar flare events, the result shows increased in average amplitude value. For the average signal strength from VTX4, India, the value is 181.94 and the average amplitude value for the signal from NWC, Australia station us 179.18 as stated in Table 4. The average percentage difference calculates that the value increased by 12% during solar flare events.

Table 4: The average amplitude value of VLF signals during quiet days and on the occurrence of solar flares.

Receiver Stations	Average Quiet Days	Average Solar Flare Days
VTX4	144.12	181.94
NWC	139.54	179.18

A sudden, rapid and intense phenomenon in solar activity releases a vast amount of energy up to 10^{25} Joules at the solar atmosphere. During the event of solar flares, it becomes the dominant source of ionization in the ionosphere. The X-ray flux from the Sun with a wavelength below 1 nm is capable to penetrate down to the lower ionosphere D-region and results in higher additional ionization at that particular layer. Thus, the amplitude reading of VLF signals value is higher because the signals are reflected at the lower layer. The rapid increase in the electron density at the ionosphere leads to several phenomena grouped together and known as Sudden Ionospheric Disturbance (SID) [10].

5. Conclusions

The objective of this paper was to analyze the amplitude data acquired by monitoring at UPSI, Malaysia site VLF radio signals emitted by two VLF transmitters; VTX4, India (19.2 kHz) and NWC, Australia (19.8 kHz) correspondingly during solar flare event. After the analysis was made, it is proved that solar flare event gives notable impact to VLF receiver system at UPSI, Malaysia with 12% increased in amplitude for the first solar flare detection (Class C-7.5). Apart from that, the occurrence of solar flare leads to further understanding of a lower region of ionosphere typically in Malaysian region.

Acknowledgment

The authors recognize Stanford Solar Center, Universiti Pendidikan Sultan Idris (UPSI), and University Malaya (UM) for given that technical support, data, and location. Above and beyond, highest gratitude sent to UniversitiTeknologi MARA for the management support and consideration throughout this project. This research financially supported by Fundamental Research Grant Scheme, FRGS (Code No. 600-RMI/FRGS 5/3 140/2014).

References

[1] D. Ruffolo, P. Tooprakai, M. Rujiwarodom, T. Khumlumlert, M. Wechakama, J. W. Bieber, P. Evenson, and R. Pyle 2006 Relativistic Solar Protons on 1989 October 22: Injection and Transport along Both Legs of a Closed Interplanetary Magnetic Loop *Astrophys. J.* **639** 1186-1205

[2] Thomson, N.R., Rodger, C.J., Clilverd, M.A. 2005 Large Solar Flares and their Ionosperic D Region Enhancements *J. Geophys. Res.* **110** 1-10

[3] Ananthakrishnan, S., Abdu, M.A., Piazza, L.R. 1973 D-region Recombination Coefficients and the Short Wavelengths X-ray Flux During a solar flare *Planet. Space Sci.* **21** 367–375

[4] Thomson, N.R., Clilverd, M.A. 2001 Solar flare Induced Ionospheric D-region Enhancements from VLF Amplitude Observations *J. Atmos. Solar Terr. Phys.* **63** 1729–1737.

[5] Singh, A.K., Singh, A.K., Singh, R., Singh, R.P. 2014 Solar flare Induced D-region Ionospheric Perturbations Evaluated from VLF Measurements *Astrophys. Space Sci.* **350** 1–9.

[6] Zigman, V., Grubor, D., Suli´c, D. 2007 D-region Electron Density Evaluated from VLF Amplitude during X-ray solar flares. J. Atmos. Solar Terr. Phys. **69** 775–792.

[7] Wait, J.R., Spies, K.P., 1964. Characteristics of the earth ionosphere wave guide for VLF radio waves. National Bureau of Standards, NBS technical note U.S. Government printing office.

[8] Ghanbari, K., Ghomi, M.K., Mohammadi, M., Marbouti, M., Tan, L.M., Modelling the Variations of Reflection Coefficiemt of Earth's Lower Ionosphere Using Very Low Frequency Radio Wave Data by Artificial Neural Network *Adv. Space Res.*

[9] Anh, Vo, Wanliss, James Watson, Stephen, & Yu, Zu-Guo 2006 Chaos Game Representation of The Dst Index and Prediction of Geomagnetic Storm Events *Chaos, Solitons & Fractals* **31** 736-746.

[10] D.M. Sulic, V.A. Sreckovic and A.A. Mihajlov 2016 A study of VLF Signals Variations Associated with The Changes of Ionization Level in The D-region in Consequence of Solar Conditions *Adv. Space Res.* **57** 1029-1043.

IOP Conf. Series: Journal of Physics: Conf. Series **1152** (2019) 012011 doi:10.1088/1742-6596/1152/1/012011

Comparison subpolar mesopause OH(6-2) temperature over Yakutia with the calculated data of NRLMSISE-00 model for 1999-2015

A.M. Ammosova, G.A. Gavrilyeva, P.P. Ammosov, I.I. Koltovskoi, V.I. Sivtseva

Shafer Institute of Cosmophysical Research and Aeronomy SB RAS

e-mail: ammosovaam@mail.ru

Abstract. The analysis of changes in temperature of mesopause region based on fluctuation measurements of rotational temperature and intensity of molecular emission of hydroxyl OH (6-2) excited at ~87 km altitude is presented. Observations were carried out at the Maimaga station (63°N, 129.5°E), at a distance of 120 km to the north from Yakutsk. Measurements were conducted with a digital infrared spectrograph. The temperature was determined from the distribution of the emission intensity in the different branches of the molecular hydroxyl band. Comparison OH(6-2) temperature with the calculated data of NRLMSISE-00 model was defined based on the data received from 1999 to 2015. 2320 midnight temperature values of hydroxyl OH (6-2) suitable for comparison were obtained at the Maimaga station over 16 years of measurements. The NRLMSISE-00 model describes temperature changes at the height of the mesopause region from October to April within the experimental fluctuations with a seasonal temperature variation of ~ 35 K, deviations from the experimental values are no more than 7 ± 4 K. A correlation analysis was performed to compare the rotational temperature OH (6-2) and the NRLMSISE-00 model calculations. When the number of observations at the station Maimaga more than 180 days (about six months) correlation coefficient R> 0.7. This means that the NRLMSISE-00 model accurately describes the temperature changes of the subauroral mesopause.

1. Introduction

The optical remote measurement of the temperature of the upper atmosphere is based on a ground-based recording of the intensity of the intrinsic radiation of the night sky. The rotational temperature, determined from the intensity distribution of the hydroxyl (OH) band, corresponds to the kinetic temperature of the neutral gas at the radiation height. The radiating layer of OH according to numerous rocket measurements is quite thin (7-10 km) and is located at the level of the mesopause (about 87 km). The mesopause region is a boundary region (80-100 km) between the mesosphere and the thermosphere, in which the temperature minimum is located. Despite variations in the height of the emission layer, this method of measuring the temperature of the mesopause region is the most convenient way of studying seasonal variations, the response of temperature to solar activity, and the long-term changes in the temperature of the upper atmosphere.

Various empirical atmospheric models, such as MSIS-90, have been constructed based on experimental data obtained by both terrestrial and satellite methods. Comparison with measurements taken at different latitudes, even after their creation, shows that there are significant deviations in the behavior of the temperature regime of the mesopause region from model ones [1].

The aim of this paper is to compare the temperature of the subpolar mesopause with the empirical model NRLMSISE-00 calculations. The study is based on the mesopause temperature data during 16-year observations of OH hydroxyl radiation (6-2) obtained at Maimaga station (63°N, 129.5°E).

2. Infrared digital spectrograph and data processing method

The results of the analysis of the mesopause temperature data obtained by an infrared digital spectrograph (ICS) made based on the SP–50 spectrograph are presented. The device is installed at the optical station Maimaga (63°N, 129.5°E), which is located 120 km north of Yakutsk. Observations were made at night, at an angle of immersion of the Sun> 9° in cloudless weather. For the analysis, the data obtained in the moonless time and in the absence of auroras were selected. From 1999 to the present, a digital CCD camera is used to record the spectrum at the output of the spectrograph, which makes it possible to use the whole spectrum for the estimation of the rotational temperature [2]. The method for estimating the rotational temperature of molecular emission is based on fitting the model spectra constructed with allowance for the apparatus function of the instrument for various predetermined temperatures to the actually measured spectrum by the method of least squares. The model spectrum whose deviation from the real does not exceed the recording noise is considered the most appropriate one and that the rotational temperature of the hydroxyl molecule, which is determined from the spectrum, corresponds to the temperature at the height of the mesopause. As estimates show, random errors in temperature measurement lie within the range of 2-5 K, depending on the level of the signal-to-noise ratio. The infrared spectrograph is calibrated twice a year using a gas discharge lamp (Ne).

Recording of the spectrum of the night sky is conducted from the middle of August to the middle of May. The longest night rows of data are recorded in the winter months. The number of measurements per month varies from 10 to 25 nights. To compare the temperature of the subauroral mesopause with empirical models, the mean nighttime temperatures of the hydroxyl were calculated. In total, from August 1999 to April 2015, there were 2320 average night values of OH (6-2) meeting the above sampling criteria. They are shown in Figure 1.

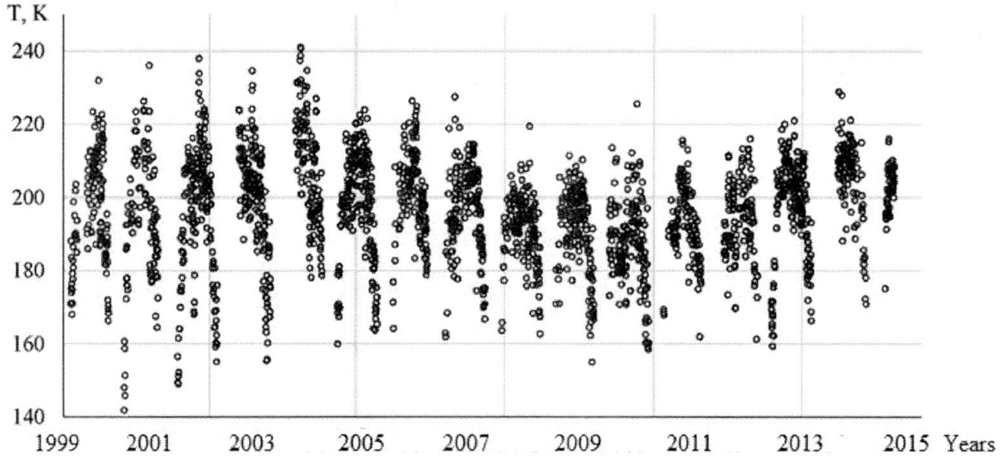

Figure 1. The OH (6-2) temperature series obtained at the Maimaga station from 1999 to 2015.

3. Comparison with the empirical NRLMSIS-00 model

To calculate the temperature and chemical composition of the atmosphere, the MSIS (Mass-Spectrometer-Incoherent-Scatter) model is widely used. The first version of this model appeared in 1977. As a result, of the accumulation of experimental atmospheric data updated versions of this model were created: MSIS-83, MSIS-86, MSIS-90, NRLMSISE-00 [3]. At the moment, the latest version of the model is NRLMSISE-00 based on the MSISE90 model. NRLMSISE-00 uses rocket, satellite and non-coherent radar data. The model calculates the concentrations of He, O, N2, O_2, Ar, H, N, total mass density, neutral temperature at altitudes from the Earth's surface to 1000 km.

The model uses the temperature profile, as a function of the geopotential height for the upper thermosphere, for the lower thermosphere is the inverse polynomial of the same variable. Atmospheric characteristics are expressed as functions of geographic and solar / magnetic parameters. To describe the main variations of the atmosphere, including latitudinal, annual, and semiannual and longitude, spherical harmonics of low orders are used.

MSIS-90 and NRLMSISE-00 models allows calculating the parameters of a neutral medium in magnetic storms. The NRLMSISE-00 model makes it possible to calculate the concentration of excited atomic oxygen. The input parameters of the NRLMSISE-00 model are the day of the year, altitude, latitude and longitude, UT local or global time. Also this model contains averaged solar activity index F10.7, index F10.7 on the day preceding the simulation day, a set of magnetic index values Ap. More precisely the average values of the index Ap on the simulation day, the three-hour values for 3, 6, 9 hours before the time of calculating the model values, as well as the average values of the index from 12 to 33 hours and from 36 to 57 hours to the specified time.

For comparison with the hydroxyl temperature OH(6-2) obtained at Maimaga station (63°N, 129.5°E) from August 1999 to May 2015 with the model NRLMSISE-00 (http://ccmc.gsfc.nasa.gov/ modelweb / models / nrlmsise00.php) latitude data were entered: 63°N, longitude: 130°E, and altitude: 87 km. Since the emissions of OH(6-2) temperature are measured at night, for the NRLMSISE-00 model, the UT global time of 15 hours was used as the input parameter, which corresponds to midnight for our geographical location.

Figure 2 shows the temperature values from the NRLMSISE-00 model calculations and the hydroxyl band (6-2) with standard errors obtained at the Maimaga station averaged over a period of 16 years. The comparison was made for all months, except for June and July, at which time observations at the station are not conducted due to geographical location. The measurements at the Maimaga station are carried out from the 20th of August to the 10th of May, therefore in these months there are very few measurements suitable for conducting the comparison.

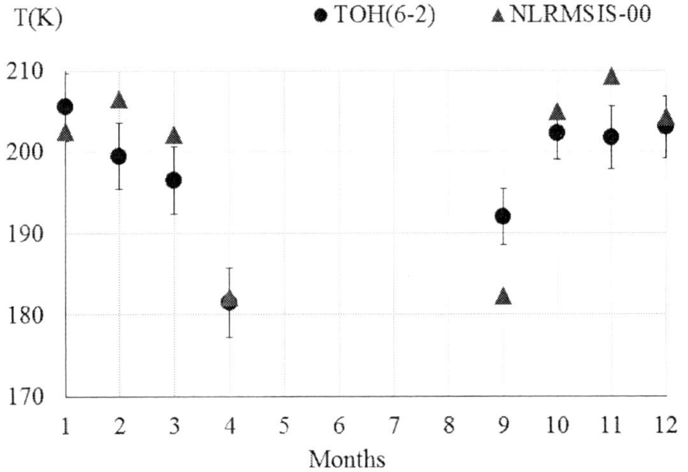

Figure 2. Average monthly NRLMSISE-00 model and OH(6-2) temperatures for 16 years.

There is a sufficient agreement between the values of OH (6-2) and NRLMSISE-00 from October to April, the difference is not more than 7 ± 4 K. In September, the temperature differences are greatest, with the hydroxyl temperature being higher, than the data of model NRLMSISE-00. This can be explained by a smaller amount of experimental data in September compared to other months, since in the autumn months there are more cloud nights. Such features of the distribution of experimental data can affect the evaluation when compared with model calculations. It should be taken into account that the subpolar mesopause is characterized by a larger temperature variation in comparison with the models during the transition from winter to summer, which is associated with seasonal circulation of the atmosphere.

To determine the dependence of the accuracy of the NRLMSISE-00 model on the variation of the temperature of the mesopause region from each year, a correlation analysis was performed, in which model calculations were made only on the days when measurements were taken at the Maimaga station. The results of the correlation analysis are presented in Table 1 in the form of the correlation coefficient R by years and the number of experimental measurements of the rotational temperature OH (6-2) for each individual year. As can be seen from the table, the number of measurements at the Maimaga station varies depending on the year. When the number of observations at the Maimaga station is more than 180 days (approximately six months), the correlation coefficient R is greater than 0.7. This means that the NRLMSISE-00 model accurately describes the temperature changes of the subauroral mesopause.

Table 1. R calculated by years the number of experimental measurements of the temperature OH (6-2)

Years	Number of measurements	R	Years	Number of measurements	R
1999-2000	139	0.7	2007-2008	173	0.64
2000-2001	109	0.65	2008-2009	169	0.68
2001-2002	180	0.78	2009-2010	168	0.19
2002-2003	158	0.65	2010-2011	122	0.55
2003-2004	133	0.38	2011-2012	123	0.45
2004-2005	190	0.71	2012-2013	183	0.82
2005-2006	144	0.47	2013-2014	100	0.71
2006-2007	167	0.62	2014-2015	62	0.61

4. Conclusions

To compare the temperature of the subauroral mesopause with the empirical model NRLMSISE-00, from August 1999 to April 2015, 2320 average night values of OH (6-2) were obtained that satisfy the above-described sampling criteria. Difference between OH(6-2) temperature and NRLMSISE-00 model data from October to April is not more than 7 ± 4 K. To determine the dependence of the accuracy of the NRLMSISE-00 model on the variation of the temperature of the mesopause region from each year, a correlation analysis was performed, in which model calculations were made only on the days when measurements were taken at the Maimaga station. NRLMSISE-00 model accurately describes the changes in subpolar mesopause temperature obtained at the Maimaga station if the number of OH (6-2) measurements is more than 180 days (the correlation coefficient R is greater than 0.7).

Acknowledgments

The reported study was funded by grant of the Head of the Republic of Sakha (Yakutia) for young scientists, specialists and students and RFBR according to the research project No. 17-05-00855 A.

References

[1] Lubken F-J 1999 *J. Geophys. Res.* D8 **104** 9135–9149
[2] Ammosov P, Gavrilyeva G, Ammosova A, Koltovskoi I 2014 *Adv. Space Res.* **54** 2518-2524
[3] Picone J M, Hedin A E, Drob D P, Aikin AC 2002 *J. Geophys. Res.* A12 **107** 1468–1483

Internal gravity and tidal waves in the mesopause region in Yakutia

V I Sivtseva, P PAmmosov, G A Gavrilyeva, I IKoltovskoi, A M Ammosova

Yu G Shafer Institute of Cosmophysical Research and Aeronomy of SB RAS, Yakutsk, Russia

E-mail: verasivtseva@gmail.com, ammosov@ikfia.ysn.ru, gagavrilyeva@mail.ru, koltigor@mail.ru, ammosovaam@mail.ru

Abstract. The temperature of the mesopause region (87 km) is monitored at the Maimaga station (63.04 ° N, 129.51 ° E) using the Shamrock (Andor) spectrograph recording the hydroxyl (OH) band (3, 1). The temperature data obtained for the seasons from 2013 to 2017 are investigated. The standard deviations corresponding to the internal gravity σ_{gw} and tidal waves σ_{td} were obtained.The seasonal course of the tidal component σ_{td} varies from 2 to 5 K throughout all observation seasons. The seasonal variation of the gravitational component σ_{gw} observed at Maimaga station almost coincides for three observation seasons except for the 2014-2015 season. In this observation season, σ_{gw} has lower values in winter than in other seasons. Moreover, in this season average monthly temperatures exceed similar values in other seasons.

1. Introduction

Recently, the mesopause region has been of great interest to the study. This is due to the fact that this region actively interacts with the radiation particles coming from above, and with a wide spectrum of waves propagating upward from the lower layers of the atmosphere.

Wave activity makes a significant contribution to the temperature regime of mesopause. The tides cause adiabatic compression and expansion of the mesopause region, respectively, causing warming up or cooling of the environment [1, 2]. Internal gravity waves (IGW), propagating upward from the lower layers of the atmosphere, transfer the momentum and energy to the mesosphere and the thermosphere. At the height of the mesopause due to the wind shear, IGW undergo spectral filtering and are absorbed, causing warming up in this region [3] (thus, the seasonal course of the mesopause temperature correlates with the wave activity). The amplitude of IGW and tidal waves increases with altitude due to a decrease in atmospheric density, which makes it possible to detect them in the upper layers of the atmosphere.

A large number of studies of wave activity at the height of the mesopause were carried out at mid-latitudes [4, 5]. Therefore, the temperature behavior and the wave activity of high-latitude mesopause are of great interest. This paper presents studies of standard deviations of temperature σ from its mean values in the mesopause region (87 km) according to the measurements at the high-latitude station Maimaga (63.04 ° N, 129.51 ° E).

2. Spectrograph and data processing method

A photosensitive infrared spectrograph Shamrock SR-303i with a working wavelength range of 1490-1544 nm was used to record the OH (3, 1) hydroxyl band. The spectrograph makes it possible to measure the temperature of the mesopause with an accuracy of ~ 2 K. The software developed in this research allows the device to work completely in an autonomous mode.

Exposure time for obtaining one measurement of the hydroxyl spectrum in the OH (3, 1) band is 60 sec. Short exposure not only allows to significantly expand the range of periods of internal gravity waves under investigation, but also includes acoustic waves (3-5 min periods) in the spectrum. The data were averaged in 3 min increments to exclude acoustic waves from consideration.

The method for estimating the rotational temperature of molecular emissions is based on fitting the model spectra constructed with allowance for the spectrograph's instrument function for various predetermined temperatures, to the actually measured spectrum [6, 7]. The rotational temperature, determined from the intensity distribution in the hydroxyl (OH) band, is close to the kinetic temperature of the neutral gas at the emission height [8].The transition probabilities calculated by Mies [9] were used to estimate the rotational temperature along the hydroxyl band.

In order to exclude data with a high level of noise interference, a sampling of the spectra satisfying the signal-to-noise ratio> 20 was performed, then the temperature was averaged in steps of 3 min. The standard deviation of temperature σ from its mean value is assumed as a characteristic of the night wave activity. It is a superposition of various active waves at night and the noise of the detector dark current. According to [10], the superposition of waves can be represented as:

$$\sigma = \sqrt{\sigma_{\text{td}}^2 + \sigma_{\text{gw}}^2 + \sigma_{\text{noise}}^2} \quad (1)$$

Where σ_{td} is the standard deviation of temperature due to tidal waves, σ_{gw} is the standard deviation of temperature due to internal gravitational waves, σ_{noise} is the component of the standard deviation of temperature caused by the noise of the dark current of the detector. σ_{noise} was calculated as the arithmetic average of the errors of each individual measurement per night. The planetary waves were not taken into consideration, since their time scale is much longer than one night.

The value of σ_{td} was determined by separating the harmonics corresponding to the 24-, 12-, and 8-hour components of the daily tide from the night temperature series by the method of least squares.

$$\text{f}_{\text{td}} = \bar{T} + A_1 \cos\left(\frac{2\pi}{1440}(t - \varphi_1)\right) + A_2 \cos\left(\frac{2\pi}{720}(t - \varphi_2)\right) + A_3 \cos\left(\frac{2\pi}{480}(t - \varphi_3)\right) \quad (2)$$

f_{td} is the sum of the harmonics of the daily tide, periods are indicated in minutes.

The result of subtracting the sum of the daily tide harmonics from the nighttime temperature range corresponds to the contribution of the dark current noise and the spread of IGW to the temperature. Hence, the standard deviation σ_{gw} is calculated.

3. The results of observations and their analysis

The tidal components of the standard deviations of temperature are shown in figure 1.

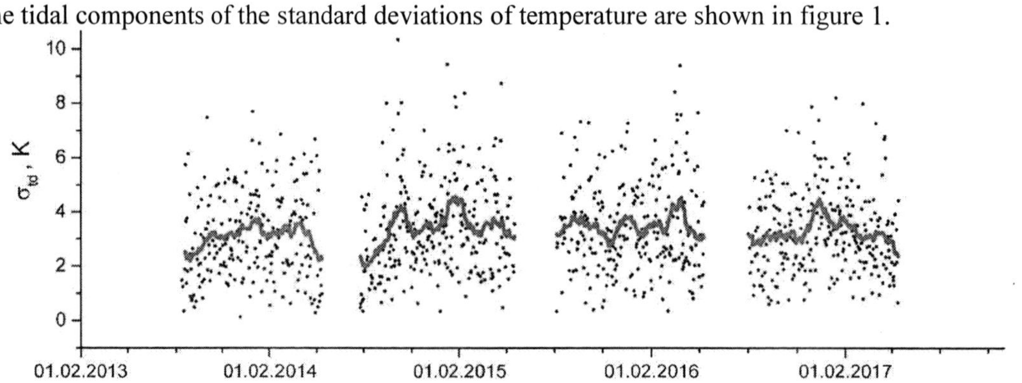

Figure 1. Seasonal variations of the tidal component of standard deviations of temperature

The red line is a moving average line with a transmission window of 30 days and represents the seasonal course of the tidal component σ_{td}. The seasonal course of the tidal component of standard deviations of temperature σ_{td} varies from 2 to 5 K throughout all observation seasons.

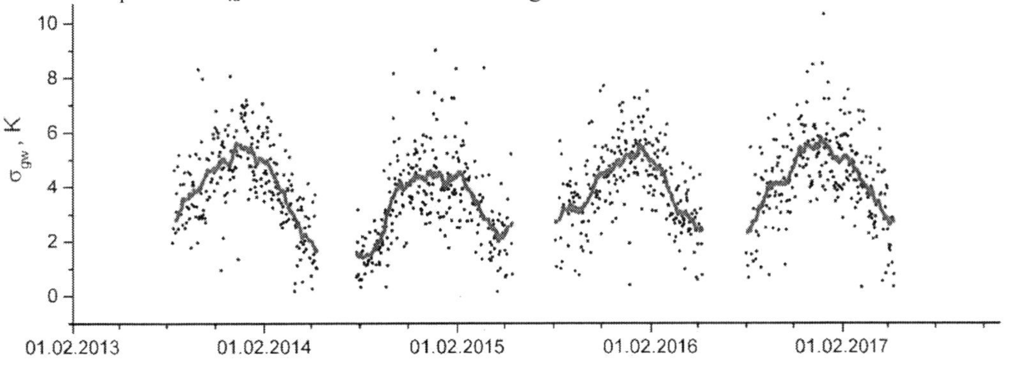

Figure 2. Standard temperature deviations caused by internal gravity waves

The standard temperature deviations corresponding to IGW are shown in figure 2, where the red line denotes a moving average with a transmission window of 30 days, representing the seasonal variation of the gravitational component σ_{gw}. The values of the gravitational component of standard deviations of temperature σ_{gw} observed at Maimaga station and their seasonal variations almost coincide for three observation seasons except one. In the observation season 2014-2015, the gravitational component of standard deviations of temperature σ_{gw} has lower values in the winter period than in the other observation seasons. In addition, during this season the average monthly temperatures of the winter mesopause, on the contrary, exceed the values in other seasons by an average of 8.3 K (figure 3). The seasonal variation of the gravitational component varies from 2 to 6 K, and in the 2014-2015 season from 1.5 to 5 K.

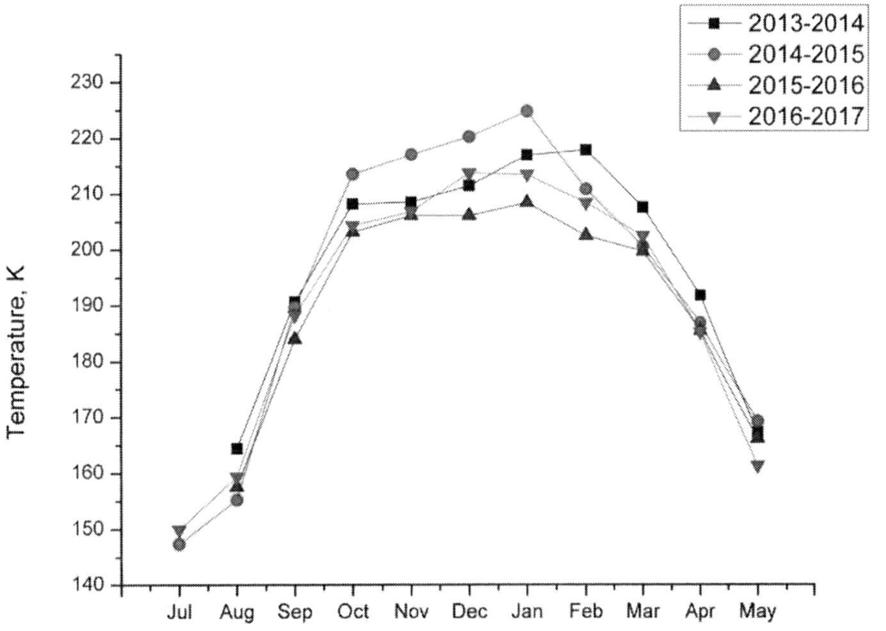

Figure 3. Mean monthly rotational temperatures of hydroxyl

The lower IGW activity in the winter period of the 2014-2015 season can be explained by the fact that it is possible that during this season a significant part of the IGW energy was absorbed at a height close to the height of the emission layer, which is confirmed by the observed elevated mean monthly temperature during this period.

4. Conclusions

The standard deviations of temperature σ from its mean values in the mesopause region were investigated according to measurements at the Maimaga station (63.04 ° N, 129.51 ° E). The spectra are recorded using a Shamrock (Andor) photosensitive infrared spectrograph recording the OH band (3, 1) in the far infrared region (about 1.5 μm). The data for the seasons from 2013 to 2017 were studied.

The standard deviations corresponding to the internal gravity σ_{gw} and tidal waves σ_{td} were obtained. The seasonal course of the tidal component σ_{td} varies from 2 to 5 K throughout all observation seasons.The seasonal variation of the gravitational component σ_{gw} observed at Maimaga station almost coincides for three observation seasons except for the 2014-2015 season. In this observation season, the gravitational component σ_{gw} has lower values in the winter period than in the other observation seasons. In addition, during this season, the average monthly temperatures of winter mesopause exceed similar values in other seasons.

Acknowledgments

The research was supported by “Scientific and Educational Foundation for Young Scientists of Republic of Sakha (Yakutia) wthin the Project №17-2-009477 “Scientific and Educational Foundation as an instrument for young scientists to develop their professional competence and science popularization” 20171201005-2, RFBR No. 17-05-00855 A and grant of the Head of the Republic of Sakha (Yakutia) for young scientists, specialists and students.

References

[1] Chapman S and Lindzen R 1972 *Atmospheric tides* (Moscow: Mir Publishers) p 295
[2] Brasseur G and Solomon S 1987 *Aeronomy of the middle atmosphere* (Leningrad: Gidrometeoizdat Publishers) p 413
[3] Hines C O 1974 *The upper atmosphere in motion* (Washington D C: AGU) p 1027
[4] Offermann D, Wintel J, Kalicinsky C, Knieling P, Koppmann R, Steinbrecht W 2011 *J. Geophys. Res.* **116** D00P07
[5] Perminov V I, Semenov A I, Medvedeva I V, Pertsev N N 2014 *Geomagn. Aeron.* 2 **54** 230–9
[6] Ammosov P P and Gavrilyeva G A 2000 *Prib. Tekh. Eksp.* 6 **43** 792–7
[7] Gavrilyeva G A and Ammosov P P 2002 *Geomagn. Aeron.* 2 **42** 267–71
[8] Shefov N N, Semenov A I, Khomich V Yu 2006 *Studying the Upper Atmosphere as an Indicator of Its Structure and Dynamics* ed M A Yatsenko (Moscow: GEOS) p 741
[9] Mies F H 1974 *J. Mol. Spectrosc.* 2 53 150–80
[10] Offermann D, Gusev O, Donner M, Forbes J M, Hagan M, Mlynczak M G, Oberheide J, Preusse P, Schmidt H, Russell J M III 2009 *J. Geophys. Res.* **114** D06110

Seasonal Variations of Equatorial Anomaly Crest using GPS Ionospheric Tomography

Siti Syukriah KHAMDAN[1], Tajul Ariffin MUSA[1] and Suhaila M BUHARI[1,2]

[1]Geomatics Innovation Research Group, Faculty of Geoinformation and Real Estate, UTM Johor, Malaysia.
[2]Department of Physics, Faculty of Science, UTM Johor, Malaysia

syukriahkhamdan91@gmail.com

Abstract. The seasonal variation of the electron density profile in equatorial anomaly region is investigated using Global Positioning System (GPS) ionospheric tomography method from a dense network of Malaysian Real-Time Kinematic Network (MyRTKnet). In this study, the monthly variation of the electron density is reconstructed to study the development of the equatorial anomaly crest over the study area. It is found that the equatorial anomaly crest appears during the noon and last for a few hours before disappeared. The result also shows that the anomaly crest show strong appearance during both equinoxes, where March to May shows stronger crest compared to October to November month-to-month variations.

1. Introduction

The equatorial regions is common with the occurrence of ionospheric special features such as equatorial electrojet (EEJ), equatorial plasma bubble (EPB), equatorial spread-F (ESF) and also equatorial ionization anomaly (EIA). Particularly at this region, the ionosphere structure are dominantly controlled by the fountain effects that caused by the eastward electric field [1]. The perpendicular east-west electric field and north-south geomagnetic field lifted the upward electrodynamic (ExB) drift of plasma, where E and B represent the electric and magnetic field, respectively, to a higher altitude during the daytime. Then, it diffuses downward along the geomagnetic field lines toward the higher latitudes under influence of gravity and pressure gradients and produces the EIA.

The EIA structures shows a decreasing structure of electron density that called as trough in F-region at the geomagnetic equator and two crests structure that are formed on the either side within ±15° magnetic equator. The strength of the anomalies are varies with the geophysical conditions of the days, seasonal and solar activity [2, 3, 4]. This phenomenon occurs in daytime near the equatorial region and it is one of the main factor that contributes to the enhancement of the scintillation effects from the spread-F or the plasma bubbles irregularities [1,5]. Numbers of study has been carried out to study the structure of the EIA.

Previously, [6] has been investigated the total electron content (TEC) variability near the EIA crest and the results is compared with the IRI model. The seasonal variation of the ionospheric total electron content (TEC) in Asian equatorial anomaly region also has been investigated by [7]. Later, [8] has been study on the solar activity dependence of the electron density in the equatorial anomaly region using CHAMP data. They used the tomography imaging to detect the tilting of the ionization crest and also found the changes of the ionospheric thickness over latitude. They found that the both northern and southern anomaly crest are fully developed during midday in winter, post-noon in equinoxes and late afternoon in summer. Recently, [4] present the study on the variability of the EIA that focused on the crest-to-trough TEC ratio in East African anomaly region.

As the GPS is low cost in maintenance and capable to improve the spatial coverage compared to other instruments, the GPS has been widely used as a tools in monitoring the ionospheric structures and its behaviors [9, 10, 11]. With the dynamics structure of the ionosphere, the tomography method has been used to reconstruct the ionospheric profile from GPS measurements as its capability in imaging the ionospheric structure and also improving the local ionospheric modelling [9,10]. In this study, we reconstructed the electron density profile using GPS tomography methods to study the ionospheric variations over the equatorial anomaly region. We focused on the seasonal variation of the electron density profile and its diurnal variations to observe the effects towards the development of the equatorial anomaly crest.

2. Data and Measurements

A dense GPS network of MyRTKnet stations, which consists of 50 GPS CORS, that covered the whole Peninsular Malaysia is used in this study. It is located at the geomagnetic belt, within 1N to 7N (geomagnetic latitude: 8.825S to 2.95S) in geographical latitude and 99E to 104E in geographical longitude. Figure 1 shows the maps of the study area based on its geomagnetic location.

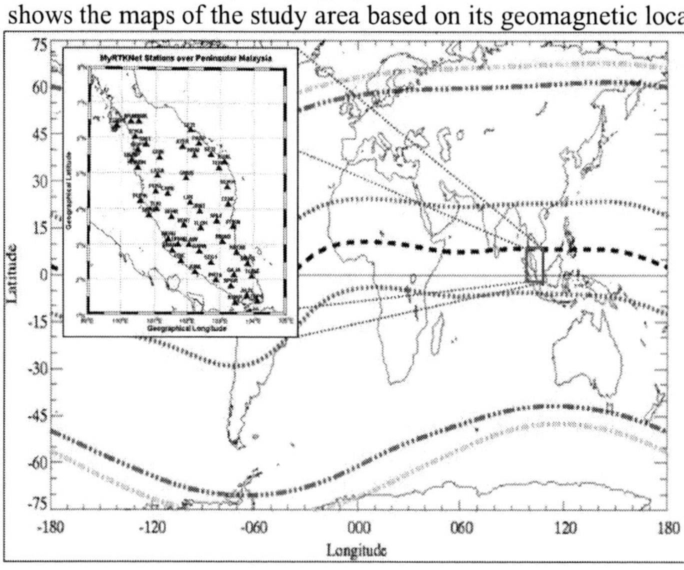

Figure 1. The geomagnetic location of Peninsular Malaysia.

To reconstruct the electron density profile of ionosphere, the slant TEC (STEC) is derived from the GPS measurements. The STEC represents the ionospheric TEC line-of-sight direction along the ray path from satellite to GPS receiver and it can be expressed mathematically as follow.

$$STEC = \frac{1}{40.3}\left(\frac{f_1 f_2}{f_1 - f_2}\right)[(P_1 - P_2) - c(DCB_r - DCB^s)] \quad (1)$$

where f_1 and f_2 are the GPS signal frequencies, P_1, P_2 represent the GPS pseudorange measurements, c is the speed of light and DCB_r, DCB^s represent the differential code bias for receiver and satellite respectively. Then, the electron density profile can be estimated using the general equation of the GPS tomography as follow.

$$Y = Ax + E \quad (2)$$

where Y represent the matrix of the observed STEC, A consists of the matrix of the length of the ray path of the satellite to receiver, x represent the matric of the electron densities that will be estimated and E is the errors during the approximation.

3. Results and Discussion

The ionospheric profile is reconstructed over the study area is reconstructed for one year data during year 2010. The month-to-month variation of the electron density profile during year 2010 is presented in Figure 2. It is noted that, even though the occurrence of the geomagnetic activities may affect the variation of the ionosphere, the results are still valid to describe the variation of the ionosphere [7].

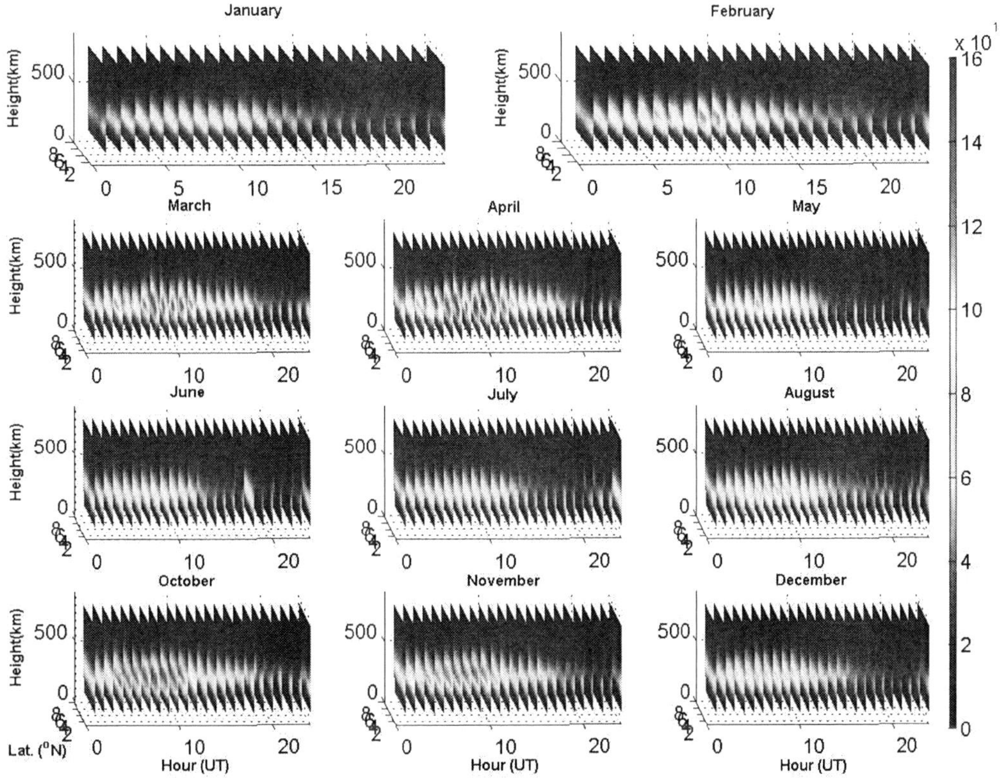

Figure 2. The month-to-month variation of the electron density profile. The horizontal axis represents the geographical latitude (x-axis) and time in universal time (UT)(y-axis) while the vertical axis represent the height of the profile in km. .

The result shows that the formation of the equatorial anomaly crest is begin at the morning local time for the whole month. It reaches maximum at post-noon, within 06 UT to 08 UT and the features decreases and disappear as the daytime passes. The same trend of formation of the equatorial anomaly crest is found by [8] that used the CHAMP electron density data, where they shows that the IRI-2001 reproduces the formation of EIA better around noon. Since the TEC is directly related with the maximum electron density of the ionosphere, [6] also explained that the increasing and decreasing of the TEC that is related with time is also depend of the production rate of ionization. During sunrise, the intensive ionization increases the electron concentration and as the temperature increase with the daytime passes, the loss rate increases and overcome the production rate and will results in gradually decreases of TEC in the evening.

Other than that, it is found that the equatorial anomaly crest is found maximum during both equinoxes, where it is higher during March to May month-to-month variations compared to October to November month-to-month variations. It is also found that the crest is minima during January monthly variations. The difference may cause by the differences of the solar activity that occurred during both equinoxes. This finding also presented by [8] that shows the difference in both equinoxes as the solar activities occurrences during March to May equinoxes are 60% stronger compared to October to November equinox.

4. Conclusion

In this study, we presented the tomography imaging of the seasonal variation of the electron density profile over the equatorial anomaly regions. A month-to-month variation is presented to observe the equatorial anomaly crest in the study area. It is found that the formation of the equatorial anomaly crest is begin at the morning and achieved maximum at the pass non local time. The feature disappears as the daytime passes. The findings also show the crest during both equinoxes is higher, where the March to May month-to-month variation is appear stronger compared to October to November variations. The difference also shown by previous study as it is triggered by the solar activity occurrences is higher during March to May month.

Acknowledgement

We would like to thank to Department of Survey and Mapping Malaysia (DSMM) for providing the MyRTKnet observation data. This work has been supported by the Research Grant University (Grant Number: Q.J130000.2526.19H39)

References

[1] Aggarwal M. TEC variability near northern EIA crest and comparison with IRI model. Advances in Space Research. 2011 Oct 1;48(7):1221-1231.

[2] Balan N, Souza J, Baily GJ. Recent developments in understanding of equatorial ionization anomaly: A review. Journal of Atmospheric and Solar Terrestrial Physics. 2017 Jul 1.

[3] Luo W, Zhu Z, Xiong C, Chang S. The Response of Equatorial Ionization Anomaly in 120°E to the Geomagnetic Storm of 18 August 2003 at Different Altitudes From Multiple Satellite Observations. 2017 Dec 1;15(12):1588-1601.

[4] Oryema B, Jurua E, Ssebiyonga N. Variations of Crest-to-Trough TEC Ratio of the East African Equatorial Anomaly Region. International Journal of Astrophysics and Space Science. 2015 Mac 2;4(1):12-20.

[5] Abdu MA. Equatorial ionosphere thermosphere system: electrodynamics and irregularities. Advances in Space Research. 2005 Jan 1;35(5):771-787.

[6] Andreeva ES, Franke SJ, Yeh KC. Some feature of the equatorial anomaly revealed by ionospheric tomography. Geophysical Research Letters. 2000 Aug 15;27(16):2465-2468.

[7] Tsai HF, Liu JY, Tsai WH, Liu CH, Tseng CL, Wu CC. Seasonal variations of the ionospheric total electron content in Asian equatorial anomaly regions. Journal of Geophysical Research: Space Physics. 2001 Dec 1;106(A12):30363-30369.

[8] Liu HX, Stolle C, Forster M, Watanabe S. Solar activity dependence of the electron density in the equatorial anomaly regions observed by CHAMP. Journal of Geophysical Research: Space Physics. 2007 Nov 1;112(A11).

[9] Chen CH, Saito A, Lin CH, Yamamoto M, Suzuki S, Seemala GK. Medium-scale travelling ionospheric disturbances by three-dimensional ionospheric GPS tomography. Earth, Planets and Space. 2016 Dec;68(1):32.

[10] Ssessanga N, Kim YH, Kim E. Vertical structure of medium-scale travelling ionospheric disturbances. Geophysical Research Letters. 2015 Nov 16;42(21):9156-9165.

[11] Hong J, Kim YH, Chung JK, Ssessanga N, Kwak YS. Tomography Reconstruction of Ionospheric Electron Density with Empirical Orthonormal Functions Using Korea GNSS Network. Journal of Astronomy and Space Sciences. 2017;34(1):7-17.

IOP Conf. Series: Journal of Physics: Conf. Series **1152** (2019) 012014 doi:10.1088/1742-6596/1152/1/012014

Comparison of equatorial plasma bubble zonal drifts and neutral winds velocity over Southeast Asia

Idahwati Sarudin[1], Nurul Shazana Abdul Hamid[1*], Mardina Abdullah[2,3], and Suhaila M Buhari[4]

[1]School of Applied Physics, Faculty of Science and Technology,
[2]Space Science Centre (ANGKASA), Institute of Climate Change,
[3]Centre of Advanced Electronic and Communication Engineering (PAKET), Universiti Kebangsaan Malaysia, 43600 UKM, Bangi, Selangor, Malaysia.
[4]Department of Physics, Faculty of Science, Universiti Teknologi Malaysia, 81310 UTM, Skudai, Johor, Malaysia.
*Corresponding author: shazana.ukm@gmail.com

Abstract. The purpose of the presented study is to investigate the zonal drifts of Equatorial Plasma Bubbles (EPB) which applied two different ground-based instruments. Malaysia Real–Time Kinematics GNSS Network (MyRTKnet), consists more than 78 GPS receivers was used to observe EPB along 96° East to 120° East longitudes. Subsequently, zonal drift of EPBs derived from GPS-ROTI is compared with neutral winds velocity from Fabry-Perot interferometer (FPI). On 10 April 2013, we founded the maximum drift is about 194.4 m/s at 1430 UT to 1500 UT whereas the minimum drift is 111.1 m/s at 1330 UT to 1400 UT. The results illustrate that temporal variations of EPB zonal drift velocities are consistent with the neutral winds.

1. Introduction

The equatorial plasma bubbles (EPB) is defined as depletion of total electron content (TEC) in the ionosphere. The investigations of EPB have been done before by using ground based data such as imaging technique and radar measurement. The first observation of EPB over Southeast Asia using GPS data was carried out by Buhari et al. [1]. One of the characteristics of EPB which have been commonly reported in the past years using various techniques was including zonal drift velocity. The study of this characteristic is a vital in maintaining a stable communication between satellite and ground. Most of the previous study observed velocities of EPB using ground based instruments such as imaging techniques [2]. They reported that the velocity of EPB were decreased with time. Apart from that, the past studied believed that the zonal drift velocity were related to the zonal neutral winds. There are numbers of discussion regarding the relation between the zonal drifts of EPB with the neutral winds. For instance, a study from Chapagain et al. [3,4] obtained a positive correlation in the variations of zonal neutral winds and the EPB zonal drift velocity. The present study focuses on the EPB zonal drift velocity estimated from high density of GPS receivers over Southeast Asia (SEA) on night-time of 10 April 2013 and compared with neutral winds velocity from Fabry-Perot interferometer (FPI).

2. Method and analysis

2.1. GPS-ROTI MEASUREMENT

The EPB zonal drifts were calculated using rate of TEC change index, ROTI obtained by high density GPS in the Southeast Asia sector [1,5]. Department of Survey and Mapping Malaysia, JUPEM was installed more than 78 receiver stations called as Malaysia Real-Time Kinematics GNSS Networks, MyRTKnet. These GPS data is in the format of Receiver Independent Exchange, RINEX were provided by JUPEM for this study.

In order to present the variations in terms of longitudes with times, the structure of EPB was extended in north-south directions. The east to west cross section of ROTI at five degree north in each five minutes was considered to plot keogram. Based on keogram, after 1200 UT about thousands kilometre scale of irregularities can be shown. According to past studies, the first EPBs structure are found to be initiate at 104°E longitude at 1230 UT and it developed faster against times [1,6]. The EPB are believed to moving in eastward direction [7,8]. In this keogram, we can showed that every EPB moves in eastward direction and disappeared almost after 1800 UT.

We only considered 1-day of ROTI-keogram that generated from MyRTKnet to determine EPB in Malaysia sites. The occurrence of EPB was clarified with ROTI values is higher than 0.1 TECU/min on the keogram. Thus, the zonal drifts of EPB can be estimated from the change of longitude with time [5]. Zonal drifts was calculated due to the initial largest ROTI value to the last largest ROTI value as can be shown on the keogram. In this case, we estimate zonal drifts of one EPB selected with range times from 1230 UT to 1530 UT.

2.2. Fabry-Perot Interferometer

The FPI was used to measure the neutral winds motions through 630-nm airglow emission lines observations of the Doppler shift with two interference fringes: inner and outer by 15 min of the time resolution [8]. FP03 code shows that this FPI is in operation at Kototabang, Indonesia. The code for each FPI will be different due to the location of their operation. This instrument was a part of the Optical Mesosphere Thermosphere Imagers (OMTIs) [9] and operated by the Space-Earth Environmental Research (ISEE), Nagoya University.

3. Results and discussions

In previous section, the method to estimate the zonal velocity drift on the night of 10 April 2013 have been discussed. By approaching the GPS-ROTI data that covered SEA, the temporal and dimensional (spatial) variations of EPB can be observed.

Figure 1 depict the keogram formed from 2-D map of ROTI with the variations of longitude. This plot gives a clear feature of EPB characteristics, where it is a cross section of ROTI that have been chosen the horizontal profiles of the ROTI at 5 degree north latitude. The white gap in Figure 1 shows the missing data during certain period. We further select one single EPB from the keogram to investigate the zonal drifts denoted by red arrow in Figure 1. This EPB are chosen based on the feature that we can fully observed from the ROTI keogram.

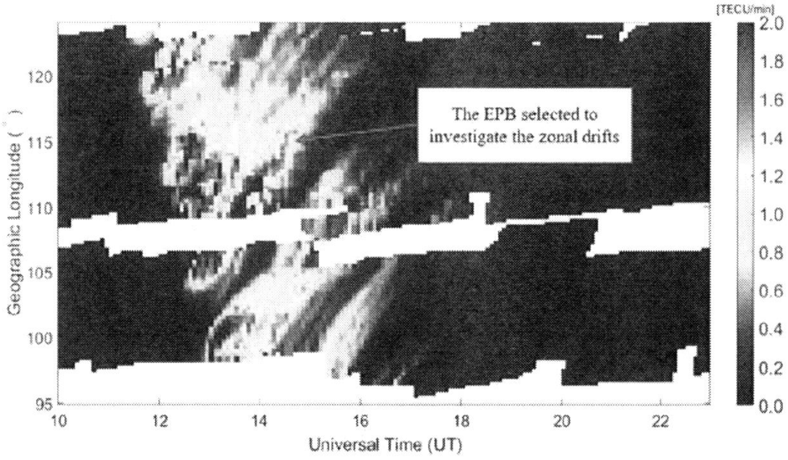

Figure 1. ROTI keogram at five degree north latitude acquired from GPS networks in SEA from 1000 UT to 2230 UT on night of 10 April 2013. The red arrow represent the EPB selected to be investigate.

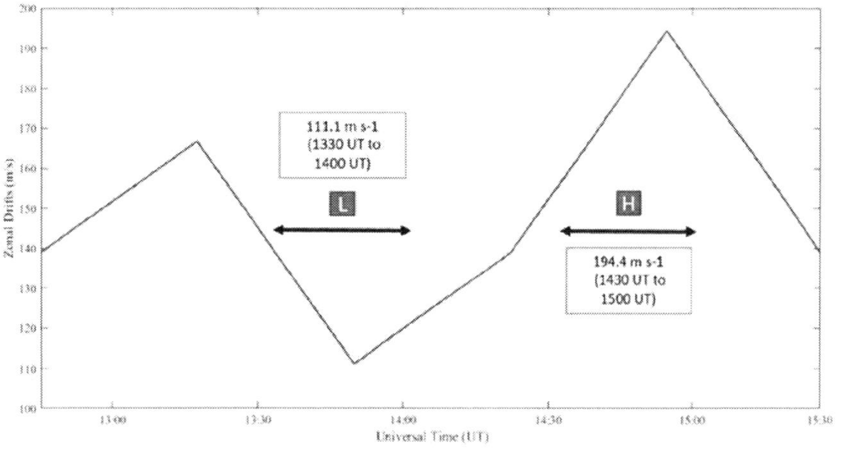

Figure 2. The zonal drifts of EPB calculated for red arrow shown in Figure 1.

Figure 2 shows plots of EPB zonal drifts obtained from GPS ROTI keogram. We can see the trend were decreased from 1230 UT to 1530 UT. Our results correlated with several past studies by [4,5,6,7] where drift velocity of EPB gently decreasing against time. The zonal drifts of the EPB shows a significant different during period of 1300 UT to 1500 UT. The maximum drifts (H) is found as 194.4 m/s at 1430 UT to 1500 UT whereas the minimum drifts (L) is 111.1 m/s at 1330 UT to 1400 UT.

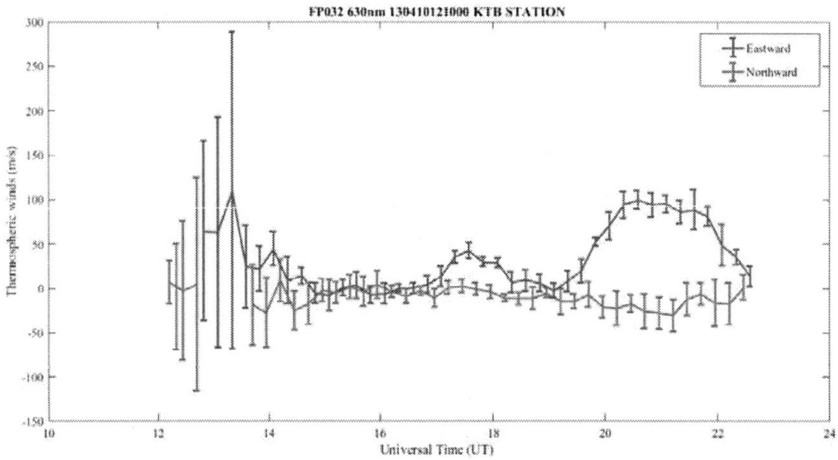

Figure 3. Temporal variation of thermospheric neutral winds observed by FPI at KTB station.

In Figure 3, we observed the thermospheric winds using FPI at KTB station. The positive values of neutral winds represent eastward velocity while negative values show winds to be northward. On this night, the estimated neutral winds peak at ~100 m/s in eastward and ~ -40 m/s in northward. The eastward winds show higher value than that of northward winds. This explained that the eastward winds play an important role as an indicator in drifting of the EPB which is moving in eastward direction. Apart of that, we can see from both Figure 2 and Figure 3, the zonal drift velocity variations of EPB and eastward winds have similar variations. At the beginning (1230 to 1300 UT), the zonal drift velocity is increased, but after 1300 UT they gradually decreased until reached at a certain time the zonal drift velocity were formed a peak and suddenly dropped.

4. Conclusion

In this paper we have exposed zonal drift of EPB from high density GPS receivers over SEA on 10th April 2013 and compared with neutral winds velocity from Fabry-Perot interferometer (FPI). The results illustrate that temporal variations of EPB zonal drift velocities are consistent with the neutral winds.

Acknowledgments:
The GPS data are obtained through http://www.rtknet.gov.my/ and also from Department of Survey and Mapping Malaysia, JUPEM data policy. This work was supported by the FRGS/1/2016/WAB08/UKM/01/1 grant from the Ministry of Education Malaysia. We would like to thank to Prof. Kazuo Shiokawa from Space-Earth Environmental Research (ISEE), Nagoya University for FPI data provided.

References

[1] Buhari S M, Abdullah M, Hasbi A M, Osuka, Yokoyama T, Nishioka M, Tsugawa T 2014 Continuous generation and two-dimensional structure of equatorial plasma bubbles observed by high-density GPS receivers in Southeast Asia *Journal of Geophysical Research* **119** doi: 10.1002/2014JA020433

[2] Fukushima D, Shiokawa K, Otsuka Y, Kubota M, Tsugawa T, Nagatsuma T 2015 Geomagnetically conjugate observation of plasma bubbles and thermosheric neutral winds at low latitudes *Journal of Geophysical Research*

[3] Chapagain N P, Makela J J, Meriwether J W, Fisher D J, Buriti R A, and Medeiros A F 2012

Comparison of nighttime zonal neutral winds and equatorial plasma bubble drift over Brazil *Journal of Geophysical Research* **117** A06309 doi:10.1029/2012JA017620

[4] Chapagain N P, Fisher D J, Meriwether J W, Chau J L and Makela J J 2013 Comparison of zonal neutral winds with equatorial plasma bubble and plasma drift velocities *Journal of Geophys. Research* **118** doi:10.1002/jgra.50238

[5] Sarudin I, Hamid N S A, Abdullah M, and Buhari S M 2017 Investigation of Zonal Velocity of Equatorial Plasma Bubbles (EPBs) by using GPS data *Journal of Physics: Conference series*

[6] Yokoyama T, Fukao S 2006 Upwelling backscatter plumes in growth phase of equatorial spread F observed with the Equatorial Atmosphere Radar *Journal of Geophysical Research* **33**

[7] Igo Paulino, Amauri Fragoso de Medeiros, Ricardo Arlen Buriti, Hisao Takahashi, Jose Humberto Andrade Sobral, Delano Gobbi 2011 Plasma bubble zonal drift characteristics observed by airglow images over Brazilian tropical region Brazilian *Journal of Geophysics* 29(2) 239-246

[8] Chapagain N P, Taylor M J, Nielsen K, Jarvis M J 2011 Airglow observations and modelling of F region depletion zonal velocities over Christmas Island *Journal of Geophysical Research* **116** A02301 doi:10.1029/2010JA015958

[9] Shiokawa K, Kadota T, Otsuka Y, Ogawa T, Nakamura T, Fukao S 2003 A two-channel Fabry-Perot interferometer with measurement in the upper atmosphere, *Earth Planet Space* 55:271-275

[10] Shiokawa K, Otsuka Y, Oyama S, Nozawa S, Satoh M, Katoh Y, Hamaguchi Y, Yamamoto Y, and Meriwether J W 2012 Development of low-cost sky-scanning Fabry-Perot interferometer for airglow and auroral studies *Earth, Planets and Space*, doi: 10.5047/eps.2012.05.004

[11] Chapagain N P Ionospheric plasma drift and neutral winds modelling, *Res. J. Physical Science*

[12] De Paula E R, Kantor I J, Sobral J H A, Takahashi H, Santana D C, Gobbi D, Medeiros A F, Limiro L A T, Kil H, Kintner P M, Taylor M J 2016 Ionospheric irregularity zonal velocities over Cachoeira Paulista *Journal of Atmospheric Solar Terrestrial Physics*, 64 1511-1516

SEAIONO: A Windows software tool to extract ionogram dataset from South East Asia Low-Latitude Ionospheric Observation Network

S A Bello[1, 2], M Abdullah[2, 3], N S A Hamid[4*] and T Yokoyama[5]

[1]Department of Physics, Faculty of Physical Sciences, University of Ilorin, Nigeria
[2]Space Science Centre (ANGKASA), Institute of Climate Change, Universiti Kebangsaan Malaysia, Malaysia
[3]Centre of Advanced Electronic and Communication Engineering, Faculty of Engineering and Built Environment, Universiti Kebangsaan Malaysia
[4]School of Applied Physics, Faculty of Science and Technology, Universiti Kebangsaan Malaysia, 43600 UKM Bangi, Selangor, Malaysia
[5]National Institute of Information and Communication Technology, Tokyo, Japan.

*shazana.ukm@gmail.com

Abstract. Analysis of the ionospheric parameters on the ionogram is important for the study of ionospheric physics and related field. This paper introduces an application program tool used for extracting ionospheric profile parameters on the ionogram from the South East Asia Low-latitude Ionospheric Network FMCW ionosonde measurement. The software runs under Windows operating system and has a user-friendly graphical user interface to accept alphanumeric input entry. In addition to the parameters on the ionogram dataset, the SEAIONO reader program is capable of computing the maximum height of the ionospheric F2 layer and the bottomside ionospheric F2 layer (B0).

1. Introduction

The ionosphere is a region of the Earth's upper atmosphere that affects the propagation of radio signal [1]. For many decades, the ionospheric observations are explored by high frequency radar known as ionosonde. The ionosonde typically works by sending pulses of a radio signal into the ionosphere. The reflected pulses that return to the ground are recorded by measuring the time delay between the transmission and reception of the sent radio pulses. By observing the fluctuating carrier frequency of short pulses of radio energy from 1 to 20 MHz, the time delay at different frequencies is recorded as an ionogram [2]. The ionogram graphs the virtual height of the reflected frequency following the equation:

$$\Delta t = \frac{2}{c} h' \quad (1)$$

where Δt is the time delay, h' is the virtual height and c is the speed of light in space of the electromagnetic field.

With the advent of digital techniques, the automation of the scaling of the ionogram and characterization of the uncertainty in determining the confidence scale of the autoscaled results are now incorporated in the latest digital ionosonde [3] (e.g. digisonde DPS-4 sounder). The challenges of analyzing ionogram from an analogue ionosonde are cumbersome and complex due to the need for manual scaling in order to extract the details of ionospheric parameters. Increasing efforts to facilitate the automatic scaling of ionogram from analogue ionosonde as resulted to a new algorithm for an ionogram generation and analysis using periodograms that are filtered and fused, utilizing an "automatic edge-detection-based time-frequency detector"[4]. In another development, Pezzopane, (2004) [2], developed a semiautomatic ionogram scaling Windows software. In recent work of Jiang et al. (2017) [5], ionogram scaler ("ionoScaler") software was developed that perform both manual and automatic ionogram scaling routines which uses the Simulate Annealing (SA) inversion algorithm that has been applied to Quasi-Parabolic Segments (QPS) model. Similarly, a software application has also been developed by Ivanov et al. (2018) [6] for ionogram image processing which implements an algorithm for determining a signal-to-noise ratio for an oblique ionosonde. Most of these ionogram-scaling approaches are designed for a specific type or brand of ionosonde. Hence, scalability of these software applications is another existing challenge.

The current paper describes an application program named SEAIONO- South East Asia IONOgram reader developed for extracting ionospheric parameters from an ionogram dataset for the ionosondes installed at the South East Asia Low-latitude Ionospheric Network (SEALION). The ionosonde of the Frequency Modulated Continuous Wave (FMCW) type is a portable analogue ionosonde that enables low transmitting power for ionospheric observation. The full description of FMCW ionosonde designed for SEALION has been reported in the work of Nozaki, 2009 [7]. The developed SEAIONO software is mainly designed for the analysis of an ionogram dataset where manually scaled ionospheric parameters extraction and computation are executed.

2. SEAIONO: Description of the Software program

The structure of SEAIONO program is in three main sections namely; input graphical user interface section, ionogram parameter extraction algorithm and output parameters. The process flow of the program is shown in figure 1. The section 1 developed using MATLAB GUI development environment first read a file named "SEAIONOReader.m" to initialize the necessary user input needed by the ionogram parameter extraction algorithm. The visualized program GUI (see figure 2) can be modified to accept user date (start and end) of the ionogram dataset, station code, and specified path to the dataset directory.

The inputted information is handled by the "SeaLionFileReader.m" function that uses a combination of looping and "datenum.m" function in generating a dummy filename that the program lookup in the specified directory. If data file exists, the program goes to the next section where the file is opened and ionospheric parameter on the ionogram can then be extracted. However, if the filenames do not exist, the program terminates and gives a "NaN" (not a number) output file.

Ionogram parameters extraction algorithm coded as "SeaLionReader.m" function is the core of the program describing the section 2 of the program flow process. Using the predefined filenames addresses, the "fopen.m" function reads files into the program as a cell array. The locations of the ionospheric parameters on the dataset (see figure 3) are tabulated in table 1. A total number of lines for complete ionogram dataset is 1536 lines. Since the FMCW ionosonde typically sounds every fifteen minutes, each sounding is thus classified as a block containing 16 lines with a distinct header showing the timestamp. The location of each parameter is constant in each block (see table 1 and figure 3) and repeats every 16 lines. Hence, a sequence of extracting the parameter is determined using the syntax expressed as:

$$\mathrm{p}\{1,\mathrm{i}\} = \mathrm{Data}\{(\mathrm{i}-1) * 16 + \mathrm{n}, 1\} \tag{2}$$

where p is the ionospheric parameters, i gives the number of iteration and n is the line number of the parameter and j is a property of the cell array to indicate the number of column (j = 1). Note that the curly parentheses used in equation (2) indicate the output file is in a form of the cell array. The numbers of iterations are calculated by dividing the total number of lines on the dataset by 16 (lines in a block).

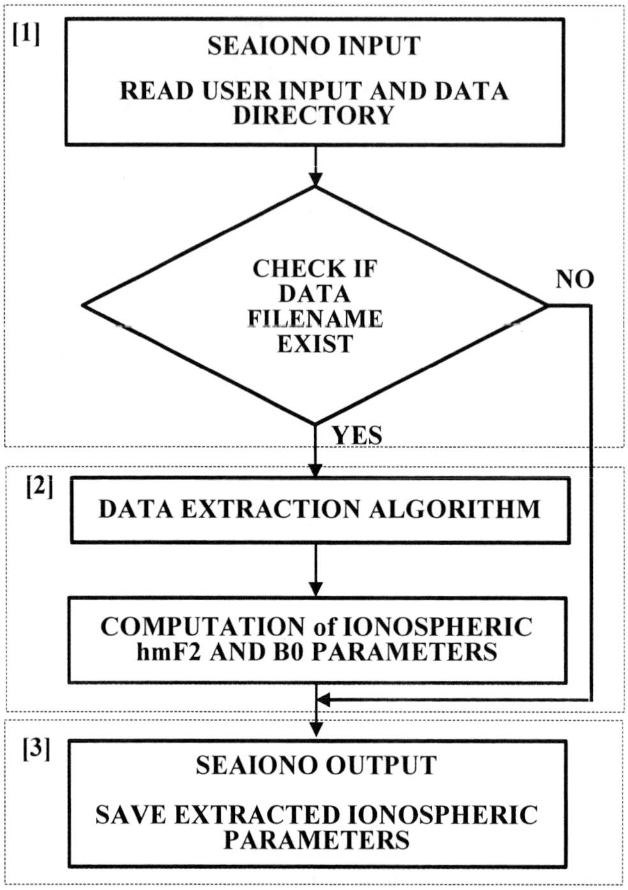

Figure 1. Flow process of the SEAIONO reader application program.

Table 1. Ionospheric parameters and their line position on the ionogram dataset.

S/N	Parameter	Symbol	Line Number	Decimal Number Correction
1.	Max. frequency of E-layer	foE	4	x 10^{-2}
2.	Virtual height of F-layer	h'F	9	
3.	Max. frequency of F2-layer	foF2	10	x 10^{-2}
4.	Maximum usable frequency	MUF(3000)F2	11	x 10^{-1}

After a set of iterations over the number of blocks existing on the dataset, the cell array is split to get the equivalent reading of the parameter. Appropriate variable type conversions are made to get the final data matrix. The data matrix can be saved directly and as well serves as an input variable for the computation of the maximum height of ionospheric F2-layer (hmF2) and ionospheric thickness parameter (B0). It is worthy to note that B0 is a parameter introduced by the International Reference Ionosphere (IRI) model [8] to describe the thickness of electron density below the ionospheric F2 layer height.

IOP Conf. Series: Journal of Physics: Conf. Series **1152** (2019) 012015 doi:10.1088/1742-6596/1152/1/012015

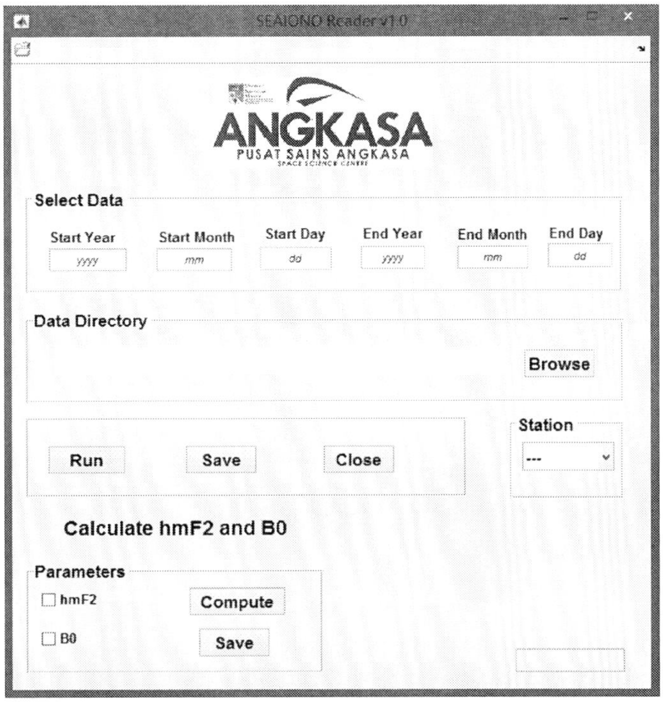

Figure 2. GUI of SEAIONO application main-menu.

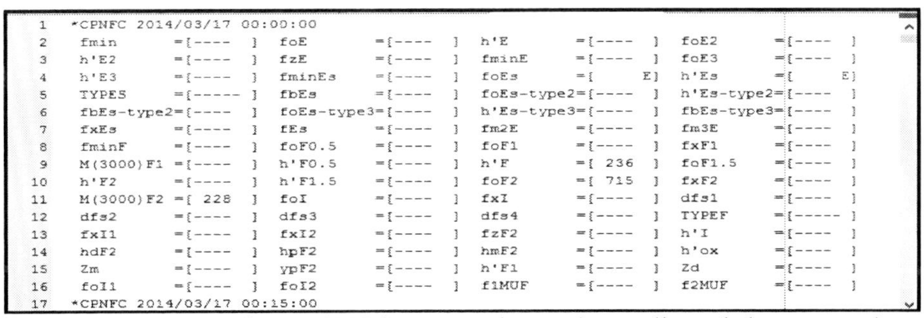

```
*CPNFC 2014/03/17 00:00:00
 fmin        =[----  ]  foE         =[----  ]  h'E         =[----  ]  foE2        =[----  ]
 h'E2        =[----  ]  fzE         =[----  ]  fminE       =[----  ]  foE3        =[----  ]
 h'E3        =[----  ]  fminEs      =[----  ]  foEs        =[     E]  h'Es        =[     E]
 TYPES       =[----- ]  fbEs        =[----  ]  foEs-type2=[----  ]  h'Es-type2=[----  ]
 fbEs-type2=[----  ]  foEs-type3=[----  ]  h'Es-type3=[----  ]  fbEs-type3=[----  ]
 fxEs        =[----  ]  fEs         =[----  ]  fm2E        =[----  ]  fm3E        =[----  ]
 fminF       =[----  ]  foF0.5      =[----  ]  foF1        =[----  ]  fxF1        =[----  ]
 M(3000)F1 =[----  ]  h'F0.5      =[----  ]  h'F         =[ 236  ]  foF1.5      =[----  ]
 h'F2        =[----  ]  h'F1.5      =[----  ]  foF2        =[ 715  ]  fxF2        =[----  ]
 M(3000)F2 =[ 228  ]  foI         =[----  ]  fxI         =[----  ]  dfs1        =[----  ]
 dfs2        =[----  ]  dfs3        =[----  ]  dfs4        =[----  ]  TYPEF       =[----- ]
 fxI1        =[----  ]  fxI2        =[----  ]  fzF2        =[----  ]  h'I         =[----  ]
 hdF2        =[----  ]  hpF2        =[----  ]  hmF2        =[----  ]  h'ox        =[----  ]
 Zm          =[----  ]  ypF2        =[----  ]  h'F1        =[----  ]  Zd          =[----  ]
 foI1        =[----  ]  foI2        =[----  ]  f1MUF       =[----  ]  f2MUF       =[----  ]
*CPNFC 2014/03/17 00:15:00
```

Figure 3. Extract of SEAIONO ionogram dataset (manually scaled parameters).

In order to calculate hmF2, "hmF2Cal.m" function is created using a set of equations which input is from the extracted parameters (foF2, foE and MUF(3000)F2) on the ionogram dataset. The computation of hmF2 involves the use of Shimazaki (1955) [9] formula given by:

$$\mathrm{hmF2} = \frac{1490}{\mathrm{M(3000)F2}} - 176 \tag{3}$$

where M(3000)F2 is the propagation factor defined as

$$\mathrm{M(3000)F2} = \frac{\mathrm{MUF(3000)F2}}{\mathrm{foF2}} \tag{4}$$

The "hmF2Cal.m" function computes hmF2 using an improved version of the Shimazaki (1955) [9] formula developed by Bilitza et al. (1979) [10] given by:

$$\mathrm{hmF2} = \frac{1490\mathrm{MF}}{\mathrm{M(3000)F2} + \Delta\mathrm{M}} - 176 \tag{5}$$

where ΔM is the correction term for the underlying ionization and expressed as

$$\Delta M = \frac{f1 \text{ x } f2}{(f5 - f3) + f4} \tag{6}$$

$$f1 = 0.00232 \text{ x } R_{12} + 0.222 \tag{7}$$

$$f2 = 1 - \frac{Rz_{12}}{150} \text{ x } \exp\left(-\left(\frac{\Psi}{40°}\right)^2\right) \tag{8}$$

$$f3 = 1.2 - 0.116 \text{ x } \exp\left(\frac{R_{12}}{41.84}\right) \tag{9}$$

$$f4 = \frac{0.096 \text{ x } (Rz_{12} - 25)}{150} \tag{10}$$

$$f5 = \begin{cases} fof2/foE & \\ -0.012 & \text{if } foE = 0, \end{cases} \tag{11}$$

where Rz_{12} is the annual average of the sunspot number and Ψ is the magnetic dip latitude. The condition of equation (11) for foE was applied from the work of Nava et al. (2008) [11]. These conditions become necessary due to multiple data gaps for the scaled foE data on the SEALION ionogram dataset. Days with sporadic E (Es) are removed from foE dataset using values > 3.6 MHz as inferred from the study of Wongcharoen et al. (2015) [12].

The thickness parameter (B0) is computed using the bottomside NeQuick 2 model and scripted as "B0Cal.m" function. The NeQuick 2 model reproduces the electron density profile of the ionosphere and has become a widely used empirical model for estimating ionospheric electron density over a region [13]. A semi-Epstein layer [14] analytically defines the bottomside of the F2 layer and the inflexion point of the electron density profile describes the B0 parameter:

$$B0 = 0.385 \frac{NmF2}{(dN/dh)_{max}} \tag{12}$$

where NmF2 is the maximum electron density of F2 layer and $(dN/dh)_{max}$ is the maximum value of the derivative of the height of the electron density which is computed using the empirical relation [15] given as

$$\ln\left(\left(\frac{dN}{dh}\right)_{max}\right) = -3.467 + 1.714\ln(foF2) + 2.202\ln(M(3000)F2) \tag{13}$$

Section 3 entails the saving of the extracted and computed ionospheric profile parameters into three file formats; *.mat (MATLAB MAT-file), *.txt (text file) and *.csv (comma-separated values). The overall ionospheric parameters processed by the SEAIONO program are six and they are the foE, foF2, h'F2, MUF(3000)F2, hmF2 and B0.

3. Software Packaging and Installation

The execution of the function scripts described in the preceding section is done by the use of a graphical user interface (GUI) to support "point-and-click" control of the designed software application. The final function scripts are packaged as a single installer file using MATLAB application compiler that collect all the necessary libraries to install and run the software. The installation (or execution) of the SEAIONO application program relieve user the need of using MATLAB software to run the ionogram extraction toolbox. The SEAIONO program is written to run under on Windows operating system. The minimum system requirements are 240 MHz processors speed with at least 200 MB of RAM, VGA graphics of 1024 x 768 system resolution. In the specified dataset directory defined by the user, the shortcut of the "SEAIONOReader.exe" must be present.

4. Conclusion

In this paper, an application software toolbox packaged to run under Windows operating system, named SEAIONO, capable of extracting ionospheric profile parameters on the SEALION ionogram dataset was described. The ionogram extraction algorithm and the application GUI was coded in MATLAB. However, the execution the SEAIONO software is independent of MATLAB. The GUI of the software enables alphanumerical input data entered by the user and gives the option to compute hmF2 and B0 parameters. The output files are essentially save using different file formats (e.g. text format, MATLAB format and spreadsheet file format).

Acknowledgement

The authors are grateful to the National Institute of Information and Communications Technology (NICT), Japan for providing the ionogram dataset. This work is supported by the grant GUP-2016-016 of Universiti Kebangsaan Malaysia.

References

[1] Bello SA, Abdullah M, Hamid NSA. Investigation of Ionospheric Minimum Frequency Near Dip Equator. Advanced Science Letters. 2017;23(2):1329-32.

[2] Pezzopane M. Interpre: a Windows software for semiautomatic scaling of ionospheric parameters from ionograms. Computers & Geosciences. 2004;30(1):125-30.

[3] Reinisch B, Huang X, Galkin I, Paznukhov V, Kozlov A. Recent advances in real-time analysis of ionograms and ionospheric drift measurements with digisondes. Journal of Atmospheric and Solar-Terrestrial Physics. 2005;67(12):1054-62.

[4] Arıkan F, Arıkan O, Salous S. A new algorithm for high-quality ionogram generation and analysis. Radio Science. 2002;37(1):1-11.

[5] Jiang C, Yang G, Zhou Y, Zhu P, Lan T, Zhao Z, et al. Software for scaling and analysis of vertical incidence ionograms-ionoScaler. Advances in Space Research. 2017;59(4):968-79.

[6] Ivanov V, Ivanov D, Ryabova N, Ryabova M, Chernov A, Elsukov A. Computer filtering of images of ionograms of HF ionospheric radio communication paths and an algorithm for determining signal-to-noise ratio. Journal of Applied Engineering Science. 2018;16(1):116-24.

[7] Nozaki K. FMCW lonosonde for the SEALION Project. Journal of the National Institute of Information and Communications Technology. 2009;56(1-4):287-98.

[8] Bello S, Abdullah M, Hamid NSA, Yoshikawa A, Olawepo AO. Variations of B0 and B1 with the solar quiet Sq-current system and comparison with IRI-2012 model at Ilorin. Advances in Space Research. 2017;60:307–16.

[9] Shimazaki T. World wide daily variations in the height of the maximum electron density in the ionospheric F2 layer. J Radio Res Labs, Japan. 1955;2(7):86-97.

[10] Bilitza D, Eyfrig R, Sheikh N. A global model for the height of the F2-peak using M3000 values from the CCIR numerical map. ITU Telecommunication Journal. 1979;46:549-53.

[11] Nava B, Coisson P, Radicella S. A new version of the NeQuick ionosphere electron density model. Journal of Atmospheric and Solar-Terrestrial Physics. 2008;70(15):1856-62.

[12] Wongcharoen P, Kenpankho P, Supnithi P, Ishii M, Tsugawa T. Comparison of E layer critical frequency over the Thai station Chumphon with IRI. Advances in Space Research. 2015;55(8):2131-8.

[13] Alazo-Cuartas K, Radicella S. An improved empirical formulation of an ionosphere bottomside electron density profile thickness parameter. Advances in Space Research. 2017;60(8):1725-31.

[14] Rawer K. Replacement of the present sub-peak plasma density profile by a unique expression. Advances in Space Research. 1982;2(10):183-90.

[15] Mosert de Gonzales M, Radicella SM. On a characteristic point at the base of F2 layer in the ionosphere. Advances in Space Research. 1990;10(11):17-25.

IOP Conf. Series: Journal of Physics: Conf. Series **1152** (2019) 012016 doi:10.1088/1742-6596/1152/1/012016

Near real time ionospheric monitoring system over Malaysia using GPS Data: My-Iono Service

Nyanasegari Bhoo Pathy[1-2], Tajul Ariffin Musa[1], Mhd Fairos Asillam[3], Wan Anom Wan Aris[1], Siti Syukriah Khamdan[1]

[1]UTM-GNSS & Geodynamics Research Group, Faculty of Geoinformation and Real Estate, Universiti Teknologi Malaysia, 81310 UTM Skudai, Johor, Malaysia.

[2]Space Science Division, National Space Agency (ANGKASA), Ministry of Science, Technology and Innovation, Jalan Turi, 42700 Banting, Selangor.

[3]National Planetarium, Ministry of Science, Technology and Innovation, 53, Jalan Perdana, 50480 Kuala Lumpur

ssegary@gmail.com

Abstract. Recently, real time information on the local ionospheric condition is becoming an important need for space and ground based technological systems which are prone to be affected by the local ionospheric state. Knowledge on the impact from the sun activity to the equatorial and low latitude ionosphere is crucial for research and development purpose in all developing nations which are relying on space-based technology systems such as the Global Navigation Satellite System (GNSS). As the first action step to achieve deeper understanding and hands on experience on real time ionospheric monitoring, the National Space Agency of Malaysia (ANGKASA) in collaboration with Universiti Teknologi Malaysia (UTM) has developed the first near real time ionospheric and space weather monitoring system for Malaysia (My-Iono) in year 2015. This web-based platform operates based on the National R&D GPS Continuously operating reference station network (NRC-net). The system runs based on a locally derived algorithm called Equatorial Ionosphere Index (EIX). The EIX was formulated based on 10 years (2004 to 2013) GPS derived empirical Total Electron Content (TEC) data obtained frpm 78 My-RTKnet stations around Malaysia. The fundamental of My-Iono Service consists of vertical TEC maps with a latency of ~ 2 mins over Malaysia and information on the current ionospheric status; Normal, Medium and Severe, time series of Mean VTEC for station specific, time series of RAte of Change of tEc (RACE) for station specific and time series of Ionospheric Zenith (Iz) Delay for station specific estimated in near real time using GPS observations. This paper presents on the development of My-Iono Service in terms of system design and architecture.

1. Introduction

Space weather refers to the conditions of the sun and in the space environment. Knowledge of the state of the ionosphere plays a crucial role in this chaotic weather system as space weather induces severe ionospheric disturbances that can cause damages to space-borne and ground-based technological systems, as well as endanger human life or health. The main industries whose operations can be severely affected by extreme space weather are electric power, space crafts, aviation and GPS/GNSS-based positioning and navigation industries (Fisher, 2011). Ionospheric disturbances are strongly affected by the number of the sunspots. Leong et al. (2009) stated that the ionospheric free electrons, known as TEC are strongly affected by the number of sunspots. It is affected the ionosphere due to the areas around the spots emit greater amounts of the ultraviolet radiation. The solar flare and coronal mass ejection (CME) are originated in magnetically active regions around the visible sunspot grouping.

Since the geomagnetic field and the ionosphere are linked in complex ways, a disturbance in the geomagnetic field often causes a disturbance in the ionosphere through fluctuations in electron density. This process can cause strong scintillation effects and large rapid changes in the ionospheric delay for GPS/GNSS signals, within time periods of about one minute (Janssen, 2012). The travelling ionospheric disturbances (TIDs) are moving ionospheric structures that may affect the use of accurate differential ionospheric refraction values for GPS applications. TIDs are also understood as plasma density fluctuations that propagate through the ionosphere over a wide range of velocities and frequencies, which can be seen in GPS TEC observations. Various studies have demonstrated that TIDs are a manifestation of the interaction between ionospheric and atmospheric gravity waves (Hines, 1960; Memarzadeh, 2009). Therefore, it would be of great advantage that the perturbations in the ionosphere are effectively quantified, translated into indices and disseminated to users as alert/warning

The common practice for quantifying ionospheric perturbations is to utilize geomagnetic indices such as Kp and disturbance storm time index, Dst (Menvielle and Berthelier, 1991). The Kp index quantifies the disturbances in horizontal component of geomagnetic field based on three hours observation of magnetometer. This index used by Space Weather Prediction Center (SWPC) for geomagnetic alerts and warnings that need to be issued to users. The Dst index is derived from a network of near-equatorial geomagnetic observatories that measures the intensity of globally symmetrical equatorial electrojet.

The geomagnetic indices have dual purposes which are (i) to give information on the geomagnetic on the geomagnetic activity level and (ii) to study the geomagnetic activity itself and its response to the various parameters such as influence of solar wind parameter and interplanetary magnetic field (IMF) (Svalgaard, 1997; Maezawa, 1978; Berthelier, 1979). These global indices are well suited to represent global phenomena so an offset from local ionospheric behavior must be expected. However, the occurrence and behavior of local ionospheric disturbances do not fully correlate with the characteristics shown in global/planetary indices (Jakowski et al., 2006). These scenarios are worse in the equatorial region as equatorial ionosphere effects is profoundly severe and much less understood than in other regions. The ionospheric TEC exhibits highly dynamic spatial and temporal morphological characteristics (Leong *et al.,* 2011). There is also a lack of ionosphere index to estimate the level of disturbances in Malaysia. To better classify, a local ionosphere index which will more closely describe the ionospheric disturbances over Malaysia must be obtained.

In this frame, the National Space Agency of Malaysia (ANGKASA) in collaboration with Universiti Teknologi Malaysia (UTM) has developed the first near real time ionospheric and space weather monitoring system for Malaysia (My-Iono Service) in year 2015. This web-based platform operates based on the National R&D GPS Continuously operating reference station network (NRC-net). The system runs based on a locally derived algorithm called Equatorial Ionosphere Index (EIX). The EIX was formulated based on 10 years (2004 to 2013) GPS derived empirical Total Electron Content (TEC) data obtained frpm 78 My-RTKnet stations around Malaysia. The details on the system design and architecture of the My-Iono Service will be presented in section 2 below.

2. System Design and Architecture of My-Iono Service Platform

The My-Iono Service design and architecture is illustrated in Figure 1. The system comprised with four (4) major sub-systems. As illustrated in Figure 1; *Subsystem 1* is data collection from NRC-net;

Subsystem 2: My-Iono Processing Centre and near real-time Ionosperic Maping; *Subsystem 3:* Data-based; and finally, *Subsystem 4:* Publicly web-based of My-Iono. Detail of each sub-systems will be described detail in section 2.1 until 2.5.

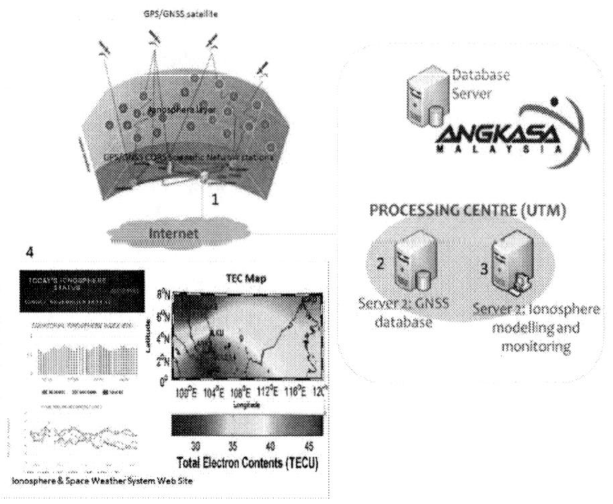

Figure 1: General My-Iono Service with four Subsystems.

2.1. Development of My-Iono Service

The development of My-Iono Service consists of three main steps.

i. Near real-time TEC derivation

The near real-time TEC value is derived from dual-frequency GPS measurement from each CORS station. The TEC real-time readings from CORS are recorded.

ii. Ionospheric perturbation Threshold

In this study, two types of threshold is used which are yearly limit that acquired empirical data from year 2004 to 2013 and monthly running threshold limit. The TEC_{lim} denotes an empirically defined threshold limit of ionospheric perturbation suited for Malaysia.

iii. Formulation of Equatorial Ionosphere Index (EIX)

The EIX is calculated based on gradient (North-South) by using linear regression of regional VTEC spatial variation which can be expressed mathematically as:

$$VTEC^{t0}_{j\left(\overline{Northing}\right)} = \alpha + \left(\beta \times \left(\overline{Northing}\right)\right) \tag{1}$$

where

$$\alpha = \overline{VTEC} - \left(\beta \times \left(\overline{Northing}\right)\right) \tag{2}$$

$$\beta_n = \frac{\sum_{j=1}^{n}(Northing_j - \overline{Northing})(VTEC_j - \overline{VTEC})}{\sum_{j=1}^{n}(Northing_j - \overline{Northing})^2} \tag{3}$$

where n is the number of line grid in latitudinal direction, j is point grid in line n, $Northing_j$ is the coordinate of j in northing component, $\overline{Northing}$ is the average value of $Northing_j$, $VTEC_j$ is the VTEC for each point grid in line n, $\overline{VTEC}$ is the average $VTEC_j$ in line n, $VTEC^{t0}_{j(\overline{Northing})}$ is the mean VTEC over the region, and β_n is the slope $\overline{VTEC}$ for each line. From equation (3), the EIX is computed which can be expressed mathematically as:

$$EIX = \left[\left(VTEC^{t0}_{j\left(\overline{Northing}\right)}\right) \times \left(\frac{1}{VTEC_{lim}}\right) \times 10\right] \tag{4}$$

where $VTEC_{lim}$ is twice the averaged VTEC over 10 years empirical TEC data. In this research project, the average VTEC from the 10 years Bernese post-processing is 74.4 TECU. The EIX is scaled from 1 to 10 to act as indicator to describe the conditions of the ionosphere in the area. Figure 2 shows the level of indicator that is used in My-Iono Service.

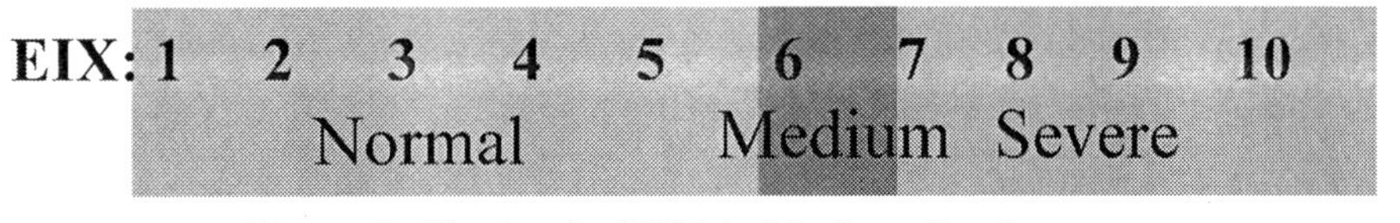

Figure 2: The level of EIX in My-Iono Service.

2.2. Analysis and performance assessment

This phase consists of three main steps.

i. ***EIX assessment***
The EIX is assessed by calculating the correlation coefficient between EIX and globally/planetary ionospheric disturbances indices.

ii. ***EIX correlation analysis***
The correlation between EIX and global/planetary ionospheric disturbance indicator is analyzed for the degree of agreement and suitability.

iii. ***Equatorial Ionosphere Analysis***
In this section, the equatorial ionospheric phenomena is recorded and probed. The regional to small scale ionospheric disturbances in monitored.

2.3. Correction in Differential Code Bias

The differential code bias (DCB), as inner delay difference between the two frequencies, must be considered when estimating the TEC. Several meters of error, which equal up to $\pm$20 to 40 TECU can occurred if the effect of the DCB is ignored. As a result, the TEC estimation can even turn to negative (Jin *et al.*, 2012; Sardon *et al.*, 1994). Based on the assessment that is carried out during analysis, the

IOP Conf. Series: Journal of Physics: Conf. Series **1152** (2019) 012016 doi:10.1088/1742-6596/1152/1/012016

results turn out to give a wrong estimation due to the neglected estimation of DCB. Therefore, the DCB should be estimated and removed during TEC estimation in My-Iono System.

The DCB consists of receiver and satellite hardware delays that affect the accuracy of estimation of TEC (Jin *et al.,* 2012; Arikan *et al.,* 2008). The crucial part in DCB estimation is to estimate the receiver DCB which are not available for all analysis center (Jin *et al.,* 2012). However, various studies have been carried out to estimate and developed the suitable algorithm and approaches to estimate the DCB values. In My-Iono Service, the DCB is estimated by adopted method from Jin *et al.* (2012) in Figure 3.

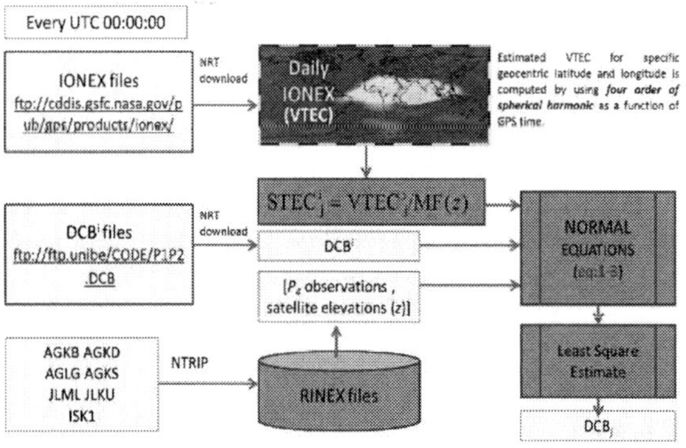

Figure 3: The processing strategy of near real time DCB in My-Iono Service.

2.4 Customization of My-Iono Service and Output

The My-Iono Service provides the near real time information such as:

a. Time measurement in local time;
b. The current ionopheric status; Normal, Medium and Severe.
c. A bar chart of the indicator,
d. Time series of mean VTEC for station specific,
e. Time series of Rate of Change of TEC (RACE) for station specific; and
f. Time series of Ionospheric Zenith (Iz) delay for station specific.

All of the above information's are displayed and saved per-epoch in form JPEG file. The system also provides the status of the program progress and problem trouble shooting such as internet connection status or system blackout where no data is provided from the server. Moreover, the near real-time TEC, RACE and Iz for each station are provided in form of numerical values and automatically written and saved in daily log file.

2.5. *My-Iono Service Webpage*

A webpage, My-Iono Service, is used in giving out alert/warning and information regarding current conditions and status from My-Iono Service to users. The My-Iono Service webpage can be access through www.myiono.angkasa.gov.my in Figure 4.

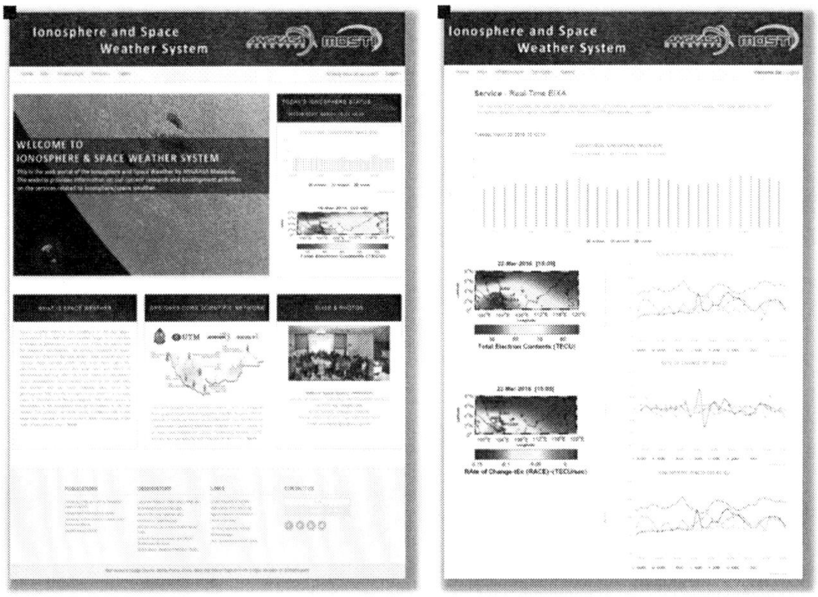

Figure 4: The layout of My-Iono Service webpage

3. Summary and outlook

This paper presents on the development of My-Iono Service in terms of system design and architecture. The My-Iono Service that has been developed has shown a good correlation with the global index and global TEC maps provided by other agencies. Moreover, this system able to trigger warning during the medium and severe conditions of ionosphere in the area during the performance assesment. Work is currently on going to address the performance of the system in regards with solar events in year 2015, 2016 and 2017.

References

Fisher, G. (2011), Satellite navigation & space weather: Understanding the vulnerability & building resilience, American Meteorological Society Policy Program, American Meteorological Society.

Hines, C. O. (1960), Internal atmospheric gravity waves at ionospheric heights, Canadian Journal of Physics Research, 38, 1441-1481.

Jakowski, N., S.M. Stankov, S. Schlueter and D. Klaehn (2006), On developing a new ionospheric perturbation index for space weather operations, Advances in Space Research, 38, 2596-2600, doi:10.1016/j.asr.2005.07.043.

Janssen, V. (2012), Likely impact of the approaching solar maximum on GNSS surveys: Be alert but not alarmed, Proceedings of the 17th Association of Public Authority Surveyors Conference (APAS2012), Wollongong, New South Wales, Australia.

Jin R, Jin S, Luo OF, Park JU (2012) MDCB: Matlab code for estimating GNSS satellite and receiver differential code biases. GPS Solut 16:541–548

Leong, S.K., T. A. Musa, K. A. Abdullah, R. Othman and choiC. Rizos (2009), GPS-derived local TEC mapping over peninsular Malaysia during solar minimum of sunspot cycle 24, Proceeding of the 2009 International Symposium on GPS/GNSS. 4-6 November. Jeju Self-Governing Province, Korea.

Leong, S. K., T. A. Musa, and K. A. Abdullah (2011), Spatial and temporal variations of GPS-derived TEC over Malaysia from 2003 to 2009, Proceedings of 10th International Symposium & Exhibition on Geoinformation (ISG2011) & ISPRS Commission II/5 & II/7 Conference, Shah Alam Convention Centre, Shah Alam, Malaysia.

Memarzadeh, Y. (2009), Ionospheric modeling for precise GNSS applications, Doctor of Philosophy. Delft University of Technology, The Netherlands.

Menvielle, M. and A. Berthelier (1991), The K-derived planetary indices: Description and availability, Reviews of Geophysics, 29 (3), 415-432, doi:10.1029/91RG00994.

Svalgaard, L., Geomagnetic activity: Dependence on solar wind parameters, in Coronal Holes and High Speed Solar Wind Streams, edited by J. B. Zirker, pp. 371–441, 1977F

IOP Conf. Series: Journal of Physics: Conf. Series **1152** (2019) 012017 doi:10.1088/1742-6596/1152/1/012017

Cross-calibration of ground-based measurement of rotational temperature of OH(3-1) at altitude ~87 km in Maimaga (63° N, λ = 129.5° E) and SABER/TIMED satellite data

P P Ammosov, G A Gavrilyeva, I I Koltovskoi, A M Ammosova and V I Sivtseva
Yu.G. Shafer Institute of Cosmophysical Research and Aeronomy of SB RAS, 31 Ave. Lenin, Yakutsk, Russian Federation
E-mail: p.p.ammosov@ikfia.ysn.ru

Abstract. Maimaga (63° N, λ = 129.5° E) is a base point in the meridional station network that we create in East Siberia to monitor changes in the upper mesosphere. Therefore, the comparison of measurements obtained by ground-based instruments and satellite data are relevant. It is known that the rotational temperature of hydroxyl (OH) corresponds to the temperature of the neutral atmosphere at the height of its radiation (~87 km). The night spectrum of OH(3-1) is measured by a new Shamrock 303i spectrograph since January 2013. The rotational temperature, estimated from the ratio of the P - branches, is compared with the kinetic temperature of the atmosphere at an altitude of 87 km measured with the SABER radiometer installed aboard the TIMED satellite. The v2.0 version of the SABER temperature profiles was used (http://saber.larc.nasa.gov). The SABER measurements obtained within the area (55°N-70°N)×(115°E-135°E) have been selected. The angle of view of the ground-based spectrograph is almost in the center of the selected area. The satellite overpasses the selected area only in September-November and January-March. The spectrograph measurement is considered if the difference in time from the satellite measurement does not exceed a half of hour. The analysis of 328 cases of coincident measurements has shown that the difference in temperatures obtained from the satellite and ground-based spectrograph lies within the errors of both instruments. The correlation coefficient between two databases is equal 0.8. Based on the performed analysis, it has been concluded that a series of hydroxyl rotational temperatures can be used to study temperature variations of different time scales including long-term trends at the OH emission altitude (~87 km).

1. Introduction

One of the available and inexpensive methods allowing tracking climatic changes in the upper atmosphere is a measurement of temperature using the emission of nightglow. A mesopause is the atmosphere region where the mesosphere borders the thermosphere (80-100 km), in which the temperature minimum is located. The radiating layer of excited hydroxyl (OH) is located at a distance of about 87 km in the mesopause region. The activated hydroxyl molecule executes $2 \cdot 104\ s^{-1}$ collisions before the radiation which is sufficient for thermalization with the surrounding medium. Therefore, the OH rotational temperature calculated from the night sky spectrum indicates to the neutral atmosphere temperature at the radiation height [1]. Recently, with the development of satellite methods, it has become possible to compare ground-based measurements of upper atmosphere temperature with satellite temperature measurements [2, 3]. As noted by Scheer et al. in [4], a comparison of ground-based instruments scattered around the globe with data from a single instrument with known characteristics located aboard a satellite can facilitate mutual calibration of the entire world network of instruments. Since January 2002, the TIMED satellite with a polar orbit equipped with a complex of instruments for studying the state of the atmosphere is successfully operating. One of the instruments is the SABER radiometer [5]. Among the many parameters measured with the radiometer there are vertical profiles of the kinetic temperature and radiation at wavelengths of 1.6 μm and 2.0 μm.

In this paper, the rotational temperatures OH (3-1) measured from January 2012 to May 2017 with the infrared Shamrock 303i spectrograph installed at the high-latitude Maimaga optical station (φ = 63 ° N, λ = 129.5 ° E) are compared with the kinetic temperature at a distant of an altitude of ~87 km obtained by the SABER.

2. Instruments

The infrared digital spectrograph consists of the Shamrock SR-303i monochromator equipped with the infrared iDus InGaAs detector. The spectrograph records the brightest emission OH(3-1) occurring near 1.5 μm wavelength. In this paper, we compare the rotational temperature OH(3-1) with the kinetic temperature of the atmosphere at the height of the hydroxyl emission measured from the satellite. The angle of view of the spectrograph is ~ 14° and is directed to the west at a zenith angle of 49° (the field of view of the spectrograph at an altitude of 87 km is an ellipsoid with size 38x32 km). The transfer function of spectrograph has an approximately Gaussian form, and its half-width is ~ 1 nm. The method for estimating the rotational temperature of molecular emissions is based on fitting of the model spectrum, convoluted with a transfer function of spectrograph for various predetermined temperatures, to the actually measured spectrum using the least-squares method [6]. The model spectrum, whose deviation from the real one does not exceed the noise of registration, is considered to be the most appropriate to the reality and that its rotational temperature corresponds to the temperature at the height of the mesopause. As the estimations show, a random error in the temperature measurement does not exceed 2 K if the signal-to-noise ratio is greater than 20. The transition probabilities calculated by Mies [7] were used to derive the rotational temperature. Observations were made at night, when a zenith angle of the Sun was > 9°. The geographical location of station makes it possible to conduct observations only from the end of July up to the middle of May, because in the summer at the latitude of Maimaga the mesopause is constantly illuminated by the Sun. For the analysis, the data obtained during the moonless time without aurorae have been selected. The spectrum of OH(3-1) was exposed with a time resolution of 1 min.

SABER (Sounding of the Atmosphere using Broadband Emission Radiometry) is an infrared radiometer probing the atmosphere from the troposphere up to 350 km. It is installed aboard the NASA's TIMED satellite [5]. The vertical profile of kinetic temperature is calculated using the emission of 15 μm CO_2 under the condition of non-local thermodynamic equilibrium (non-LTE) in this altitude region. The algorithm for calculating of the temperature profile, which takes into account the non-LTE effect, including all collisional and radiative processes, is described in detail in [8]. In our work, the v2.0 version of the SABER database was used (http//saber.larc.nasa.gov). The random error in measuring the temperature with SABER at an altitude of 87 km is 5 K; the systematic error is estimated by 2 K.

In this work, we have selected those SABER data that were obtained when scanning the emitting layer of OH within the area (55°N-70°N) × (115°E-135°E) with the radiometer. The ground-based spectrograph's sight line intersects almost the center of this area. Because of the orbital characteristics, the SABER overpasses the station Maimaga only in September-November and January-March.

3. Results and discussion

The ground-based spectrograph integrates the radiation along the field of view and the rotational temperature reflects the average atmosphere temperature weighted in thickness of the emission layer. It is necessary to distinguish that part of the SABER vertical temperature profile that would be equivalent to the hydroxyl rotational temperature. The convolution of the vertical profile of volume emission of OH with the kinetic temperature profile is equivalent to the temperature measured with the ground-based instrument. The SABER radiometer two channels are intended to carry out a vertical scanning of airglow: in the 1.6 μm wavelength region, where the OH(5-3) and OH(4-2) bands are emitted, and in the 2.0 μm wavelength region where the OH(9-7) and OH(8-6) band are emitted. the radiation vertical profile in the 1.6 μm region has been chosen as a weight function since OH(3-1) radiates approximately at the same wavelength. To reduce the effect of solar illumination, the data obtained at the zenith angle of the Sun of less than 115 degrees, we have excluded these data from analysis. For comparison, the time difference between SABER and the ground-based measurements should not to be greater than 1 hour. Since 17 January 2013, 328 nights of measurements coinciding in time and space were found. During night, the SABER radiometer scanned the limb above Maimaga three–four times, and the scanning results have been averaged. A scatter plot of OH(3-1) rotational temperatures versus SABER kinetic temperature is shown in Figure 1 a. A correlation coefficient between two data sets is equal 0.8.

The time series of OH(3-1) rotational temperatures and the equivalent SABER temperatures are shown in Figure 1 b. The open blue squares and red triangles are the nighttime rotation temperature of OH(3-1) and the SABER temperature, respectively. The solid large black squares and triangles connected with lines correspond to the average temperature during the observation period from September to March, except December, when the SABER data are absent. As can be seen, the temporal temperature changes in the measurements of two independent devices are in good agreement. However, the mean rotational temperature is approximately 3 - 4 K higher than the mean temperature measured with the SABER radiometer.

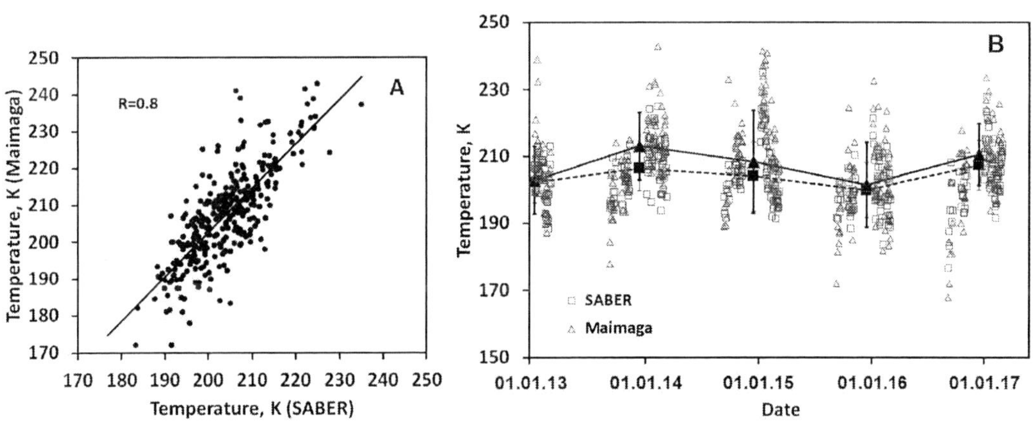

Figure 1. Left panel – the comparison of OH(3-1) rotational temperature measured at st. Maimaga with the equivalent kinetic temperature obtained with the SABER. Right panel - the OH(3-1) temperature series (red triangles), and SABER temperatures (blue squares) measured from 2013 to 2017. The solid large black squares and triangles are the annual mean temperatures. Vertical bars show the standard deviations.

The difference in the measurements of two independent devices can depend on various reasons. The first works devoted to inter-comparison of ground-based and satellite-borne measurements of the mesospheric temperature indicated that such a comparison is associated with some difficulties. A ground-based spectrograph receives emission from the entire emitting layer, the half-width of which is about 8–10 km. There is usually no information on the change of the height of emission layer and shape of the layer itself. The satellite device scanning the limb integrates the light horizontally at a certain fixed height. As a result, both instruments measure the temperature of different volume of air, even if the angles coincide precisely. Under quiet conditions, the measured temperatures should be close, however, with strong wave disturbances, the measured temperatures can differ greatly. On the other hand, the spectrograph and the radiometer measure the temperature at the radiation height of the hydroxyl bands from different vibrational levels. The difference in altitude between them is about 2 km and the average temperature can be different. Finally, a systematic error can consist of systematic errors with two different devices. Nevertheless, it can be concluded that a series of rotational temperatures can be used to study temperature changes at different time scales at the altitude of OH emissions (~ 87 km), including long-term trends.

Acknowledgments
This work was supported by the Russian Foundation for Basic Research (Grant No. 15-17-10001), and partly by Program of State Academies of Science (Project II.16.1.7. 0375-2016-0004).

References
[1] Khomich V Yu, Semenov A I and Shefov N N 2008 *Airglow as an Indicator of Upper Atmospheric Structure and Dynamics* (Berlin: Springer-Verlag) p 740
[2] von Savigny C et al. 2004 First near global retrievals of OH rotational temperatures from satellite based Meinel band emission measurements *Geophys. Res. Lett.* **31** L15111
[3] Oberheide J, Offermann D, Russell III J M and Mlynczak M G 2006 Intercomparison of kinetic temperature from 15 μM CO2 limb emissions and OH(3,1) rotational temperature in nearly coincident air masses: SABER, GRIPS *Geophys. Res. Lett.* **33** L14811
[4] Scheer J, Reisin E R, Gusev O A, French W J, Hernandez G, Huppi R, Ammosov P P, Gavrilyeva G A and Offermann D 2006 Use of CRISTA mesopause region temperatures for the intercalibration of ground-based instruments *J. Atmos. Sol.–Terr. Phys.* **68** pp 1698–1708
[5] Mlynczak M G 1997 Energetics of the mesosphere and lower thermosphere and the SABER experiment *Adv. Space Res.* **20** pp 1177–83
[6] Ammosov P P and Gavrilyeva G A 2000 Infrared digital spectrograph for hydroxyl rotational temperature measurements *Instruments and Experimental Techniques* **43** pp 792–7
[7] Mies F H 1974 Calculated vibrational transition probabilities of OH ($X^2\Pi$) *J. Mol. Spectrosc.* **53** pp 150–88
[8] Garcıa-Comas M et al. 2008 Errors in sounding of the atmosphere using broadband emission radiometry (SABER) kinetic temperature caused by non-local-thermodynamic-equilibrium model parameters *J. Geophys. Res.* **113** D24106

Statistical Analysis of Solar Radio Burst Type III and Type IV with Relation to Solar Activities

***Z.S.Hamidi[1,4], N. Ramli[2], N.N.M.Shariff[3]**
[1] Institute of Science, Universiti Teknologi MARA, Shah Alam, Selangor, Malaysia
[2]Department of Physics, Faculty of Science, University of Malaya, 50603, Kuala Lumpur, Malaysia
[3]Academy of Islamic and Contemporary Studies, Universiti Teknologi MARA, Shah Alam, Selangor, Malaysia
[4] Faculty of Applied Sciences, Universiti Teknologi MARA, Shah Alam, Selangor, Malaysia

Email: zetysh@salam.uitm.edu.my

Abstract. Solar activities are natural phenomena that occur at the solar surface by ejecting energy into interplanetary space and consequently hitting Earth's atmosphere where the commencement of geomagnetic storms. Generally, solar radio bursts originated at the same level as the solar flares, Coronal Mass Ejections (CMEs) and shock at solar atmosphere which the Sun project the radio energy into the interstellar medium. This is a novel study to investigate the correlation between solar burst type IV and type III based on their properties. The analysis is based on the statistical data of type IV and type III solar bursts from 2006 – 2011. The statistics of solar burst type IV associated with type III were obtained through e-Callisto (Compact Astronomical Low-frequency Low-cost Instrument for Spectroscopy Transportable Observatory) Spectrometer in the years 2006 until 2011. The aim of this study is to find out whether type III bursts preserved energy to solar bursts type IV thus led to the formation of geomagnetic storm. Statistical data from this study showed that most of the burst type IV were preceded and followed by type III burst. The total number of type IV events studied was 37, with 20 preceded by type III and the remainder with other types. Predominantly, solar bursts events are accompanied with great solar flares (class X and M) producing major storms. Although the study did not show much correlation between bursts type IV and type III in producing these storms, there are several possible explanations for this result. It may be the case that solar events such as solar flare, CMEs and active region ejected from the solar disc is not heading towards Earth and the energy has eventually dispersed into the interplanetary medium.

Introduction

Solar radio burst type III solar radio burst generated from the fast electron particle ejected from the Sun. One of the characteristic type III burst is the rapid drift of frequency from high to low that attribute the decreasing of electron plasma frequency . Besides, type III burst is usually long lasting, intense bursts seen in the low-frequency observations made from the space. It caused by streams of electron travelling along open field lines from flaring regions near the Sun into the interplanetary medium [1]. Type III bursts have consistently the fastest drift rates of bursts at metric wavelengths and was created by an electron beam through plasma emission. The radio emission tracks the electron beam as it travels through the decreasing plasma density of the solar corona and solar wind [2]. Meanwhile, the type III solar burst is a common signature of near-relativistic electrons streaming through the background plasma of the solar corona and interplanetary space, offering a means to remotely trace these electrons

[13]. This burst shows the coalescence of electron plasma waves from a non-thermal distribution with electron plasma waves from the distribution of thermal charge fluctuations [14, 15]. It is estimated that the total number of electrons $\gtrsim$ 22 keV required to produce a type III burst is $\lesssim 10^{34}$ [16, 17]. This burst can occur singularly, in groups, or storm. This fast drift burst is the most common of the meter wavelength bursts [18]. This main purposed of this study was intending to see the relationship of solar radio burst type IV in the presence of type III burst that associated with solar activity. Thus, led to the occurrence of a geomagnetic storm that affected the Earth magnetosphere. However, this study not focused on solar cycle whether it happens at the solar minimum or solar maximum.

These are from electron beams escaping along open magnetic field lines. Type III bursts typically start in the corona at frequencies of order 100 MHz, and then drift downwards in frequency as the driving electrons move out into the increasingly dilute plasma of the solar wind. Interplanetary type III bursts almost never show fundamental-harmonic structure. It is well accepted that type III bursts are associated with electron beams, which develop as faster electrons outrun slower electrons to form a localized hump in the electron distribution at velocities parallel to the magnetic field which are much larger than the electron thermal speed in the solar wind. Type IV are broadband quasi-continuum features associated with the decay phase of solar flares. Type IV emission has a duration of a few minutes to a few hours, covers a broad range of frequencies, varies smoothly in intensity. It generally believed to be synchrotron radiation from relativistic electrons spiralling in magnetic fields and results from trapped electron populations. There are several subclasses of type IV bursts, the type most tightly coupled to flares is called type IV continuum. The centimeter-wavelength bursts, particularly the intense ones, are closely associated with solar flares. Solar burst type IV always related to the development of sunspot groups. It has been well established that type IV bursts have a high probability of being followed by geomagnetic disturbance.

During solar flares, the X-ray flux received at the Earth increases dramatically, often within a few minutes, and then decays again in times ranging from a few tens of minutes to several hours [3]. The X-rays radiation from solar flares ionize the neutral atmosphere at D region heights greatly increasing the electron densities there and thus markedly lowering the effective VLF reflection height [4]. The energy of solar flare dependence of the decrease may go inversely with particle rigidity, or it may be flatter, depending on the details of the magnetic configuration in and around the blast wave [5, 6]. Detailed analysis on the temperature of the underlying loops increased toward higher altitudes, while the temperature of the coronal source increased toward lower altitudes [7]. The previous results of the analysis indicate that (1) most of the flare plasma was at temperatures between 3 and 10 million degrees; (2) the peak temperature decreased with time from about 8 x 10^6 K to 5 x 10^6 K over a period of 3.5 hours; (3) the differential emission measure steadily decreased with time at nearly all temperatures; (4) both radiation and conduction were important cooling mechanisms for the plasma at $T>10^5$ K; and (5) a substantial amount of energy, of the order of 3 x 10^{31} ergs [8]. A time-intensity profiles for solar proton events are analyzed with respect to the acceleration of energetic particles from the ambient solar wind by an interplanetary shock [9]. The uncertainty in the determination of the level of the primary electron bremsstrahlung as well as the lack of measurements on the γ -ray emission above 100 MeV combine to allow rather a wide range of energy distribution parameters [10]. Some of these variations were irreversible, occurred in the vicinity of magnetic neutral lines, and likely were related to magnetic energy release in the flare [11]. It is shown that Sweet's mechanism is much more effective in a highly compressible medium if the merging magnetic fields are exactly antiparallel [12].

Methodology

A statistical study of solar radio burst type III and type IV were performed using the data from Compound Astronomical Low cost Low frequency Instrument for Spectroscopy and Transportable Observatory (CALLISTO) from 2006 until 2011. Year 2006 until 2011 was chosen to examine the properties of solar radio burst type III and type IV because of the availability of archived complete list of solar radio burst data provided by e-CALLISTO website.

We use the Compound Astronomical Low cost, Low frequency Instrument for Spectroscopy and Transportable Observatory (CALLISTO) as a main system of solar radio bursts and RFI-monitoring for astronomical science and education [19-21]. The e-CALLISTO is the worldwide network connected via internet [22, 23]. This system can cover 24 hour data of solar burst, which operates identical spectrograph in different locations around the world. It is more than 133 instruments in more than 67 locations with users from more than 132 countries [24, 25]. However, it solar burst can be detected only after they have travelled far enough from the Sun [15, 26]. This system has an individual channel has 300 kHz bandwidth and can be tuned by the controlling software in steps of 62.5 kHz [27].

Results and Discussion

It is widely accepted that magnetic reconnection plays an important role in releasing the energy stored in sheared magnetic fields on the surface. In a magnetic reconnection process, oppositely directed field lines come closer and join, resulting in the release of magnetic energy in the form of thermal energy and particle acceleration (eruption of a plasma blob, associated M-class flare, and large-scale EUV wave observed by SDO. Magnetism in the Sun plays a key role in the dynamics of its surface. It is in part responsible for eruptions called coronal mass ejections that release high amounts of energy into space. Magnetism is phenomenon that arises out of the movement of electric charge, a fundamental property of matter. It creates a magnetic force, a "push" or "pull", on objects with moving electric charge. The electrons within the magnets are thought to spin in the same direction, resulting in a magnetic field. As a result of this self-created field, each magnet has a polarity, or two poles, north and south. When a pole of a magnet is brought to the same pole of the other, the magnets repel. When the poles are different, magnets attract. As a pair magnet can show, the movement of electric charge is the driving force of magnetism. The Sun has a very weak overall magnetic field (average dipole field). However, the solar surface has very strong and tremendously complicated magnetic fields.

As it shown in the Table 1, the occurrence of type III burst more frequent compared to the type IV burst, it was recorded with 320 and 37 events respectively. The events increasingly by year as the sites of CALLISTO also increasing. The data of solar burst increased drastically in 2011 due to an additional site to detect and monitor the Sun with a longer period of observation. This is because of the increment of the CALLISTO spectrometer all over the world, thus can monitor the Sun with 24 hours per days.

Year	Solar Radio Burst Type IV	Solar Radio Burst Type III
2006	1	15
2007	0	19

2008	3	10
2009	1	31
2010	8	45
2011	24	200
Total	37	320

In 2006, only a few sets of CALLISTO were installed and not cover the 24 hours Sun monitoring. In the beginning of the installation of CALLISTO spectrometer, its only have 12 sites only. Other than that, both burst rising from 2010 to 2011 because of solar activities that increase during that time which is from minimum solar cycle to maximum solar cycle.

Table 2 present the distribution of solar radio burst type IV events associated with others solar radio burst for 2006-2011. From 37 events of type IV, 20 events were preceded by type III burst and 22 events was followed by type III and the rest with other types and no burst detected. Type IV burst was preceded by 20 bursts of type III, 5 by DCIM, 2 from type II burst, 5 from type I burst and the rest of 5 with no burst detected at all. Other than that, type IV burst with 18 events was preceded and followed by both type III burst.

	Type IV		
	Preceded	Followed	Both
Type III	20	22	18
Other Type	12	12	7
No burst detected	5	3	-
TOTAL		37	

Table1: Distribution of solar radio burst type IV events associated with others solar radio burst for 2006-2011.

From this table it appears that the occurrence of solar radio burst type IV mostly associated with type III solar radio burst. As this study focused only on solar radio burst type III associated with type IV, the preceded of solar radio burst type III only included based on the durations not exceeding than 24 hours before type IV occurred. Type III emission that commenced after type IV was not focused in this study. However, there were 22 events listed in the e-Calisto archived in which type III followed type IV bursts.

As a comparison, the classification is by the presence or absence of an associated type III solar radio burst. In order to examine their possible relation, the set of 37 events was divided into two groups by two classifications, one depends on the presence of type III burst that preceded type IV burst and the other on type IV burst only. Figure 2 shows plot of KP index, an index of geomagnetic variability, for several days after each of the 37 type IV events at 3-hourly intervals.

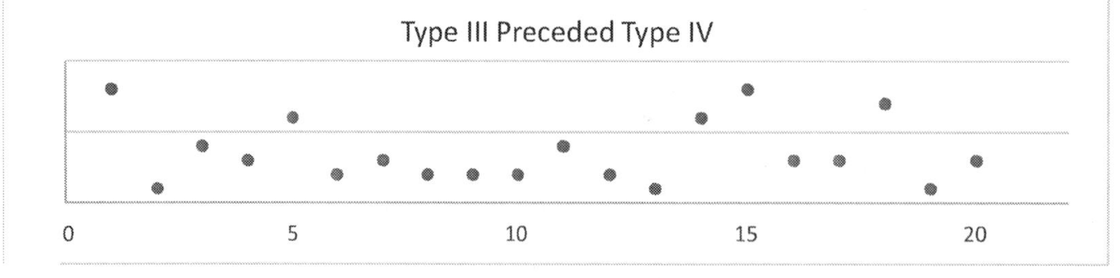

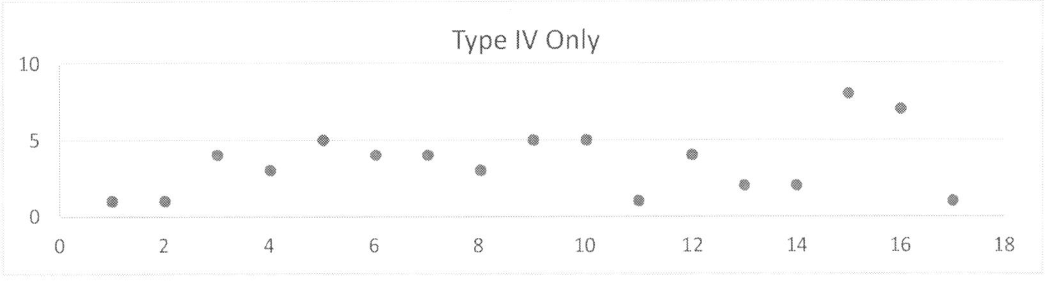

Figure 1. Distribution of geomagnetic storm activity as a function of events: (above) type III preceding type IV burst with 20 events; (below) type IV only without type III with 17 events.

Similar trends are clear among both type IV burst in the presence of type III burst and type IV burst only. With 20 events of solar radio burst type IV preceded by type III solar radio burst, there are 5 events associated with great geomagnetic storms from 6 to 8 Kp index. 2 events with geomagnetic disturbance and the rest of 13 events are geomagnetic calm from 1 to 3. While, for type IV burst only without associated with type III burst, 5 events associated with geomagnetic storm from 5 to 8 Kp index, with three of its at 5 Kp index. 4 events of geomagnetic disturbances and 8 events of geomagnetic calm.

As can be seen, with the presence of the type III burst before type IV burst event, the association of geomagnetic storm was great compared to type IV burst only. Even though type IV burst only also have geomagnetic storm, but it occurred at the index of 5 and only two events at 7 and 8 indexes, compared to type IV that have been preceded by type III burst where it associated with 6 to 8 Kp index.

This association between type III and type IV solar radio burst can be explained by the fact that both of this solar radio burst are generated by fast electrons. The disturbance which gives rise to type III bursts have high velocities and may be either streams of electrons or fast, transverse shock. While, the electrons which give rise to type IV are principally have a similar velocity range.

Although not showing an outstanding difference in the presence and absence of type III solar radio burst as shown in Figure 2, this might be due to some other factors. Among the factors that affect the geomagnetic storm are solar flare and solar wind. Table 6.3 compares the various categories of all type IV events in the presence of type III events and type IV events only with respect to association with powerful solar flares, great geomagnetic storm, speed of solar wind and from which active region the events was release.

As shown in figure 6.3, all type IV storms are found to be associated with very large solar flares. Predominantly, solar bursts events are accompanied with solar flares (from class C to M) producing major storms.

For type IV events preceded by type III burst, about half (9 events) of the events are found to be associated with major flare of X-class and M-class of solar flares. It clearly shown that 6 out of 9 events associated with great flare have sunspot that are directed toward Earth and the rest are not Earth directed or near to east or west limb. However, it can be seen that only 2 events with high speed of solar wind more than 500 km/s. Vice versa for type IV bursts only, it absolutely showing that 10 events associated with powerful solar flare have a slower solar wind speed which is below than 500 km/s.

No.	Type IV With Type III					Type IV Only				
	Date	Solar Flare	Geomagnetic Storm Index	Solar Wind, km/s	Active Region	Date	Solar Flare	Geomagnetic Storm Index	Solar Wind, km/s	Active Region
1	13-Dec-06	X3.4	8	645	930	2-Nov-08	Class B	1	401.2	1007
2	3-Nov-08	Class B	1	379.9	1007	11-Dec-08	Class B	1	418.1	1009
3	13-Feb-10	C9.6	4	308.2	1048	12-Feb-09	Class B	4	339.7	1012
4	13-Jun-10	M1.0	3	400.4	1079	12-Feb-10	M8.3	3	340.9	1046
5	1-Aug-10	C3.2	6	476.4	1092	16-Feb-11	M1.6	5	407	1158
6	5-Aug-10	C1.3	2	518.7	1093	3-Mar-11	C5.4	4	630.8	1164
7	14-Aug-10	C4.4	3	408.7	1099	7-Mar-11	M3.7	4	389.9	1164
8	18-Oct-10	C2.5	2	379.5	1112	30-May-11	C7.0	3	566.2	1226
9	19-Oct-10	C1.3	2	431.2	1112	2-Jun-11	C3.7	5	417.3	1226
10	22-Jan-11	C2.4	2	372.3	1149	7-Jun-11	M2.5	3	467.2	1226
11	14-Feb-11	M2.2	4	400.9	1158	29-Jun-11	Class B	1	339.1	1242
12	15-Feb-11	X2.2	2	460.2	1158	2-Aug-11	M1.4	4	405.8	1261
13	24-Feb-11	M3.5	1	325.2	1163	22-Sep-11	X1.4	2	405	1302
14	1-Mar-11	C6.0	6	621.8	1164	23-Sep-11	M1.9	2	373.4	1302
15	3-Aug-11	M6.0	8	356	1261	24-Sep-11	X1.9	8	326.3	1302
16	9-Aug-11	X6.9	3	551.5	1263	25-Sep-11	M7.4	7	347	1302
17	11-Aug-11	C6.2	3	413.6	1263	25-Dec-11	M4.0	1	368.8	1387
18	1-Oct-11	M1.2	7	532.7	1305					
19	9-Nov-11	M1.1	1	353.3	1343					
20	13-Nov-11	C2.6	3	366.7	1347					

Table 2: Association of solar radio burst events with solar flares, geomagnetic storm, solar wind and active region.

Despite it is associated with great solar flare but still have lower geomagnetic index, this is might be from which active region releases and the solar wind speed. From which part of solar disc also take in consideration whether it was ejected from the sunspot that is directed to Earth or not. If not directed towards Earth, the energy that disturb Earth's magnetic field is too small thus led to minor geomagnetic storm only. Compared to other energy that Earth directed which is, of course, will give extra amount of energy to magnetic field. As is known, the occurrence of solar radio burst type IV with geomagnetic storm have an outstanding relation. This was proven in this study, 12 events of the great geomagnetic storm (indexes from 5-8) associated with powerful flare. Out of 12 events, 3 events were C class flare and the remainder are M-class and X-class flare.

Conclusion

This clarify that, even though it has high solar flare ejected from the active region that Earth directed, but the solar wind speed is low thus do not have enough energy to eject the electron to the interplanetary medium to disturbing Earth's magnetosphere. High speed winds bring geomagnetic storms while slow speed winds bring calm space weather. Other than that, there also have some events which have slow solar wind speed but have higher index of geomagnetic storm and great solar flare. This happen due to the accompanying CME to the events thus propelled more energy to the magnetic field. The association of CME in producing geomagnetic storm will be discussed in selected events.

References

[1] R. Lin, H. Hudson, Non-thermal processes in large solar flares, Solar Physics, 50 (1976) 153-178.
[2] B.R. Dennis, Solar hard X-ray bursts, Solar Physics, 100 (1985) 465-490.
[3] Z.S. Hamid, N. Shariff, Z. Ibrahim, N. Ramli, Investigation of the statistical properties of solar radio burst type II and III, Space Science and Communication (IconSpace), 2015 International Conference on, IEEE, 2015, pp. 335-338.
[4] N.R. Thomson, C.J. Rodger, M.A. Clilverd, Large solar flares and their ionospheric D region enhancements, Journal of Geophysical Research: Space Physics, 110 (2005).
[5] E. Parker, Sudden Expansion of the Corona Following a Large Solar Flare and the Attendant Magnetic Field and Cosmic-Ray Effects, The Astrophysical Journal, 133 (1961) 1014.
[6] Z. Hamidi, N. Shariff, Z. Ibrahim, C. Monstein, W.W. Zulkifli, M. Ibrahim, N. Arifin, N. Amran, Magnetic Reconnection of Solar Flare Detected by Solar Radio Burst Type III, Journal of Physics: Conference Series, IOP Publishing, 2014, pp. 012006.
[7] L. Sui, G.D. Holman, Evidence for the Formation of a Large-Scale Current Sheet in a Solar Flare.
[8] G. Withbroe, The thermal phase of a large solar flare, Astrophysical Journal, 225 (1978) 641-649.
[9] D. Reames, Acceleration of energetic particles by shock waves from large solar flares, Astrophysical Journal, 358 (1990).
[10] N. Vilmer, A. MacKinnon, G. Trottet, C. Barat, High energy particles accelerated during the large solar flare of 1990 May 24: X/$\gamma $-ray observations, Astronomy & Astrophysics, 412 (2003) 865-874.
[11] A. Kosovichev, V. Zharkova, Magnetic energy release and transients in the solar flare of 2000 July 14, The Astrophysical Journal Letters, 550 (2001) L105.
[12] E. Parker, The Solar-Flare Phenomenon And The Theory Of Reconnection And Annihilation Of Magnetic Fields, Astrophysical Journal, Supplement Series (US), 8 (1963).
[13] H.A.S. Reid, H. Ratcliffe, A review of solar type III radio bursts, Research in Astronomy and Astrophysics, 14 (2014) 773.
[14] D. Melrose, On the theory of type II and type III solar radio bursts. I. The impossibility of nonthermal emission due to combination scattering off thermal fluctuations, Australian Journal of Physics, 23 (1970) 871-884.
[15] Z. Hamidi, N. Shariff, Automatic Solar Radio Burst Type II and III Image Processing Tracking by Using CALLISTO System, Industrial Engineering, Management Science and Application (ICIMSA), 2017 International Conference on, IEEE, 2017, pp. 1-4.
[16] S. Kane, Evidence for a common origin of the electrons responsible for the impulsive X-ray and type III radio bursts, Solar Physics, 27 (1972) 174-181.
[17] Z. Hamidi, N. Anim, N. Shariff, Z. Abidin, Z. Ibrahim, C. Monstein, Dynamical structure of solar radio burst type III as evidence of energy of solar flares, AIP Conference Proceedings, AIP, 2013, pp. 11-15.
[18] M. Ali, S. Sabri, Z. Hamidi, N. Husien, N. Shariff, N. Zainol, M. Faid, C. Monstein, e-CALLISTO Network System and the Observation of Structure of Solar Radio Burst Type III, Industrial Engineering,

Management Science and Application (ICIMSA), 2016 International Conference on, IEEE, 2016, pp. 1-5.
[19] Z. Hamidi, Z. Abidin, Z. Ibrahim, N. Shariff, Indication of radio frequency interference (RFI) sources for solar burst monitoring in Malaysia, ICPAP 2011, AIP Conference Proceedings-American Institute of Physics, 1454 (2012).
[20] Roslan Umar, Zamri Zainal Abidin, Zainol Abidin Ibrahim, Mohd Saiful Rizal Hassan, Zulfazli Rosli, Z.S.Hamidi, Population density effect on radio frequencies interference (RFI) in radio astronomy, ICPAP 2012, AIP Conference Proceedings, Bandung Indonesia, 2012, pp. 4.
[21] Z.Z. Abidin, N.M. Anim, Z.S. Hamidi, C. Monstein, Z.A. Ibrahim, R. Umar, N.N.M. Shariff, N. Ramli, N.A.I. Aziz, I. Sukma, Radio frequency interference in solar monitoring using CALLISTO, New Astronomy Reviews, 67 (2015) 18-33.
[22] S. Sabri, N. Zainol, M. Ali, N. Shariff, N. Hussien, M. Faid, Z. Hamidi, C. Monstein, The Dependence of Log Periodic Dipole Antenna (LPDA) and e-CALLISTO Software to Determine the Type of Solar Radio Burst (I-V), Industrial Engineering, Management Science and Application (ICIMSA), 2016 International Conference on, IEEE, 2016, pp. 1-5.
[23] Z.H. NM Anim, ZZ Abidin, C Monstein, NS Rohizat, Radio frequency interference affecting type III solar burst observations, in: R.S. et.al (Ed.) PERFIK 2012, American Institute of Physics, Malaysia, 2013, pp. 5.
[24] Z. Hamidi, N. Zainol, M. Ali, S. Sabri, N. Shariff, M. Faid, N. Husien, C. Monstein, Signal Detection of the Solar Radio Burst Type III Based on the CALLISTO System Project Management, Industrial Engineering, Management Science and Application (ICIMSA), 2016 International Conference on, IEEE, 2016, pp. 1-4.
[25] N. Zainol, S. Sabri, Z. Hamidi, M. Ali, N. Shariff, N. Husien, C. Monstein, M. Faid, Effective Data Collection and Analysis of Solar Radio Burst Type II Event Using Automated CALLISTO Network System, Industrial Engineering, Management Science and Application (ICIMSA), 2016 International Conference on, IEEE, 2016, pp. 1-5.
[26] N. Husien, N. Zainol, Z. Hamidi, S.N. Sabri, N.N. Shariff, M. Faid, M. Ali, C. Monstein, Solar radio bursts detected by CALLISTO system and their related events, 3rd International Conference on Industrial Engineering, Management Science and Applications, ICIMSA 2016, Institute of Electrical and Electronics Engineers Inc., 2016.
[27] Z. Hamidi, N. Shariff, An Evaluation Performance of Log Periodic Dipole Antenna Based on the Parameter of Flux Density of the Solar Radio Burst Event, Industrial Engineering, Management Science and Applications 2015, Springer Berlin Heidelberg2015, pp. 685-692.

Acknowledgments
We are grateful to CALLISTO network, STEREO, LASCO, SDO/AIA, NOAA and SWPC make their data available online. This work was partially supported by the UITM internal BESTARI grant, 600-IRMI/DANA 5/3/BESTARI (067/2017) from the Kementerian Pengajian Tinggi Malaysia. Special thanks to the National Space Agency and the National Space Centre for giving us a site to set up this project and support this project. Solar burst monitoring is a project of cooperation between the Institute of Astronomy, ETH Zurich, and FHNW Windisch, Switzerland, Universiti Teknologi MARA and University of Malaya. This paper also used the NOAA Space Weather Prediction Centre (SWPC) for the sunspot, radio flux and solar flare data for comparison purpose. The research has made use of the National Space Centre Facility and a part of an initiative of the International Space Weather Initiative (ISWI) program.

The Relationship between Solar Radio Burst Type III and Geomagnetic Storm on the Earth that Occurred on 4th October 2017

N A M Norsham[1,2], **Z S Hamidi**[1,2]
[1] Institute of Science, Universiti Teknologi MARA Shah Alam Selangor Malaysia

[2] Faculty of Applied Sciences, Universiti Teknologi MARA Shah Alam Selangor Malaysia

E-mail: zetysh@salam.uitm.edu.my

Abstract. In this paper, the analysis of the Sun on 4th October 2017 is been highlighted by using the data from e-CALLISTO system, SWPC, SOHO and NASA. From the e-CALLISTO data, solar radio burst type III has occurred at 10:03 UT during the low activity of the sun. There are two active regions occurred during this event which were AR2682 and AR2683. The ejection of solar flares which are correspond to the solar radio burst is caused by the active region AR2683. Further analysis showed that the energy range of the burst are between 1.325 x 10^{-26} J and 5.301 x 10^{-26} J. The main objective of this study was to analyze the relationship between solar radio burst type III with the geomagnetic storm. Hence, we found that the K-index value during that day is 3 and interplanetary magnetic field of 6.6 nT.

Keywords: Sun; active region; 2683; solar radio burst; type III; geomagnetic storm

1. Introduction

The first person that introduced that solar radio burst type III is dominant with solar flare phenomenon was Wild [1]. Solar radio burst type III can been classified by three types, which are single, group or storm burst. The occurrence of these burst type stays in frequency range of 10 kHz to 1 GHz. The characteristics of type III solar burst can be seen from the rapid drift of frequency from high to low which attributes from the decreasing of electron plasma frequency [2]. Besides that, this burst can be recognizing by the rapid fast drift, short duration and large bandwidth. Type III radio burst usually come from the active region. This type III burst event rate can been used as an additional index of solar activity [3]. By investigating the solar radio burst type III, the details about the properties and structure of active region streamers can be obtained. The active region streamers are the most notable structure in solar corona correlated to the plasma level for the observation of frequency. Besides, solar radio burst type III structure is a sign of the magnetic reconnection starting point [4].

As stated by Wild that solar radio burst type III is dominant with solar phenomena, hence, the unpredictable solar effects on the climate of the Earth could be significant. Solar phenomena such as solar flare and Coronal Mass Ejections (CMEs) can release a big amount of plasma and magnetic flux to the solar wind. Solar flares can be classified into three classes based on their energetic particle emission. The first is the small flares without the energetic phenomena in particles, radio or X-ray emission. The second class is the small flares that produce low electrons' energy, which occurred with type III and microwave radio burst, and energetic X-ray bursts. The third class is the major solar flare eruptions distinguished by the energetic production of solar proton and type II and IV radio burst which is appeared with intense microwave, relativistic energy electrons, and X-ray emission. This is related to the event that occurs when the outer atmosphere of the sun magnetically heated [5].

When in high solar activity, the magnetosphere of the Earth will be disturbed which is caused by the increase in solar wind. This will cause the production of the magnetic storm and the aurora [6]. The production of geomagnetic storm can affect the communications, navigation systems, disruption of electrical systems, caused damage to satellite hardware and radiation hazards to the human.

2. Methodology

For this study, the data was taken from the new CALLISTO configuration at station Fritz Lensch, OE3FLB, Austria. Compact Astronomical Low-Cost Frequency Instrument for Spectroscopy and Transportable Observatory (CALLISTO) is a global network spectrometer system that observes the activities of the Sun in the different region of the world [7]. At this current moment, there are more than 144 instruments at more than 80 locations worldwide. Malaysia also participated in e-Callisto spectrometer network to contribute data and monitor solar activity for space weather research with the initiative from the International Space Weather Initiative (ISWI) project [8-10]. CALLISTO is built by two systems, which are outdoor and indoor system. Outdoor system consists of CALLISTO antenna and low-noise preamplifier. The indoor system consists of indoor amplifier, CALLISTO receiver and PC. CALLISTO spectrometry has a total range of frequency from 45 MHz to 870 MHz [11].Other sources that really help in providing the details are the freely available data from Space Weather Prediction Centre and Solar and Heliospheric Observatory (SOHO), Solar Monitor and National Aeronautics and Space Administration (NASA).).

3. Result and Discussion

The analysis of the sun on 4th October 2017 have been carried out by taking the data from the e-CALLISTO system. Figure 1 showed the sun in hydrogen alpha filter with the existence of two sunspots facing the Earth, which were 2682 and 2683. The active region 2683 produced minor B-class events where the flare was originated with the value of B2.4 at 03:54 UT. This active region produces 1% of M-class and X-class within the 24 hours' period. The sunspot number during that day is 25 with the radio flux of 86 sfu. Based on that data, it can be considered that the sun is at the low activity. Besides that, the solar flare activities were also occurred at a very low level and did not poses any threat for strong solar flares. Coronal holes image was presented in Figure 2. The image showed the solar corona part where the bright part are the active solar regions and flares while the darker areas indicate the coronal holes, where there is less radiation. The coronal hole area will let the solar wind to pass through the open magnetic field lines as this area have lower temperature and densities and thus have the thin regions. As can be seen on Figure 3, there was production of coronal mass ejection on 5th October 2017. It is believed that this CME was caused by the solar wind that flowed from the sun's surface on 4th October 2017. The value of the K-index during 4th October 2017 was 3 (quiet). However, the production of the CME was not too severe and did not cause any serious effect to the Earth and interplanetary space.

Based on Table 1 below, the wind speed for the sun that day was in normal value. The higher this value is, the higher the chances for a better geomagnetic condition enhancement. The density showed the moderate value with 6.5 protons/cm^3 in the solar corona. We will get to see the aurora if there are more particles in the solar wind as the particles will collide with the Earth's magnetosphere. In addition, the magnetic field, B_z was directed to the South where the lower this value goes, the higher the chances for the auroral activity.

IOP Conf. Series: Journal of Physics: Conf. Series **1152** (2019) 012019 doi:10.1088/1742-6596/1152/1/012019

Table 1. The Condition of the Sun on 4th October 2017 (Credited to Space Weather).

Parameter	Value
Solar Wind Speed	389.4 km/sec
Density	6.5 protons/cm^3
Radio Sun	10.7 cm flux: 86 sfu
Interplanetary Magnetic Field	B_{total} : 6.6 nT Bz : -5.9 nT South

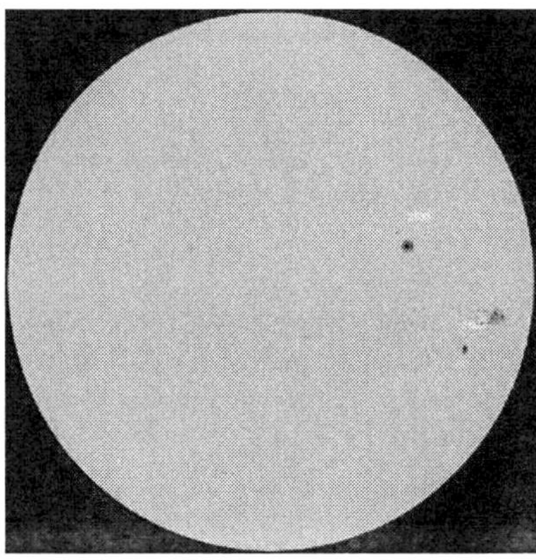

Figure 1. Active region on the sun's surface on 4 October 2017. (Credited to SDO/HMI)

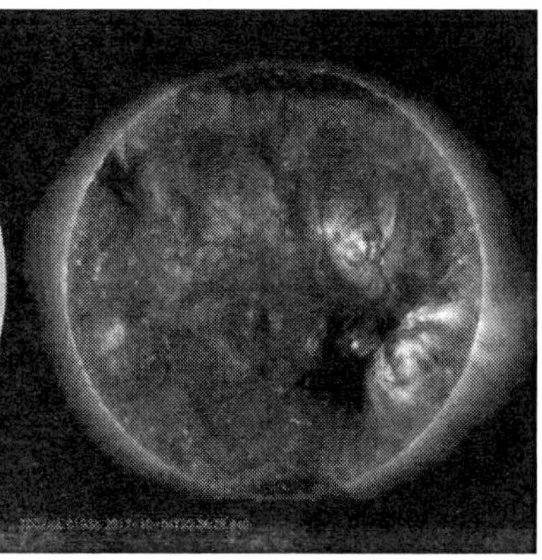

Figure 2. Coronal hole image. (Credited to SDO/AIA)

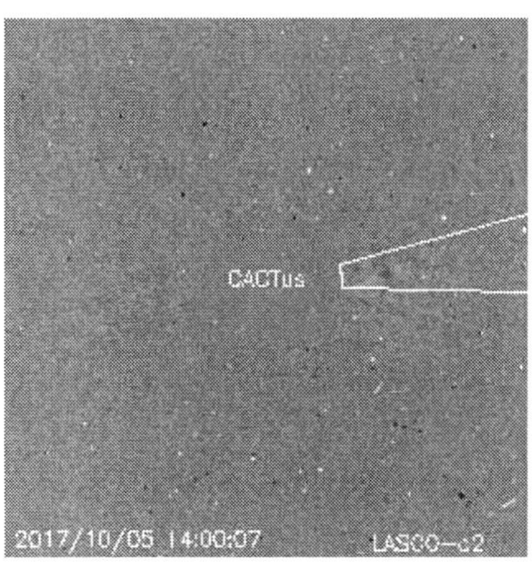

Figure 3. CME image on 5th October 2017. (Credited to LASCO-c2)

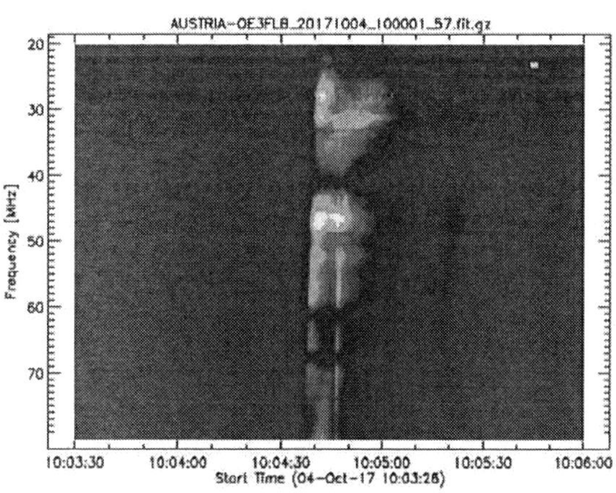

Figure 4. Solar radio burst type III on 4th October 2017. (Credited to e-CALLISTO)

Solar radio burst type III above was recorded by CALLISTO station at OE3FLB, Fritz Lensch, Austria. The event occurred at starting frequency of 20 MHz and ended at a frequency of 80 MHz. The energy range from the frequency of 20 MHz to 80 MHz was 1.325 x 10^{-26} J to 5.301 x 10^{-26} J. Solar radio burst type III above showed the group and single line type. Solar radio burst type III always has relation to the production of the solar flare.

4. Conclusion

By analyzing the data on 4th October 2017, a few conclusions can be made. We can learn to recognize type III solar radio burst characteristics, which occurred in the rapid fast drift, short duration and large bandwidth. Besides that, type III radio burst usually come from the active region and indicate that there was occurrence of solar flares. There was also CME occurred on 5th October 2017 which is believed came from the solar wind from the day before. The CME was not severe and cause no harm to the Earth. All this information will help in better understanding of the space weather; thus, any precaution can be taken in the future.

References

[1] J. Wild, S. Smerd, and A. Weiss, "Solar bursts," *Annual Review of Astronomy and Astrophysics,* vol. 1, no. 1, pp. 291-366, 1963.

[2] R. Lin, D. Potter, D. Gurnett, and F. Scarf, "Energetic electrons and plasma waves associated with a solar type III radio burst," *The Astrophysical Journal,* vol. 251, pp. 364-373, 1981.

[3] V. Lobzin, I. H. Cairns, and P. A. Robinson, "Solar cycle variations of the occurrence of Coronal Type III radio bursts and a new solar activity index," *The Astrophysical Journal Letters,* vol. 736, no. 1, p. L20, 2011.

[4] Z. Hamidi *et al.*, "Magnetic Reconnection of Solar Flare Detected by Solar Radio Burst Type III," in *Journal of Physics: Conference Series*, 2014, vol. 539, no. 1, p. 012006: IOP Publishing.

[5] J. Wild and S. Smerd, "Radio bursts from the solar corona," *Annual Review of Astronomy and Astrophysics,* vol. 10, no. 1, pp. 159-196, 1972.

[6] G. Mann, "Theory and observations of coronal shock waves," *Coronal Magnetic Energy Releases,* pp. 183-200, 1995.

[7] A. Benz *et al.*, "The background corona near solar minimum," *Solar Phys,* vol. 55, pp. 121-134, 2004.

[8] Z. Hamidi, Z. Abidin, Z. Ibrahim, and N. Shariff, "Indication of radio frequency interference (RFI) sources for solar burst monitoring in Malaysia," in *AIP Conference Proceedings*, 2012, vol. 1454, no. 1, pp. 43-46: AIP.

[9] S. Sabri *et al.*, "The Dependence of Log Periodic Dipole Antenna (LPDA) and e-CALLISTO Software to Determine the Type of Solar Radio Burst (I-V)," in *Industrial Engineering, Management Science and Application (ICIMSA), 2016 International Conference on*, 2016, pp. 1-5: IEEE.

[10] Z. S. Hamidi, N. N. M. Shariff, C. Monstein, A. Ibrahim, and M. Yusof, "The international cooperation of solar radio burst project using e-Callisto system network," in *Information Science and Applications (ICISA) 2016*: Springer, 2016, pp. 309-317.

[11] A. O. Benz, C. Monstein, and H. Meyer, "CALLISTO–a new concept for solar radio spectrometers," *Solar Physics,* vol. 226, no. 1, pp. 143-151, 2005.

Acknowledgments

We appreciate efforts of the teams CALLISTO network, STEREO, LASCO, SDO/AIA, NOAA, SWPC and CACTus CME catalog. This research was supported by UiTM under grant, 600-IRMI/DANA 5/3/BESTARI (067/2017),600-IRMI/PERDANA 5/3/MITRA (005/2018)-2,600-RMI/FRGS 5/3 (0077/2016).

Observation of Ionosphere Scintillation and Total Electron Content (TEC) Characteristic at Equatorial Region

Norsuzila Ya'acob, Noraisyah Tajudin, Mohd Shahrul Azree Remly, Darmawaty Mohd Ali, Suzi Seroja Sarnin and Nani Fadzlina Naim

Faculty of Electrical Engineering,Universiti Teknologi MARA (UiTM) Malaysia, 40450 Shah Alam, Selangor,Malaysia.
Wireless Communication Technology (WiCoT), Faculty of Electrical Engineering, Universiti Teknologi MARA, 404500 Shah Alam Selangor, Malaysia

norsuzilayaacob@yahoo.com,tnoraisyah@gmail.com,mohdshahrulazree@yahoo.com, darma504@salam.uitm.edu.my, suzis045@salam.uitm.edu.my

Abstract. Ionosphere scintillation and total electron content (TEC) measurement were investigated in terms of value at Equatorial based on GPS/GNSS measurement. This paper presents the analysis month-to-month variation of Scintillation (S4) and Total Electron Content (TEC) based on during night time and day time activity. This paper also presents the analysis of scintillation (S4) and Total Electron Content (TEC) during equinox and solstice. The analysis presented from January to December 2016. The measurement and recorded data of scintillation (S4) and Total Electron Content (TEC) was done by GPS Ionosphere Scintillation and TEC Monitoring (GISTM), installed at UKM, Malaysia (2.92°N 101.78°E). Strong Scintillation (S4) was obtained from PRN 135 with S4 ≥ 0.4 (0.430368) on 11:33 UTC (19:33 UT). The maximum Total Electron Content (TEC) was found on midday, generally the daily peak is around 5:00 to 10:00 UTC (13:00 to 18:00LT). April shows that the highest Total Electron Content (TEC) about 152 TECU. For seasonal variation, Scintillation (S4) highest on equinox and lowest on solstice, meanwhile Total Electron Content (TEC) highest on equinox and lowest on solstice. The disturbance will cause error in distance measurement for positioning and navigation.

Keywords— Scintillation, Total Electron Content, Ionosphere, GPS / GNSS, The first section in your paper

1. Introduction

Ionosphere is an upper layer of the earth's atmosphere that contains a high concentration of ions and free electrons, which can exert an effect on GPS signals. An ionosphere condition in Malaysia is located in the south geomagnetic latitude (lower latitudes) has a high electron density, thus causing potential disruption of electromagnetic waves through the ionosphere in the area. Interference caused by the activity of the ionosphere on the GPS / GNSS passing through the ionosphere could be a delay for code measurements, as well as fluctuations in amplitude and phase to phase measurement. One of the phenomena in the ionosphere irregularities that interfere with the satellite signal propagation are ionosphere scintillation. Ionosphere scintillation is the rapid fluctuations of the amplitude and phase of the satellite signal as shown in Figure 1. Satellite communication systems disrupted by ionosphere scintillation is a system that uses multiple GHz frequency or frequency band L (L-Band). GPS signals, which uses L-band frequencies, used as ionosphere scintillation monitoring and interruption of satellite communications to satellite communication systems in the L-band frequency. To determine the effects of ionosphere scintillation on satellite communications then do a comparison between the index data S4 with CNO ratio of GPS data. The results show that the greater the S4 index, the ratio CNO increasingly fluctuating. Fluctuations such signals are satellite signals are disrupt by ionosphere scintillation.

Furthermore, the TEC's study at Equatorial region is significant for unpredictable and highly variable of phenomena like fountain effect and Equatorial anomaly. TEC is refer of the total number of electrons be present along a path in between a radio transmitter to receiver. Where is the presence of electrons will be affected the radio waves. This means that the more radio signal will affect when there are many electrons along the path of the radio waves. The unit is TEC - Unit (TECU) which is equivalent to 10.000 billion electrons per square meter or 1 TEC Unit = 1016 el / m2. TEC good value will vary daily, seasonal, annual (following cycles of solar activity) even the location (latitude). In one day, TEC will be undervaluing in the morning, getting higher during the day, to reach the peak, and then decreases back toward evening. This is because the electron production is highly dependent on the level of solar radiation. In the daytime, the intensity of solar radiation received ionosphere is very high and decreases in the afternoon.

High solar activity will cause any disruption in the ionosphere higher. The effects are decreasing in number of electrons, the velocity of the iron will increase and the temperature and the plasma cut-off frequency will decrease [1]. Disturbances in the ionosphere can cause problems in the application of radio waves such as radio communications, satellite navigation systems and space weather so the study ionosphere disturbances become an important subject in the research contributions of space weather. The radio signal from the satellite as it passes through the disturbed ionosphere will cause interference with the faraday rotation signal and the signal is also fluctuated rapidly in amplitude and phase are received in the receiver due to the irregularity of the ionosphere electron density medium. These fluctuations are known as ionosphere scintillation [3], [11].

The parameters of spatial and temporal variations ionosphere at mid-latitudes are quite small comparing to the low-latitude and equatorial regions. The specific knowledge is important to determine the value of TEC for different location in different geophysical conditions. A few researches have been conducted to investigate the characteristics of short-term temporal and spatial for the condition of local ionosphere. This is to be ensuring these characteristics are well understood in Malaysia [4]. However, the coupling effect of sun earth on the equatorial ionosphere is lacking due to the barely distributed GPS stations at the equatorial region in comparison to the high latitude and mid-latitude regions [3].

The objective of this research is to investigate Ionosphere Scintillation and Total Electron Content (TEC) Characteristic at Equatorial region. This research also will evaluate the value of Ionosphere Scintillation and Total Electron Content (TEC) at Equatorial region. From this evaluation, it will verify and validate the characteristic of Ionosphere Scintillation and Total Electron Content at Equatorial region.

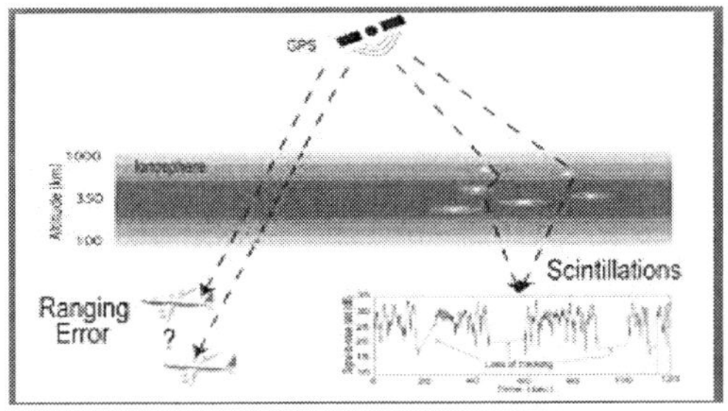

Figure 1. Schematic of ionosphere effects on GPS signal

2. Data and Methodology

This research was done by data collection of dual-frequency GSV4004B GPS Ionosphere Scintillation and TEC Monitoring (GISTM) installed at UKM, Malaysia. There were 11 GPS satellites able to track simultaneously by GPS receiver at L1 and L2; and the satellite data were converts to slant TEC (STEC). All available satellites TEC data were sampled into 60s interval. The slant TEC (TEC along the line of sight) is obtained through an excel file as shown in Figure 2.

$$STEC = TEC0 \qquad \text{(Eq. 1)}$$

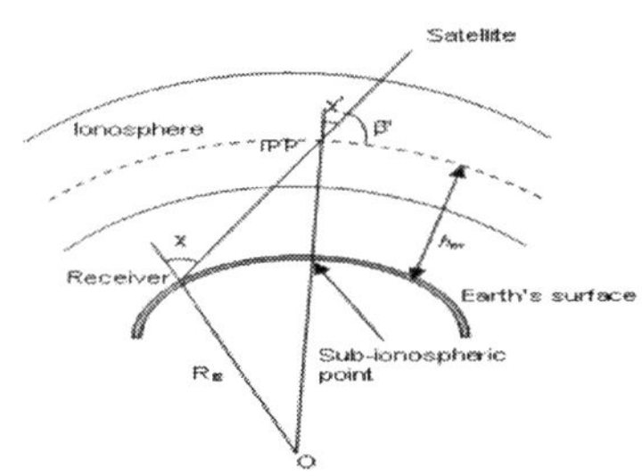

Figure 2. Ionosphere single layer model

The sampling rate of amplitude scintillation is recorded more than 60 second intervals in 50Hz. The two parameters of S4, which are the correction of S4 (S4cor) and total S4 (S4t) are compute by the receiver. Where is the total of S4 (S4t) is includes of ambient noise effect [14]. The corrected S4 index analysis was conducted without the influence of ambient noise and it is determined by using:

$$S4 = \sqrt{(S_{4t}^2 - S_{4cor}^2)}$$

Table 1. Scintillation Type and Range

Scintillation Type	Range
Negligible	S4 < 0.2
Weak	0.2 ≤ S4 ≤ 0.3
Moderate	0.3 ≤ S4 ≤ 0.4
Strong	S4 ≥ 0.4

The 12-months data analysis was collected from January to December 2016. The data was analyze based on Equinox (20 March 2016 and 22 September 2016) and Solstice (20 June 2016 and 21 December 2016) occurring. The result is present the scintillation (S4) and Total Electron Content (TEC) for January to December 2016.

3. Result and Discussion

This section present result of measurement of Scintillation (S4) and Total Electron Content (TEC) month – to – month variation based on night time and day time activity, also seasonal variation from January to December 2016.

3.1. *Month – to – month variation of Scintillation (S4) and Total Electron Content (TEC)*

Refer to Figure 3 and Figure 4 show that variation month – to – month of Scintillation and Total Electron Content (TEC) for all PRNs was monitored at UKM station from January to December 2016. Coordinate Universal Time (UTC) is equal to Local Time (LT) when plus eight (LT=UTC+8).

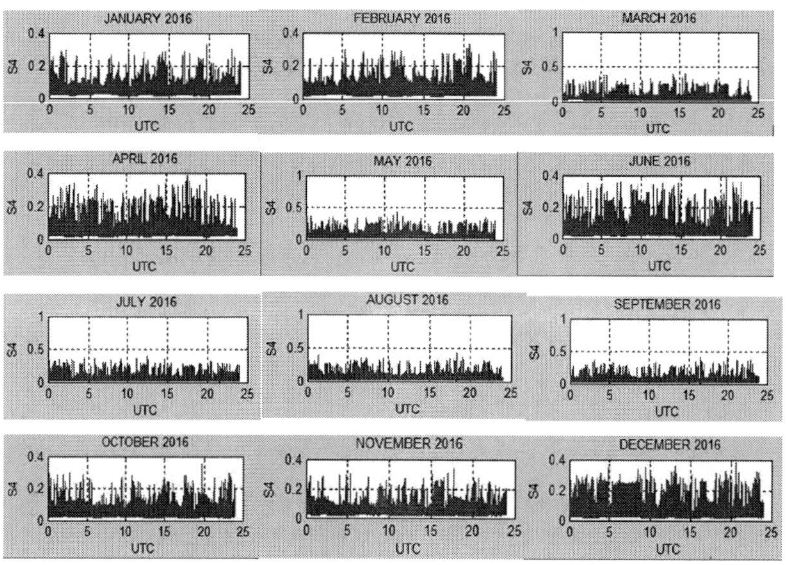

Figure 3. Month – to – month variation of Scintillation (S4)

As shown as in Figure 3, the strong Scintillation ($S4 \geq 4$) is occurred on March, May, July, August and September during the nighttime. This strong Scintillation ($S4 \geq 4$) occurred from 10:00 to 22:00 UTC (18:00 to 6:00 LT). Meanwhile on January, February, April, June, October, November and December shows that moderate Scintillation ($0.3 \leq S4 \leq 0.4$) was occurred during night time from 10:00 to 22:00 UTC (18:00 to 6:00 LT). Moderate Scintillation ($0.3 \leq S4 \leq 0.4$) was occurred during daytime from 00:00 to 10:00 UTC (8:00 to 18:00 LT) during the year of 2016. Generally, the occurrence of the ionosphere scintillation events that occurs in equatorial region is happen after local sunset and before local sunrise due to irregularities of equatorial density [14].

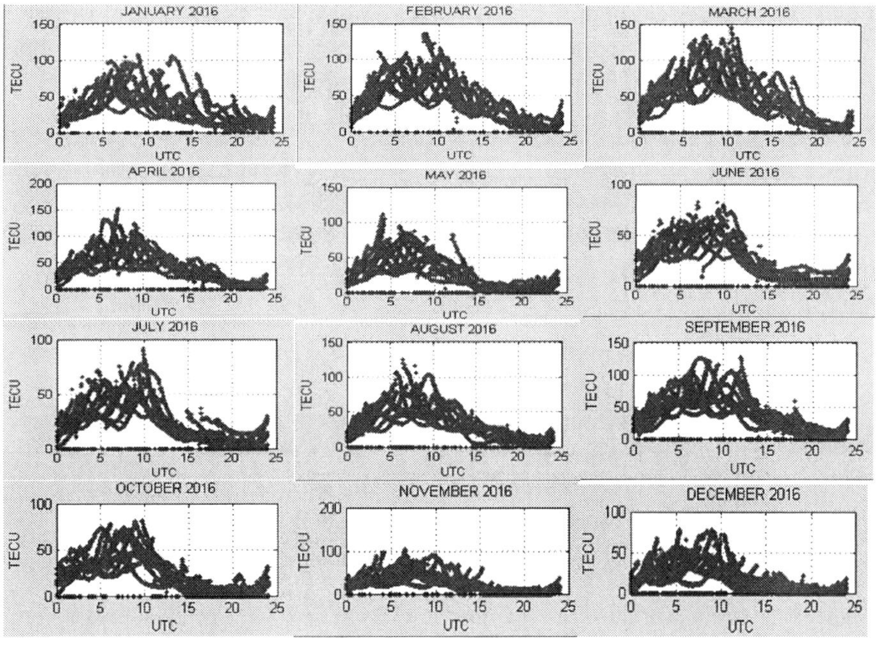

Figure 4. Month – to – month variation of Total Electron Content (TEC)

As per result in Figure 4, the graph shows the low of TEC in the early morning and will be increase with time. The post noon is recorded as the maximum TEC value and slowly decreases after sunset. We identify the pattern of gradual morning rise starting on 00:00 UTC (8:00 LT) and evening decrease of TEC is around 10:00 UTC (18:00 LT) in all month. While, around 5:00 to 10:00 UTC (13:00 to 18:00LT) is recorded as a daily peak of TEC values. April shows that the highest Total Electron Content (TEC) about 152 TECU and the lowest Total Electron Content (TEC) is on December about 77 TECU.

3.2. Scintillation (S4) and Total Electron Content (TEC) activity during night time

As per result on Figure 3, it showed that Scintillation (S4) was occurred during night time and has a strong Scintillation (S4). For example, on 30 May 2016, measurement of Scintillation (S4) was obtained from PRN 135 with S4 $\geq$ 0.4 (0.430368) on 11:33 UTC (19:33 UT). This night time ionosphere disturbance is happening may be due to large and small-scale irregularities that exist together during nighttime at UKM station, Malaysia [14].

Figure 4 shows that TEC is low during night time. The recombination between electrons and ions during night time will cause of TEC decays.

3.3. Scintillation (S4) and Total Electron Content (TEC) activity during day time

Refer to Figure 3, we can see that Scintillation (S4) was occurred in day time in moderate Scintillation (S4) type. For example, on 31 October 2016, measurement of Scintillation (S4) was obtained from PRN 9 with $0.3 \leq S4 \leq 0.4$ (0.301076) on 2:46 UTC (10:46 UT). Mostly the Daytime Scintillation (S4) is occurs without the phase scintillation occurrence. However, significant TEC variations are rarely took place throughout daytime scintillation activity. As Figure 4, the maximum TEC was found on midday, generally the daily peak is around 5:00 to 10:00 UTC (13:00 to 18:00LT). For example, TEC show the highest value on February, March and April. The TEC of daytime variation shows large variability comparing to nighttime variation. This is due of decays of TEC values because of

IOP Conf. Series: Journal of Physics: Conf. Series **1152** (2019) 012020 doi:10.1088/1742-6596/1152/1/012020

recombination between electrons and ions during nighttime. While, the electrons density is increasing during daytime due to particle emission by sun in the ionosphere. This daily variation happens because of daily earth rotation on its own axis and following the movement of the Sun.

3.4. Equinox and Solstice phenomena of Scintillation (S4) and Total Electron Content (TEC) activity
The data that are collected from January to December 2016 (12 months data) was analyzed based on Equinox (20 March 2016 and 22 September2016) and Solstice (20 June 2016 and 21 December 2016) occurring. The result is presented scintillation (S4) and Total Electron Content (TEC) based on Equinox and Solstice.

Table 2. Equinox and Solstice phenomena of Scintillation (S4) and Total Electron Content (TEC) activity

Phenomena	Month	S4	TEC
	March	0.401251	145
Equinox	**September**	0.407598	127
Solstice	**June**	0.386964	82
	December	0.382412	77

Based on Table 1, seasonal variation of Scintillation (S4) and TEC are presented. From the table it show that both Scintillation (S4) and TEC highest on equinox and lowest on solstice. The increase in the emergence of the scintillation associated with the terminator summer sun's magnetic field and meridians that form on these months. Formation of the magnetic field and the solar terminator causes the dynamo EXB drift current in the equatorial F layer strip thereby increasing plasma irregularity. This disorder is also associated with increased plasma bubbles in the coming months until the scintillation intense in these months. The neutral composition and O/N2 ratio at low latitude and equatorial will increase when there is changes in the flow of meridional winds during daytime from pole to equator region. The decrease will be maximum during the equinox. Hence, the result will be higher TEC in equinox due to higher of electron density. The ionosphere is varying with geographic location, time, certain solar and geomagnetic activities. The quantity of emission and solar activity from the Sun are very correlate with the sunspots number on its surface. In addition, the sudden sun eruption may cause the increasing activity in ionosphere and this correlated with number of sunspot.

Conclusion
Ionosphere Scintillation (S4) and Total Electron Content (TEC) were observed in terms of value at equatorial region based on GPS/GNSS measurement.
Strong Scintillation (S4) was obtained from PRN 135 with S4 $\geq$ 0.4 (0.430368) on 11:33 UTC (19:33 UT). The maximum TEC was found on midday, generally the daily peak is around 5:00 to 10:00 UTC (13:00 to 18:00 LT). April shows that the highest TEC about 152 TECU. For seasonal variation, Scintillation (S4) highest on summer and lowest on winter, meanwhile Total Electron Content (TEC) highest on equinox and lowest on summer. In order to observe the effects and characteristic of Ionosphere Scintillation (S4) and Total Electron Content (TEC) on GPS/GNSS receivers at equatorial region, the investigation of Ionosphere Scintillation (S4) and TEC activity should be made based on long-term data series.

4. Acknowledgement
The authors would like to thank Faculty of Electrical Engineering, Universiti Teknologi MARA (UiTM) for their valuable support.The authors also would like to thank to Institute of Space Science (ANGKASA),Universiti Kebangsaan Malaysia (UKM) Bangi for providing dual-frequency GSV4004B GPS Ionosphere Scintillation and TEC Monitoring (GISTM) data.

References

[1] NurHidayahBintiMohdSukaimi "Investigation of GPS Ionospheric Scintillation during Annular Solar Eclipse on 15th January 2010".

[2] M. Abdullah, A.F.M. Zain, Y.H. Ho and S. Abdullah "TEC and Scintillation Study of Equatorial Ionosphere: A Month Campaign over Sipitang and Parit Raja Stations, Malaysia" American J. of Engineering and Applied Sciences 2 (1): 44-49, 2009.

[3] Alina, M.H., Mohd, A.M.a, and Norbahiah, M., "ionospheric TEC and Scintillation during 15-16 May 2005 Major Storm over Equatorial Anomaly Region at Arau", Asia-Pasific Conference on Applied Electromagnetic Proceeding 4-6 December 2007,Melaka, Malaysia.

[4] Z.M.Radzi, M.Abdullah, A.M Hasbi, J.S. Mandeep and S.A.Bahari, "Seasonal variation of Total Electron Content at equatorial station, Langkawi,Malaysia" 2013 IEEE International Conference on Space Science and Communication (ICONSPACE), pp.186-189.

[5] Rohaida Mat Akir, Mardina Abdullah, KalaivaniChellappan, and Alina Marie Hasbi "Preliminary Vertical TEC Prediction Using Neural Network: Input Data Selection and Preparation".

[6] "Development of a Ground-Based Augmentation System (GBAS) Ionospheric Threat Model (GITM) for Singapore" The Sixth Meeting of Ionospheric Studies Task Force (ISTF/6), Bangkok, Thailand, 19 – 21 January 2016.

[7] C. N. Mitchell,L. Alfonsi, G. De Franceschi, M. Lester, V. Romano, and A. W. Wernik "GPS TEC and scintillation measurements from the polar ionosphere during the October 2003 storm" GEOPHYSICAL RESEARCH LETTERS, VOL. 32, L12S03, doi:10.1029/2004GL021644, 2005.

[8] GaoShu, "Activities for Ionospheric study for Ground Based Augmentation Systems (GBAS) in Singapore" IDCAS Workshop - SP/6, 6 May 2011.

[9] Rupesh M. Das, SnehaYadav, R. S. Dabas, Shailendra Saini and A. K. Gwal "Investigation of TEC variation and Ionospheric scintillation at the two hemispheres over the Polar region using GPS measurements" 978-1-4244-5118-0/11 ©2011 IEEE.

[10] TajulAriffin Musa, ShienKwun Leong, KhairulAnuar Abdullah, and Rusli Othman "ISKANDARnet IOMOS: Near real-time equatorial space weather monitoring and alert system in Peninsular Malaysia", SPACE WEATHER, VOL. 10, S11003, doi: 10.1029/2012SW000813, 2012.

[11] Charles S. Carrano, Atmospheric and Environmental Research Inc., Lexington, MA Keith M. Groves, Air Force Research Laboratory, Space Vehicles Directorate, Hanscom AFB, MA "TEC Gradients and Fluctuations at Low Latitudes Measured with High Data Rate GPS Receivers" ION 63rdAnnual Meeting, April 23-25, 2007, Cambridge, Massachusetts.

[12] Department Civil Aviation Malaysia (DCA),"Malaysia Performance-Based Navigation Implementation Plan", second edition 2012.

[13] International Civil Aviation Organization, "Guide for Ground Based Augmentation System Implementation", May 2013.

[14] ArameshSeif , Mardina Abdullah, Alina Marie Hasbi and YuhuaZou "Observation of GPS Ionospheric Scintillation at UKM, Malaysia", Proceeding of the 2011 IEEE International Conference on Space Science and Communication (IconSpace) 12-13 July 2011, Penang, Malaysia.

[15] RajendraMalla and Robert Fries "Multi-Layer Modeling and Simulation of the Effects of Ionospheric Scintillation on Service Availability of the GPS Augmentation Systems", 2010.

[16] YuhuaZou, Dongli Wang, QiushengShen "GPS Scintillations and TEC Monitoring at Guilin, China", 2008 International Workshop on Education Technology and Training & 2008 International Workshop on Geoscience and Remote Sensing.

IOP Conf. Series: Journal of Physics: Conf. Series **1152** (2019) 012021 doi:10.1088/1742-6596/1152/1/012021

Link Budget and Noise Calculator for Satellite Communication

NorsuzilaYa'acob,Noraisyah Tajudin, Suzi Seroja Sarnin,Siti AmalinaEnche Ab Rahim and Azrif Manut

Faculty of Electrical Engineering,UniversitiTeknologi MARA (UiTM),40450 Shah Alam,Selangor,Malaysia.
Wireless Communication Technology (WiCoT), Faculty of Electrical Engineering, Universiti Teknologi MARA, 404500 Shah Alam Selangor, Malaysia

norsuzilayaacob@yahoo.com,tnoraisyah@gmail.com,suzis045@salam.uitm.edu.my, amalinaabr@gmail.com,azrifmanut@salam.uitm.edu.my

Abstract. MATLAB is widely used in all areas of applied mathematics, in education and research at universities, and in the industry. MATLAB stands for MATrixLABoratory and the software is built up around vectors and matrices. This makes the software particularly useful for linear algebra but MATLAB is also a great tool for solving algebraic and differential equations and for numerical integration. Among MATLAB software main function is to produce a calculator to calculate the Link Budget System and the noise of the system. A link budget is accounting of all of the gains and losses from the transmitter, through the medium used (i.e. free space, cable, waveguide, and fiber) to the receiver in a telecommunication system. The purpose of calculating link budget is to investigate the system performance tied to operation threshold, to get the minimum power, Cmin that should be received at the demodulator input in order for communication to work properly with reading typically of 10 dB.By using this design, all the parameter (i.e. Effective Isotropic Radiated Power (EIRP), Path loss/ Free space path loss, Effective aperture, Flux density, Physical aperture, Transmit gain, Receive gain, Transmit power and Receive power) can be calculated. Lastly, the purpose of calculating noise in satellite link system is to get noise temperature ratio and noise density ratio. A noise temperature ratio is a figure of merit used to represent the quality of a satellite or an earth station receiver. While, noise density ratio is one of the most important and most often used parameters when evaluating a digital radio system.

Keywords-Automatic; Calculator; Space Technology; Link Budget;

1. Introduction

Link analysis is shows in detail how the difference between the transmit power and the receive power is accounted for [1]. The fundamental elements that related in communications satellite Radio Frequency (RF) or free space link are employing in link analysis. The basic link of power equation is exploited from basic transmission parameter which are antenna gain, free space path loss and beam width [2,3].The development of concept and quantify for system noise on the RF link is to define the parameters such as noise temperature , noise power, noise figure and figure of merit. The communications link design and performance defined by the carrier-to-noise ratio and related parameters, which are, based on the basic link and system noise parameters [4].

In communications, the carrier-to-noise ratio, is define as ratio of the strength of received carrier relative to the strength of the received noise and commonly it has written as CNR or C/N. The better quality reception and higher accuracy and reliability are providing by high C/N ration compare to the low C/N ratios [5, 6]. The C/N ratio between the carrier power of the desired signal and the total of received noise power are specifying in decibels (dB) [7]. If both carrier strength, Pc and the noise level, Pn in microwatts, then the carrier-to-noise ratio, C/N (dB) is given by following equation (1).

$$C/N = 10 \log 10 \, (Pc/Pn) \qquad (1)$$

The C/N ratio is measure in a similar approach to the technique of the signal-to-noise ratio (S/N) measurement and both specifications are performing as the quality of a communications channel. Nevertheless, the S/N ratio specification shows the more significant in practical situations [8]. Commonly, the C/N ratio is use in satellite communications systems to align the receiving dish; so that the best dish alignment is defining by maximize C/N ratio.

The purpose of calculating noise in satellite link system is to determine noise temperature ratio and noise density ratio. A noise temperature ratio is a figure of merit used to represent the quality of a satellite or an earth station receiver [9]. Noise density ratio is the one most important and often use parameters in evaluating a digital radio system [9]. While, Link budget analysis is need to analyze the signal loss factor during the propagation and estimate the required power in a certain transmitter to overcome noise that happen in the receiver [10].

2. Methodology

The link budget and noise calculator is design to ease user to calculate all the signal gains and losses to ensure that the signal intended for a given receiver is sufficiently strong enough to get the job done. The flowchart in Figure.1 shows the overall process of the project.

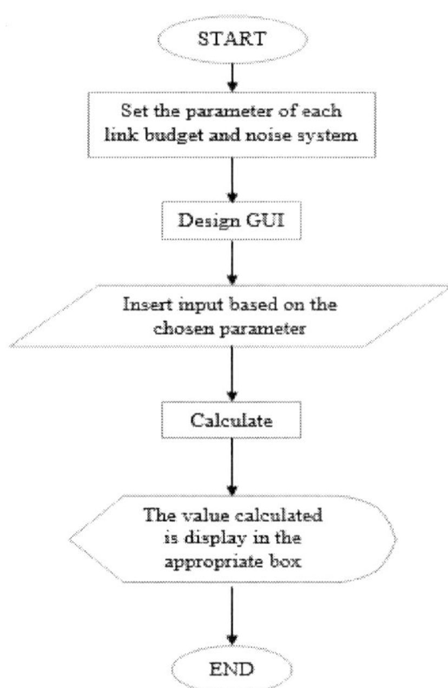

Figure.1. Flowchart of the overall project

By refer to Figure. 1 , the system starts by setting the parameter of each link budget and noise system. Next, design the GUI of the system. By successfully designing the GUI, user may now insert an input based on the chosen parameter. After that, the calculation part took place. User need to wait a while for the system to calculate. Finally, the calculated value will be display in the appropriate box. By displaying the calculated value, the system has successfully calculated the linked budget and the noise of the system.

The noise calculation involving the calculation of the carrier to noise ratio (uplink), carrier to noise ratio (downlink), total carrier to noise ratio, carrier to noise ratio density (uplink), carrier to noise ratio density (downlink) and total carrier to noise ration density. As for link budget that is in Radio Frequency (RF) is to get rough feel of viability and to find tune actual design. Figure. 2 below shows the Application of link budget in Radio Frequency.

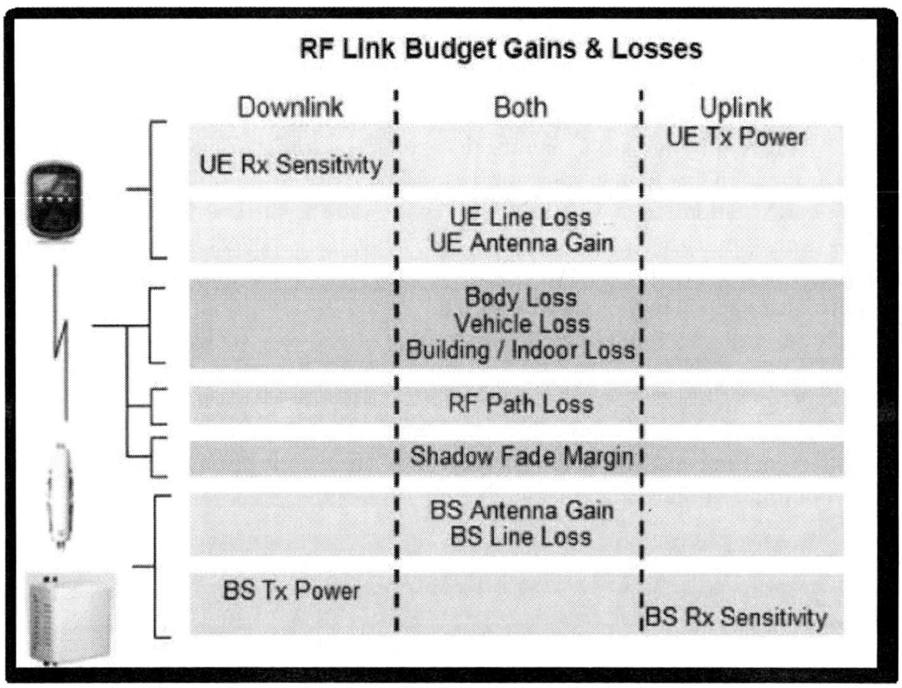

Figure. 2. Application of link budget in Radio Frequency

3. Result and Discussion

The design of GUI is show in Figure 3. The user need to insert the input values such as the speed of light (c), frequency (f), efficiency (η), effective area of isotropic antenna (A0), power transmit (Pt), transmitter gain (Gt),aperture antenna (Ar), input backoff (Boi), wavelength (λ), effective aperture (Ae), noise temperature (Tn), losses (L), distance radius (R) , Boltzmann constant (k), saturation flux density(Ψs), bandwidth (Bw), figure of merit for uplink and downlink (G/T),output back off (Boo) , diameter (D), effective isotropic radiated power (EIRP) and power receive (Pr). Number 1 until 16 is the parameter that available in link budget and noise calculator program. The formula for each parameter of the link budget has been set in the program, hence, user was only need to insert the input for the parameter that need to be calculate and click the "compute" button. Then the value will be display in the appropriate box.

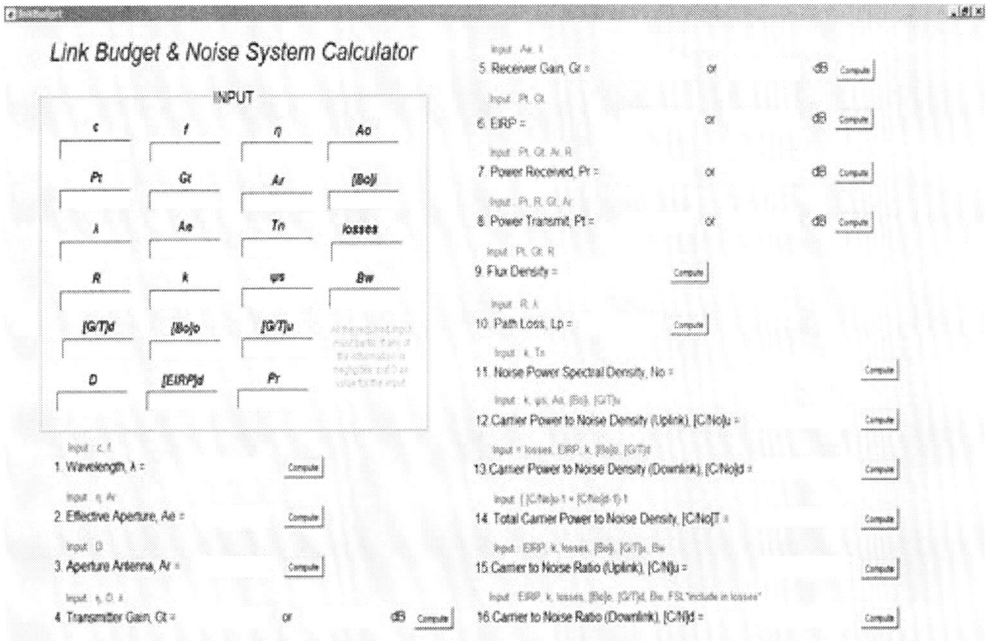

.Figure. 3. GUI for link budget and noise system calculator

Firstly, to find the wavelength and effective aperture, user must key in the value of speed of light (c), frequency (f), efficiency (η), and aperture antenna (Ar) as shown in Figure. 4 below. Here, other parameters will not give an output as the inputs are only involving the wavelength and the effective aperture only.

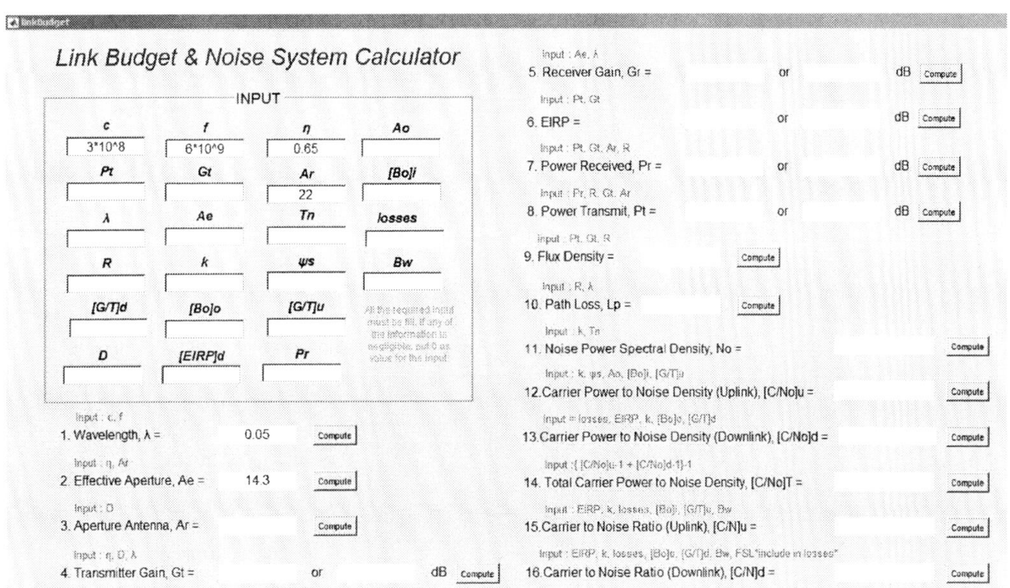

Figure. 4. Output for wavelength and effective aperture

Next, to find the path loss, user must key in the value of wavelength and distance radius (R). The wavelength value that has been calculating earlier also can be use. The interface is as shown in Figure. 5 below. The user must key in the value of Pt and Gt. to find the effective isotropic radiated power (EIRP), this calculator will calculate the value in dB as shown in Figure. 6.

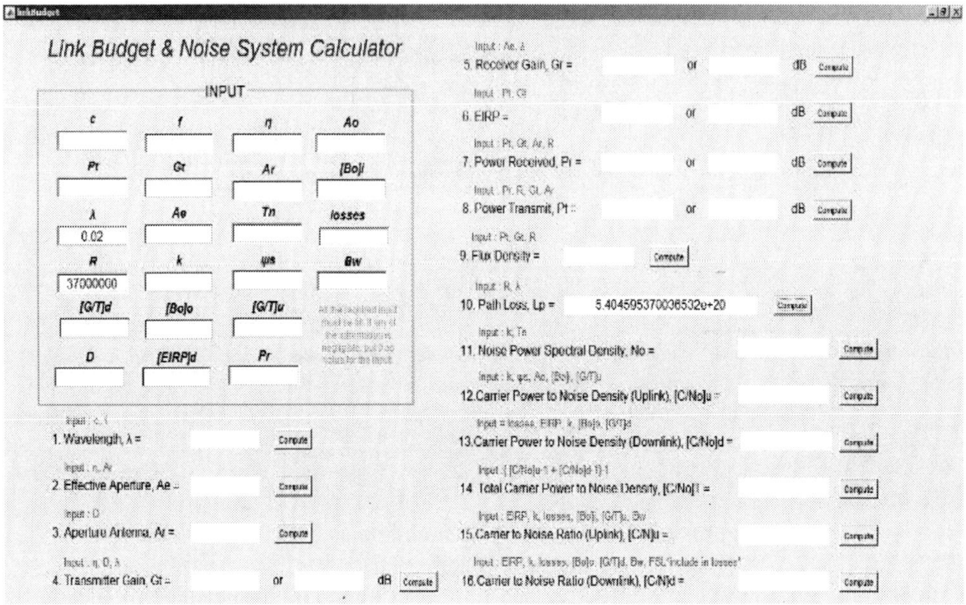

Figure. 5. Path loss Output

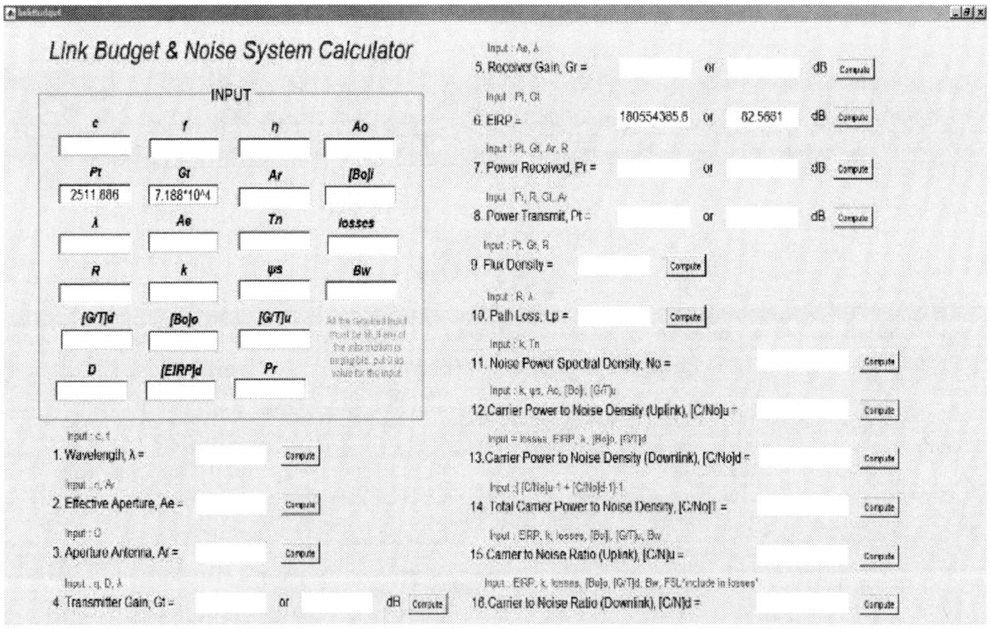

Figure. 6. EIRP Output

Next, to find the noise power spectral density (N0), the input that user needed to key in are Boltzmann constant (k) and noise temperature (Tn) as shown in Figure. 7. Lastly, refer Figure. 8 below to find the uplink, downlink and the total carrier power to noise density.

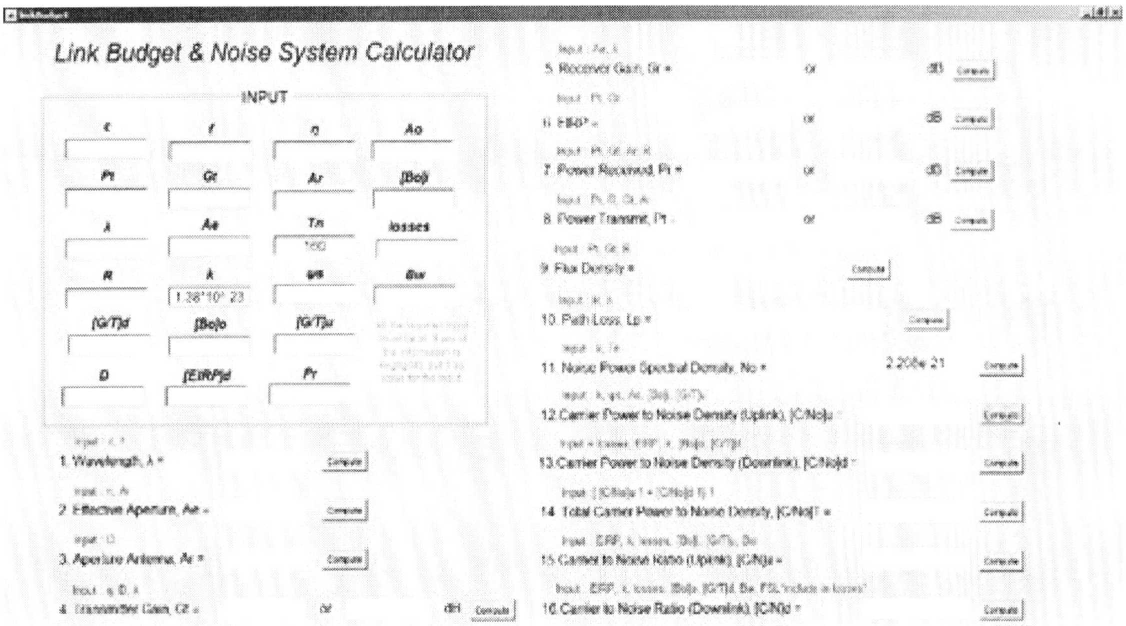

Figure. 7. Noise power spectral density Output

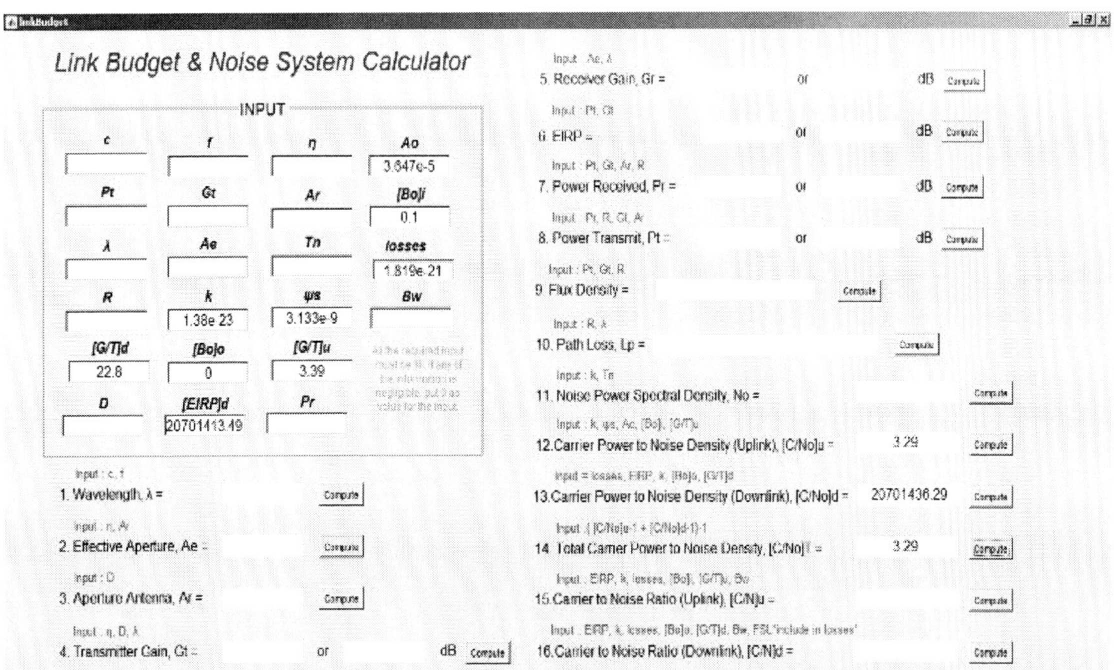

Figure. 8. Carrier Power to Noise Density Output

4. Conclusion

As a conclusion, the Link Budget and Noise System Calculator is successfully created using the interface in MATLAB software. The link budget items are including transmitter and received power, antenna gain, aperture antenna, effective area of isotropic antenna, antenna feeder losses and path losses. Whereas for noise system the items are noise temperature, figure of merit for uplink and downlink, input back off and output back off. Finally yet importantly, link budget and noise system calculator is important tool to investigate the areas, gain and losses that possibly will occur between transmitter and receiver of satellite communication system.

5. Acknowledgement
The authors would like to thank Faculty of Electrical Engineering, Universiti Teknologi MARA (UiTM) for their valuable support. This research is partly funded by the Malaysian Government through UiTM under 600-RMI/DANA5/3/ARAS (0042/2016).

References

[1] Minghui Tao; Lei Zhou; Huaiyu Zeng; Shengping Li; Xiang Liu"50-Gb/s/λ TDM-PON based on 10G DML and 10G APD supporting PR10 linkloss budget after 20-km downstream transmission in the O-band", Optical Fiber Communications Conference and Exhibition (OFC), 2017.

[2] Luis Escolar Haro; Gregorio Juliana Quirós; Álvaro Rodríguez Villalba; Ramón Martínez Rodríguez-Osorio; Pablo Honold; Antonio Arana; Alejandro Martínez; Inés Sanz"HispaSim: A web application for satellite link budget optimization and management", 11th European Conference on Antennas and Propagation (EUCAP), 2017.

[3] Ali Nauman; MoazamMaqsood"System design and performance evaluation of high altitude platform: Linkbudget and power budget", 19th International Conference on Advanced Communication Technology (ICACT), 2017.

[4] N. Yarkoni; N. Blaunstein; D. Katz "Link budget and radio coverage design for various multipath urban communication links", Radio Science, Vol. 42, Issue 02. 2007, pp. 2.

[5] Andrea Cagliero; Assunta De Vita; RossellaGaffoglio; Bruno Sacco"A New Approach to the Link Budget Concept for an OAM Communication Link", vol. 15, IEEE Antennas and Wireless Propagation Letters, 2016, pp. 568-571.

[6] B. Jeffrey; J. Wout; P. David "Comparison of the Link Budget with Experimental Performance of a WiMAX System", EURASIP Journal on Wireless Communications and Networking, 2009.

[7] N.Shkelzen; M. Artan"Link budget analysis in the network designed mobile WiMAX technology in the territory of the urban area of the city of Gjakova"IJCSI International Journal of Computer Science Issues, Vol. 9, Issue 5, No 2, September 2012.

[8] Yan Zhang; Laurence T. Yang; Jianhua Ma, "UNLICENSED MOBILE ACCESS TECHNOLOGY- Protocols, Architecture, Security, Standards and Applications" 2009.

[9] N.Makhijani "Statistical analysis of link budgets satellite communications" 2014 24th International Crimean Conference Microwave & Telecommunication Technology, 2014.

[10] R.Hui; M.Vaziri;J.Zhau"Separation of noise from distrortion for high speed system link budgeting"IEEE Photonics Technology Letters, 1999.

IOP Conf. Series: Journal of Physics: Conf. Series **1152** (2019) 012022 doi:10.1088/1742-6596/1152/1/012022

Relationship between plasma bubble and ionospheric current, equatorial electrojet, and equatorial counter electrojet

T Akiyama[1], A Yoshikawa[12], A Fujimoto[3] and T Uozumi[2]
[1] Department of Earth and Planetary Sciences, Kyushu University, 744, Motooka, Nishi-Ku, Fukuoka, Japan
[2] International Center for Space Weather Science and Education, Kyushu University, 744, Motooka, Nishi-Ku, Fukuoka, Japan
[3] Department of Systems Design and Informatics, School of Computer Science and Systems Engineering, Kyushu Institute of Technology, 680-4, Kawazu, Iizuka, Fukuoka, Japan

E-mail: 2SC17211N@s.kyushu-u.ac.jp

Abstract. In recent years, it has been clarified from previous studies that plasma bubbles and equatorial electrojets (EEJs) are related. In general, EEJs are calculated by subtracting the magnetic field H component of the magnetic equator from that at low latitude. However, in this study, EE-index data at Langkawi (magnetic equator), which includes all local current systems, were used for the analysis during the period from January 1, 2011, to November 8, 2014. By using the EE-index, it was found that plasma bubbles tend to occur for larger EEJ strengths. This result differs from the previous studies. In addition, if an equatorial counter electrojet (CEJ) occurs, it is understood that plasma bubbles will rarely occur due to the westward current; however, we found that when the lunar tidal effect is strong, plasma bubbles can occur even in conjunction with CEJs. Finally, we want to find the relationship between plasma bubbles and ionospheric current to predict them.

1. Introduction

Plasma bubbles are one of the ionospheric disturbance phenomena that occur in the equatorial F-region ionosphere after sunset, and they can cause irregularities in the plasma density. Therefore, if plasma bubbles occur, they may seriously affect communication systems such as GPS satellites. Plasma bubbles are generated by the Rayleigh–Taylor instability, and their basic characteristics are well-known; however, their dominant condition is not well-understood.

The equatorial electrojet (EEJ) is the eastward-current-flowing magnetic equatorial ionosphere during the daytime. Because of the eastward electric field generated by the E-region dynamo, the Cowling effect is enhanced near the magnetic equator (dip lat. –3°–+3°), and EEJs flow as a result. Generally, the direction of the EEJ is eastward, but the electric field sometimes switches to a westward direction before sunset. This westward electric field is called the equatorial counter electrojet (CEJ). As the EEJ and CEJ exhibit a significant variation in the horizontal component of the magnetic field, they have mainly been studied using geomagnetic data.

In the E-region ionosphere, the vertical E×B drift is upward in the daytime and downward in the nighttime. This inversion occurs around sunset, and the upward E×B drift often increases before reversal. This phenomenon is called the pre-reversal enhancement (PRE), and it is known that the F-region dynamo may be the origin. It is also known that the PRE greatly affects the growth rate of plasma bubbles.

In recent years, the relationships between the plasma bubble and EEJ strength have been reported. The intensity of EEJs at 11 Local Time (LT) was reported to have a positive correlation with the occurrence of plasma bubbles in Southeast Asia during the period from September to December 1989 [e.g., Dabas et al. 2003]. However, there was no relationship between the integrated EEJ ground strength from 07 to 17 LT and the occurrence rate of plasma bubbles during the period from November 2007 to October 2008 [e.g., Uemoto et al., 2010]. Furthermore, when the integrated EEJ ground strength during the period from 1 to 2 h prior to sunset at the E-region height was negative, which means the generation of CEJs, the generation of PRE and plasma bubbles were suppressed [e.g., Uemoto et al., 2010].

This study aims to investigating the relationship between plasma bubbles and ionospheric current.

2. Dataset and methods

When there is an ionospheric disturbance such as a plasma bubble and the electron density is unstable, the radio waves transmitted from the GPS satellite are diffracted in the irregularity and the phase and amplitude of the radio waves received on the ground fluctuates (scintillation). One way to measure the strength of the scintillation is the S4-index. In this study, the S4-index obtained from the GPS receiver installed in Kototabang, Indonesia, was used as the criterion for plasma bubble generation. In this study, the plasma bubble drifted eastward, so it was defined only when it occurred before 21 LT. If the plasma bubble occurred after 21 LT.

In general, the strength of the EEJ is calculated from the magnetic field fluctuation on the ground. EEJs can be estimated by subtracting the H component of the low-latitude station, which is outside the EEJ region from the H component of the magnetic equator station, which is inside the EEJ region. In contrast, the EE-index was used in this study. The EE-index is proposed to monitor the magnetic field fluctuations at low latitude and at magnetic equator, and it is calculated from the ground magnetic field H component [Uozumi et al., 2008]. In the EE-index, there are the two indices, EDst and EUEL. EDst represents the magnetic field fluctuation caused by the global current because of the magnetospheric disturbance (orange line in figure 1), whereas EUEL represents that caused by the local current because of the ionospheric disturbance (red dashed circle in figure 1). In this study, we used the EUEL 1 min value in Langkawi, Malaysia as the strength of the EEJ, and we used EUEL in Kototabang as the reference station. Information of the station is presented in Table 1. We investigated the relationship between the plasma bubble and the EEJ and CEJ during the period from January 1, 2011, to November 8, 2014.

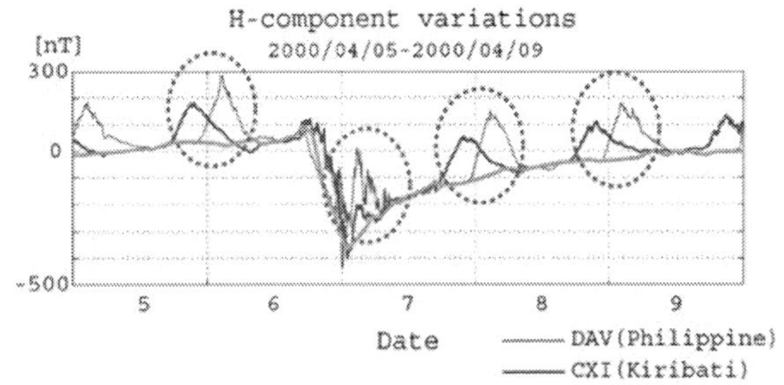

Figure 1. Global and local fluctuation of magnetic field. Y axis means the magnitude of magnetic field. The variation from the magnetosphere is EDst (orange line) and the variation from the ionosphere is EUEL (red dashed line).

Table 1. Coordinates of the station.

Station	Code	GGlon	Dip lat.
Langkawi	LKW	99.78	−1.07
Kototabang	KTB	100.32	−10.1

3. Results

By using the same method as Uemoto et al. (2010), we compared the integrated EUEL value to the occurrence ratio of plasma bubbles. The integrated EUEL strength from 07 to 17 LT is referred to as the daytime IEUEL, and that from 17 to 19 LT is referred to as the pre-sunset IEUEL. Figure 2 illustrates how the equatorial spread-F occurrences depend on the daytime IEEJ (panel a) and pre-sunset IEEJ (panel b) obtained by Uemoto et al. (2010). Figure 3 presents the relationship between the occurrence ratio of the plasma bubble and the daytime IEUEL (panel a) and the pre-sunset IEUEL (panel b) obtained by this study.

Figures 2a and 3a illustrate the dependence of the plasma bubble on the daytime EEJ strength, but different trends can be seen. This difference may be caused by the difference in the evaluation of the EEJ. In this study, as the EUEL is used for the criterion of the EEJ, the magnetic field fluctuation due to the sq current system is also included in the EEJ. It is conceivable that not only the enhancement of the EEJ but also the current such as the sq current system also contribute to the generation of plasma bubbles. Since the growth rate of plasma bubbles is influenced by the eastward electric field, sq current is considered to be important in addition to the enhancement of EEJs. However, EEJs occur in E-region height and plasma bubbles occur in F-region height, so it is difficult to consider these relationships directly. Figures 2b and 3b present the relationship between the EEJ before sunset and the plasma bubble, respectively, and they exhibit identical behavior. In Uemoto et al. (2010), it is considered that the negative value of the pre-sunset IEEJ implies the generation of CEJs, and the plasma bubble and PRE are suppressed because of the westward electric field caused by the CEJ. The result of this study can be considered using the same mechanism. In figure 3b, however, when the pre-sunset IEUEL is negative (occurrence of CEJ), it is seen that there are a few events accompanied with plasma bubbles.

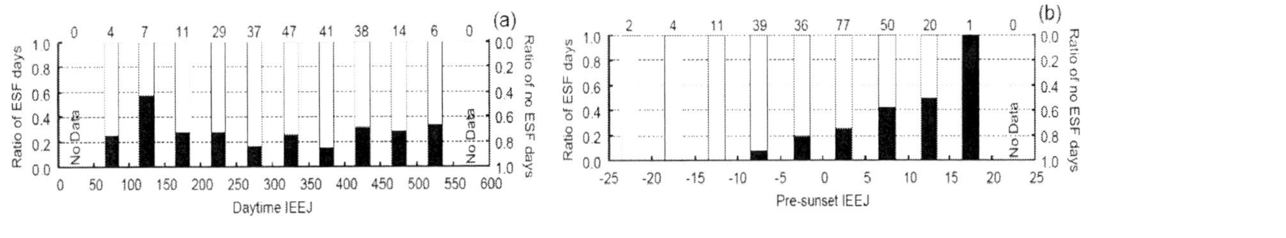

Figure 2. Dependence of equatorial spread-F (ESF) occurrence on (a) daytime and (b) pre-sunset IEEJ. The black and white bars indicate the ratio of ESF and no ESF days to the total number of days contained in each bin, respectively. The total number of days in each bin is given on the top of each bar. [Uemoto et al., 2010]

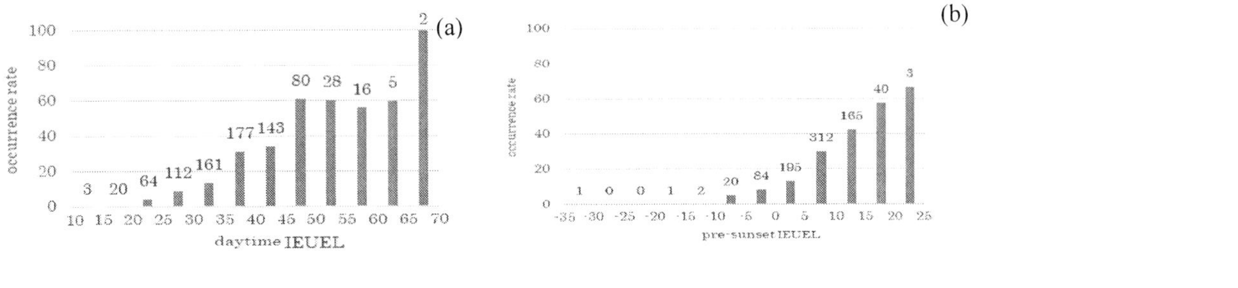

Figure 3. Relationship between the plasma bubble and (a) daytime and (b) pre-sunset IEUEL. The vertical axis shows the generation ratio of the plasma bubble. The total number of days in each bin is given on the top of each bar.

Figure 4 presents a plot of the magnetic field variations on the CEJ day, which has a peak time from 16 to 17 LT. Figure 4a shows the day when the plasma bubble occurs, while figure 4b shows the day when the plasma bubble does not occur. The red line indicates the average of each variation, and it is seen that the magnetic field variation after the CEJ tends to change from negative to positive when the plasma bubble occurs (see the enlarged view in the upper right inset of figure 4). Therefore, the magnetic field variation after CEJ occurrence is classified as follows: (a) returns to baseline, (b) remains negative until after sunset, and (c) fluctuates from negative to positive. Table 2 lists the total number of classified days. In pattern (c), plasma bubbles tend to occur even if the CEJ is generated. It is considered that the eastward electric field was created because of a phenomenon similar to PRE after the CEJ, so it is a good condition for the plasma bubble. Moreover, since there is no plasma bubble event in pattern (b), it is seen that the electric field before and after sunset is important for generation of plasma bubbles.

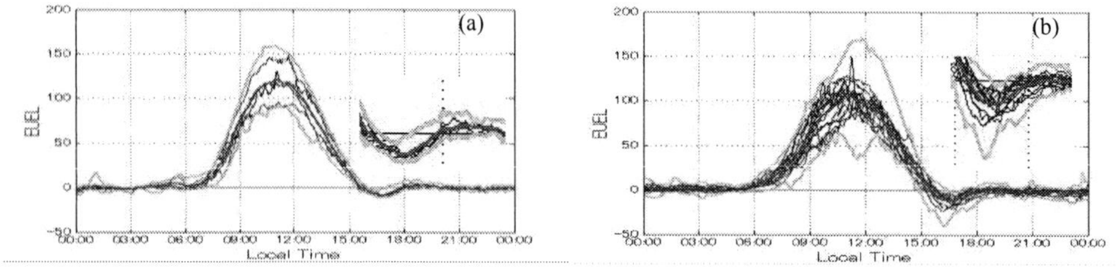

Figure 4. Magnetic field variations of LKW on CEJ day. The CEJ peak is 16–17 LT. (a) Occurrence of plasma bubble and (b) no plasma bubble. The red line represents the average for each time, while the green line represents the maximum and minimum. The upper right inset is an enlarged view of the CEJ time.

Table 2. Number of days when changes after the CEJ are classified as follows: (a) returns to baseline, (b) remains negative until after sunset, and (c) fluctuates from negative to positive.

Pattern	Bubble occurs	Bubble does not occur
a	2	12
b	0	13
c	6	26

There are some origins that cause the CEJ, such as lunar tidal effects and the disappearance of the sporadic-E layer. Thus, we examined which origin affected the CEJ. The sporadic-E layer was confirmed by the ionogram in Kototabang. The intensity of the lunar tidal effect was classified by lunar age, as listed in table 3. Table 4 presents these relationships; Table 4a shows the case when the plasma bubble occurs, while Table 4b shows that when the plasma bubble does not occur. From this table, we can see that the occurrence of the plasma bubble is influenced not by the existence of the sporadic-E layer but by the strength of the lunar tidal effect.

Table 3. Correspondence between lunar age and strength of the lunar tidal effect. Number in table means lunar age.

	High lunar tide		Low lunar tide	
	New moon	Waxing moon	Waning moon	Full moon
Lunar age	0–3.625 25.375–29	3.625–10.875	18.125–25.375	10.875–18.125

Table 4. Relationship between plasma bubble, lunar tide, and sporadic-E layer. (a) Occurrence of plasma bubble and (b) no plasma bubble. Number in table means days satisfying the condition.

(a)	Strong lunar tide	Week lunar tide
With sporadic-E layer	2	1
No sporadic-E layer	5	0

(b)	Strong lunar tide	Week lunar tide
With sporadic-E layer	8	15
No sporadic-E layer	12	16

4. Conclusion

In this study, we examine the relationship between the plasma bubble and EEJs using the EE- index at the magnetic equator, and it was revealed that plasma bubbles are more likely to occur for stronger EEJs in all seasons. Generally, it is considered that the generation of plasma bubbles is reduced by the suppression of the PRE when the CEJ occurs, but in this study, a CEJ and plasma bubble event existed on the same day. On such a day, the magnetic field tended to reverse direction from south to north after the onset of the CEJ, and it was found that phenomena such as PRE may occur even if the CEJ occurs. Although there are various phenomena that cause the CEJ, when the plasma bubble and CEJ occur on the same day, it was found that the CEJ is primarily caused by the lunar tidal effect. This study covered Southeast Asia, so other longitudes must be analyzed in order to determine whether these features are global. In this study, we did not consider solar cycle or seasonal variation, so considering this makes further understanding of plasma bubble. Furthermore, it is also necessary to construct the mechanism of plasma bubble occurrence on the CEJ day.

Acknowledgement

The magnetic field data and s4-index data at Kototabang were used from the Institute for Space-Earth Environment Research (ISEE), Nagoya University, Japan. The ionosonde at Kototabang was used from NICT, Japan.

References

[1] Dabas R S, Singh L, Lakshmi D R, Subramanyam P, Chopra P and Garg S C 2003 Evolution and dynamics of equatorial plasma bubbles: Relationships to ExB drift, postsunset total electron content enhancements, and equatorial electrojet strength *Radio Sci.* **38**(4) 1075 doi:10.1029/2001RS002586A reference

[2] Haerendel G and Eccles J V 1992 The role of the equatorial electrojet in the evening iono-sphere *J. Geophys. Res.* **97** A2 1181-92

[3] Kil H 2015 The morphology of equatorial plasma bubbles - a review *J. Astron. Space Sci.* **32**(1), 13–19,

[4] Mayaud P N 1977 The equatorial counter-electrojet – a review of its geomagnetic aspects *J. Atmos. Terrestr. Phys.* **39**(9–10), 1055–70

[5] Saito S and Maruyama T 2006 Ionospheric height variations observed by ionosondes along magnetic meridian and plasma bubble onsets *Ann. Geophys.* **24** 2991–96

[6] Saito S and Maruyama T 2007 Large-scale longitudinal variation in ionospheric height and equatorial spread F occurrences observed by ionosondes *Geophys. Res. Lett.* **34**(16) L16109 doi: 10.1029/2007GL030618

[7] Uemoto J, Maruyama T, Saito S, Ishii M and Yoshimura R 2010 Relationships between pre-sunset electrojet strength, pre-reversal enhancement and equatorial spread-F onset *Ann. Geophys.* **28**(2) 449–454

[8] Uozumi T, Yumoto K, Kitamura K, Abe S, Kakinami Y, Shinohara M, Yoshikawa A, Kawano H, Ueno T, Tokunaga T, McNamara D, Ishituka J K, Dutra S L G, Damtie B, Doumbia V, Obrou O, Rabiu A B, Adimula I A, Othman M, Fairos M, Otadoy R E S and MAGDAS Group 2008 A new index to monitor temporal and long-term variations of the equatorial electrojet by MAGDAS/CPMN real-time data: *EE*-index *Earth Planets Space* **60**(7) 785–790

[9] Yamazaki Yand Kosch M J 2014 Geomagnetic lunar and solar daily variations during the last 100 years *J. Geophys. Res.* **119** 6732-44 doi:10.1002/2014JA020203

Determining Land Use and Land Cover Changes and Predicting the Growth of Dhaka, Bangladesh Using Remote Sensing and GIS Techniques

Abstract. Developing countries are experiencing a rapid growth in the urbanization in recent decades. Urbanization yields many positive effects if happens within the appropriate limits. However, extensive urbanization mostly results in adverse effects. This study aims to evaluate and observe the various changes like vegetation, built-up and water in the urban area of the greater Dhaka area of Bangladesh using Landsat 7 ETM+ and Landsat 8 OLI images between 2011 and 2018 and to predict its future using different GIS and remote sensing techniques. The analysis shows the superabundant growth of the buildup areas as well as degradation in the vegetated and water areas of greater Dhaka, Bangladesh. The classified maps have been assessed through an accuracy assessment. Based on such, the result shows and predict the expected urban growth, vegetation areas and the water index for the year 2020. The experimental result indicates and aims to provide an idea about subsequent understanding on the urban growth of greater Dhaka, Bangladesh.

1. Introduction

Rapid changes in LULC (land use and land cover) in urban areas becoming a major problem for the mega cities like Dhaka, Bangladesh. The conversion for this rapid change are currently occurring at an unprecedented rate. Only monitoring and proper management plans can solve the emerging problems timely and effectively. By using satellite images and their data's we can measure the subsequent changes and take the necessary procedure for the upcoming changes in urban expansion.

Bangladesh has witnessed a rapid change in the urban growth in recent times. Dhaka, the capital of Bangladesh is the prime center for this change. However, with astonishing growth in population of 3 million in 1980's to over 16 million· Dhaka has become one the most densely populated cities in the world where its expansion is disordered and irregular. This growth however has placed without proper planning which is resulting in a city with inadequate urban structure, contraction of river banks, vegetated area and so on. Remote sensing and GIS (Geographic Information Systems) are powerful yet most cost-effective tools for accessing the spatial and the temporal dynamics of LULC [1]. Multi-temporal data are provided through Remote sensing which is useful for the processes and the patterns of LULC change whereas GIS is useful for mapping and analyzing those patterns [2]. Most developed countries have both recent and extensive LULC information, but such scenario is prevalent for developing countries, particularly Bangladesh. For instance. City does not have official land uses patterns or proper statistics, nor the master plans contain maps or quantitative information's on the existing pattern of land uses map in the city [3]. Therefore, this study is an attempt to identify the spatial and temporal dynamics of LULC changes of greater Dhaka to identify a scientific approach to LULC change detection and for the policy makers to access the LULC change pattern in this area of Bangladesh.

The objective for this approach is to collect, record, analyze and to explore multi-temporal data of greater Dhaka using Landsat ETM+ and OLI images ranging from 2011 to 2018 and to predict the future phases of changes in the urban expansion.

2. Study Area

The study area of Dhaka, is located at the center of Bangladesh between 23°55`N and 90°23` E and 23° 39`N and 90° 26`E respectively. The study area surrounded by four river systems: The Bur-iganga, Turag, Tongi and the Balu, which flow to the south, west, north, and east, respectively. As of World Bank data, this city has over 16 million people while the city itself has a population estimated at about 8.5 million with population growth of 4.2% annually and expected to be the third largest city in the world by 2020[4]. It is one of the mostly populated areas and one of the fastest megacities around the world carrying a density of 23,000 people living per square kilometer within a total area of approximate 300 square kilometers [5].

3. System procedure

3.1. Data acquisition

Landsat data (ETM+, OLI) were acquired to evaluate the changes in land cover and uses in the urban expansion of Dhaka city. All the data acquired by Landsat 7 from 2003 has data gaps which are known as SLC-off image. The scenes cover 78 percent of their pixels, but these are the most accurate geometric and radiometric [5] satellite data. This study kept Landsat 7 images as references and collected, recorded, Landsat 8 images to analyze the urbanization patterns in different classification method.

Table 1. Collected Landsat Images

Satellite Data	Spatial Resolution (meters)	Date of Acquired data
Landsat 7 ETM+	30	2013Dec16; 2015Dec22; 2017Dec27
Landsat 8 OLI	15	2013Dec24; 2015Dec30; 2017Dec03

3.2. Methodology

All the images are collected from United States Geological Survey (USGS). There are numerous change detection methods to assess the variations in LULC using satellite data [7]. Among them pre and post-classification comparison are the most used one's. This study followed both among which, pre-classification method includes image differencing [8], change vector analysis and vegetation index differencing [9]. Initially, an unsupervised classification was run by using the ISO data decision rule to determine "naturally" occurring spectral clusters which is subsequently reduced to defined 4 class through the construable of similar/diverse statistics. Next, upon visual assessment and LULC class types were assigned to each index. A supervised classification approach was then followed by using maximum likelihood method to relate the study area pixels. The detailed classification report for all the images are shown in table 2. A detailed workflow of the methodology has shown in the figure 1.

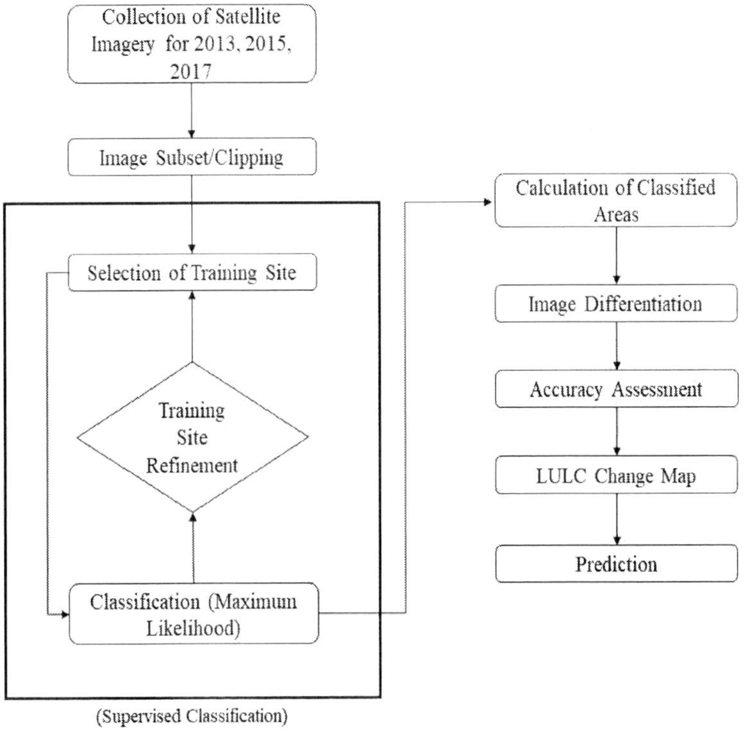

Figure 1. Methodology of the study

Supervised classification has been performed in all the obtained satellite images and the obtained classification report for all the images are shown in table 2. Classification report graph has shown in figure 2.

Table 2. Classification Report

Features	Year 2013 (Sq. km)	Year 2015 (Sq. km)	Year 2017 (Sq. km)
Water	20.18	19.09	14.31
Built-up	122.43	157.13	163.23
Bare soil	87.31	66.25	73.45
Vegetation	69.23	56.78	48.12
Total Area (Sq.km)	299.15	299.25	299.11

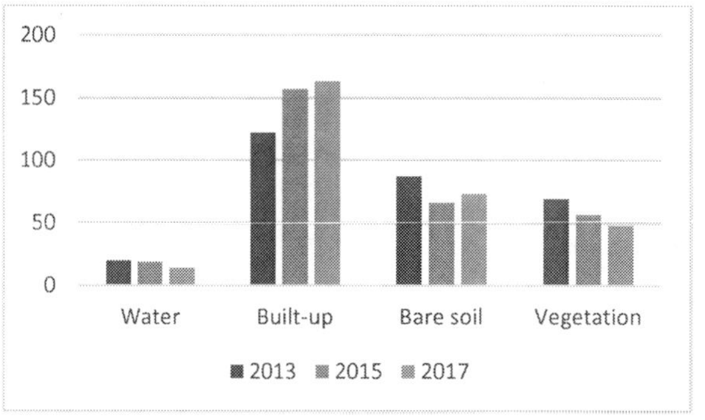

Figure 2. Classification Report for Satellite image

3.3. *Data Analysis and Result*

The classified output of 2013, 2015, 2017 are showed by figure 3, 4, 5 respectively. The supervised classification has been done based on four separate LULC types: Water index, Bare soil, Built-up and Vegetation index using maximum likelihood classification method which uses Bayes' theorem of decision making. However, several classes were incorrectly classified in classification method of LULC due to their similarities in spectral characteristics. Therefore, a post-classification method was used to improve the accuracy of the classification for its simplicity and effectiveness [6].

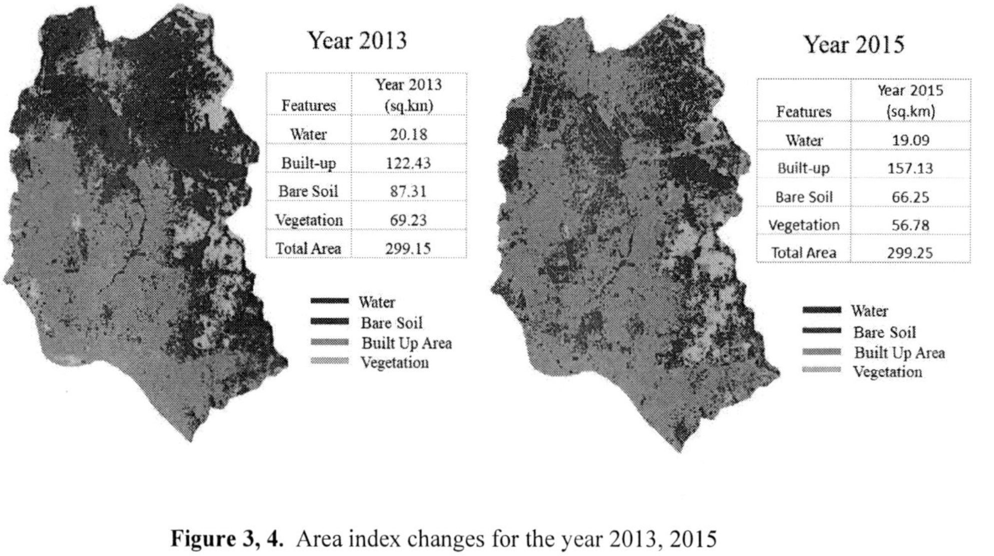

Figure 3, 4. Area index changes for the year 2013, 2015

IOP Conf. Series: Journal of Physics: Conf. Series **1152** (2019) 012023 doi:10.1088/1742-6596/1152/1/012023

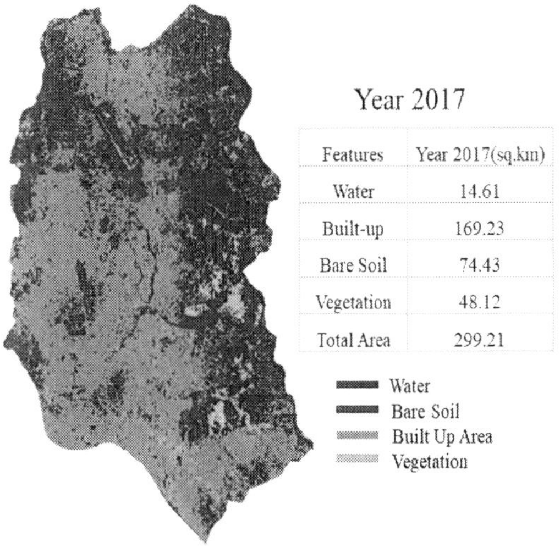

Figure 5. Area index changes for the year 2017

The area has been calculated for the respected index in each time after running post-classification method. The total area calculated is approximately 300 square kilometers. The total accuracy from the Landsat-derived LULC data was 86%, 86.7% and 85% respectively for the year 2013, 2015, 2017 verifying standard accuracy of 85-90% for LULC mapping studies as recommended by Anderson [8] and the result has been shown on table 3. It is comparing the classified image to another data source which in this case, ground truth data exerted from Google earth image. The LULC cover changes are predicted for the year 2019 by using arithmetic progression. The figure3, 4, 5 shows the respected area index changes for the year 2013, 2015, 2017 and predicted result for the expected growth for each index in 2019 are shown in table 4.

Table 3. Overall accuracy assessment

Year	2013	2015	2017
Overall accuracy	86%	86.7%	85%

Table 4. Prediction result of 2019

Features	Year 2013 (Sq. km)	Year 2015 (Sq. km)	Year 2017 (Sq. km)	Year 2019 (Sq. km)
Water	20.18	19.09	14.31	11.85
Built-up	122.43	157.13	163.23	187.33
Bare soil	87.31	66.25	73.45	62.87
Vegetation	69.23	56.78	48.12	37.08
Total Area	299.15	299.25	299.11	299.12

4. Future plan and conclusions

Satellite remote sensing and GIS technology has proven its usability in determining Land use and Land cover changes analysis. That is why this kind of study is a good way organize as well as cost and time effective for urban planners and decision makers. This study shows the changes in terms of built-up, water, bare soil and vegetation indexes of Dhaka city has experienced a massive change in LULC and risen high as it was from 2013 and predicted its growth much higher than the year 2013.As of the data shown for 2017, the total built-up increased 40.8% and the total vegetated area decreased to -21.11%

from what it was in 2013. This follow-up data will be more accurate with the combination of GIS, RS and socio-economic data to understand and predict the spatial and temporal dynamics of LULC changes. Furthermore, this study will continue to improve and analyze the given data and will apply different machine learning techniques and forecast the future growth patterns of Dhaka city using neural network techniques. As reliable and current data are not quite available for Bangladesh, the data's acquired and the calculations will contribute for both the sustainable urban land uses and forecasting the possible future changes in growth patterns.

5. References

[1] Islam, N. (2005). Dhaka now: Contemporary development. Dhaka: The Bangladesh Geographical Society.

[2] Hathout, S. 2002 The use of GIS for monitoring and predicting urban growth in East and West St Paul, Winnipeg, Manitoba, Canada. Journal of Environmental Management, 66, 229–238.

[3] Zhang, B. P., Yao, Y. H., Cheng, W. M., Zhou, C. H., Lu, Z., & Chen, X. D. (2002). Human -induced changes to biodiversity and alpine pastureland in the Bayanbulak Region of the East Tianshan Mountains. Mountain Research and Development, 22, 1–7.

[4] Islam, N. (1996). Dhaka from city to megacity: Perspectives on people, places, planning and development issues. Dhaka: Urban Studies Program, Department of Geography, University of Dhaka.

[5] The World Bank. (2007). Dhaka: Improving living conditions for the urban poor. Sustainable Development Unit, South Asia Region, Report No. 35824-BD. Toll, D. L., Royal, J., & Davis, J. B. (1980). Urban areas update procedures using Landsat data. In Proceedings of American Society of Photogrammetry. Falls Church, VA: American Society of Photogrammetry.

[6] Islam, N. (2005). Dhaka now: Contemporary development. Dhaka: The Bangladesh Geographical Society.

[7] Coppin, P., Jonckheere, I., Nackaerts, K., Muys, B., & Lambin, E. (2004). Digital change detection methods in ecosystem monitoring: a review. International Journal of Remote Sensing, 25(9), 1565–1596.

[8] Toll, D. L., Royal, J., & Davis, J. B. (1980). Urban areas update procedures using Landsat data. In Proceedings of American Society of Photogrammetry. Falls Church, VA: American Society of Photogrammetry.

[9] Townshend, J. R. G., & Justice, C. O. (1995). Spatial variability of images and the monitoring of changes in the normalized difference vegetation index. International Journal of Remote Sensing, 16, 2187–2195.

[10] Harris, P. M., & Ventura, S. J. (1995). The integration of geographic data with remotely sensed imagery to improve classification in an urban area. Photo-grammetric Engineering & Remote Sensing, 61(8), 993–998.

[11] Anderson, R., Hardy, E. E., Roach, J. T., & Witmer, R. E. (1976). A land use and land cover classification system for use with remote sensor data. USGS Professional Paper 964. Washington, DC.

IOP Conf. Series: Journal of Physics: Conf. Series **1152** (2019) 012024 doi:10.1088/1742-6596/1152/1/012024

Studying the Effects of Disturbance Torques on a 2U CubeSat in Low Earth Orbits

Asma Mohammad Nusrat Aman, Roselina Arelhi and Nafizah Khan
University of Nottingham Malaysia, Jalan Broga, Semenyih, Selangor, Malaysia

Abstract. Disturbance torques in Low Earth Orbits may present a harsh environment for the successful operation of small satellite missions. The effects of gravity-gradient and aerodynamic torques are studied individually at various altitudes for low Earth orbit. Results show that for a basic 2U CubeSat, the gravity-gradient torque is highly perturbing, whereas the effect of aerodynamic torque is established to be so dominant that it offers attitude stabilization in lower altitudes; and may be used as a means of passive attitude stabilization.

1. Introduction

CubeSats, a class of small low earth orbiting satellites, are based on standardized units of mass and volume. The basic CubeSat unit measures 10x10x10 cm^3 (often called "1U"), and weighs around 1 *kg*. A combination of two units forms a larger 2U CubeSat (20x10x10 cm^3). CubeSats typically piggyback on the launch of larger satellite missions, and are mostly deployed into Low Earth Orbits (LEO) [1]. For such small satellites, LEO presents a harsh space environment with various forces that could act as disturbing forces on the CubeSat. The most prominent disturbance torques in LEO are the magnetic disturbance torques due to interaction with Earth's magnetic field; gravity-gradient torques, which are a result of the differences in the Earth's gravitational pull on the CubeSat body; and the aerodynamic torques due to aerodynamic drag, which is predominant in lower orbits. These disturbance torques act on the CubeSat and ultimately alter its orientation. Maintaining the CubeSat's orientation in space to some desired attitude is important for it to fulfil its mission requirements. Therefore, the attitude of a CubeSat needs to be controlled or maintained in a certain manner. Passive attitude stabilization techniques such as gravity-gradient stabilization, passive magnetic control, and aerodynamic stabilization make use of these disturbance torques to control the CubeSat's attitude.

This paper studies the effects of the gravity-gradient and aerodynamic disturbance torque individually. The magnetic disturbance torques are not studied independently due to an inadequate model of the Earth's magnetic field, as only a simple dipole model was used. In reality, the Earth's magnetic field model is one that is highly variable and unpredictable. Also, magnetic torque disturbance, if modelled accurately, will vary with orbit inclination, which is not considered in this study. Therefore, effects of only gravity-gradient and aerodynamic disturbance torques are considered. The main aim is to study how each disturbance torque plays a role in influencing the CubeSat attitude stability. This may further aid in changing design parameters to implement passive attitude stabilization.

2. Modelling

2.1. 2U CubeSat Dynamic Model

A basic 2U CubeSat is modelled as a rigid body (three degrees of freedom), with its center of mass coinciding with its center of gravity. Figure 1 shows the modelled dimensions and attitude rotation of 2U CubeSat. For attitude analysis, the CubeSat's overall dynamics equation is given in Equation 1 [2].

$$\boldsymbol{I}\dot{\omega} + \omega \times (\boldsymbol{I}\omega) = \tau_{total} \tag{1}$$

where τ_{total} is the sum of all the disturbance torques in *Nm*, $\omega = [\omega_x \ \omega_y \ \omega_z]^T$ represents the angular velocities (*degrees/sec*) in roll, pitch and yaw; and $\boldsymbol{I}$ is the inertia matrix about the center of gravity.

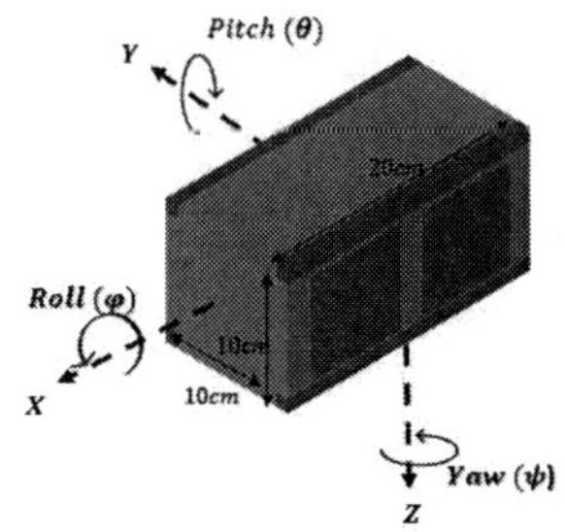

Figure 1. 2U CubeSat model showing dimensions and Euler angle representations.

2.2. Disturbance Torques

The major disturbance torques considered are: the magnetic torque, gravity-gradient torque, and aerodynamic torque. τ_{total} in Equation 1 is the sum of all these torques, which will affect the CubeSat attitude stabilization.

2.2.1. Magnetic Torque. For simplification purposes, the Earth's magnetic field is modelled as a dipole. The magnitude of the magnetic field is given in Equation 2

$$B = \frac{\mu_o}{R^3} \tag{2}$$

where B is magnetic field strength (*Tesla*), which varies with altitude and latitude, μ_o is permeability of free space ($\mu_o = 4\pi \times 10^{-7} H/m$), and R is the distance from the center of Earth to the CubeSat in meters. If D is the sum of all magnetic moments on the CubeSat, the magnetic torque acting on the CubeSat, τ_m (*Nm*), will be given by

$$\tau_m = D \times B \tag{3}$$

where τ_m is the magnetic torque (*Nm*) on the CubeSat.

2.2.2. Gravity-Gradient Torque. The gravity-gradient torque is caused by the variation in Earth's gravitational forces exerted on the CubeSat body. It is therefore in the *z*-direction, and the gravity-gradient torque is given by [3]:

$$\tau_g = 3(\frac{\mu}{R^3}) z_o \times \boldsymbol{I} \cdot z_o \tag{4}$$

where z_o is the *z*-axis in the orbit frame, as shown in figure 1.

2.2.3. Aerodynamic Torque. The expression for aerodynamic torque [2] is given as:

$$\tau_a = \rho V_{orbit}\left[V_{orbit} A_{drag} c_p \hat{V}_{orbit} - (\boldsymbol{I} + \hat{V}_{orbit}\boldsymbol{J})\omega\right] \tag{5}$$

IOP Conf. Series: Journal of Physics: Conf. Series **1152** (2019) 012024 doi:10.1088/1742-6596/1152/1/012024

where ρ is atmospheric density (*kg/m³*), V_{orbit} is orbital velocity of CubeSat (*m/s*), A_{drag} is the area perpendicular to the velocity vector (*10cm×10cm*), c_p is centre of pressure, and $\boldsymbol{J}$ is the new moment of inertia matrix for drag. The atmospheric density and orbital velocity both vary with altitude. Within the same altitude, the atmospheric density also tends to vary slightly. For simulation and analysis of worst-case scenario, the maximum atmospheric density values are considered in the calculation of aerodynamic torque.

3. Results and Discussion

According to Equation 1, all the available disturbance torques act on the CubeSat body at any given time. However, for this study, the disturbance torques are simulated separately to study their effects on the CubeSat attitude. The Earth magnetic field model is incorporated in both the gravity-gradient and aerodynamic analyses. Simulations were carried out using Matlab/Simulink as it offers a convenient graphical environment to design the dynamic models and disturbance torques. A simple detumbling controller for CubeSats was incorporated into the model, so detumbling may be achieved and the attitude stabilization analysis is improved further [4]. The effects of disturbances on the CubeSat are studied without the intervention of any attitude control. The simulation is carried out for 6 orbits. Results of the simulations presented in figures 2-7, show the Euler angles for the CubeSat attitude versus number of orbits. In order to study of the effects of individual disturbance torque, the following altitudes were considered: 400 *km*, 700 *km*, and 1000 *km*.

3.1. Effects of Gravity-Gradient Torque

Figures 2-4 are plots of simulations for attitude (roll, pitch, and yaw) in degrees of the 2U CubeSat in the presence of gravity-gradient disturbance torque, at varying altitudes.

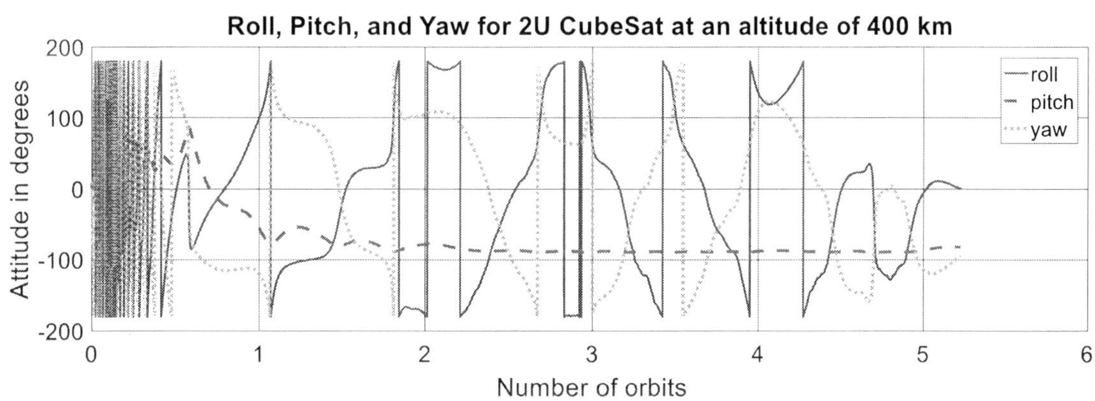

Figure 2. Plot of 2U CubeSat attitude (roll, pitch, and yaw) evaluated at altitude of 400 km, inlcuidng gravity-gradient disturbance torque.

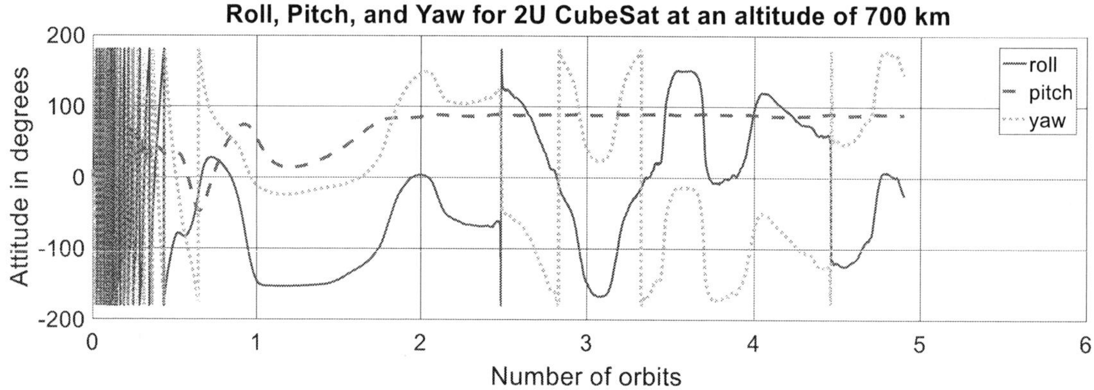

Figure 3. Plot of 2U CubeSat attitude (roll, pitch, and yaw) evaluated at altitude of 700 km, inlcuidng gravity-gradient disturbance torque.

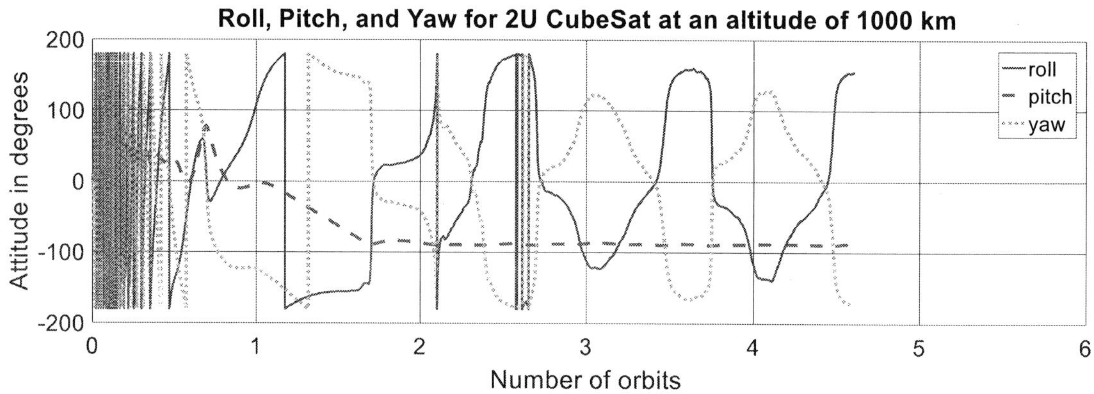

Figure 4. Plot of 2U CubeSat attitude (roll, pitch, and yaw) evaluated at altitude of 1000 km, inlcuidng gravity-gradient disturbance torque.

As can be seen in figures 2-4, there is high angular displacement in the roll and yaw axes. However, these high perturbations are not observed in the pitch axis (normal to orbit plane). This is evident due to the effect of gravity-gradient which is influenced by the configuration of the CubeSat. There were no significant changes in attitude performance for the different altitudes. This suggests that the CubeSat cannot be passively stabilized via gravity-gradient torque for the configuration considered.

3.2. Effects of Aerodynamic Torque

Figures 5-7 are plots of simulations for attitude (roll, pitch, and yaw) in degrees of the 2U CubeSat in the presence of gravity-gradient disturbance torque, at varying altitudes.

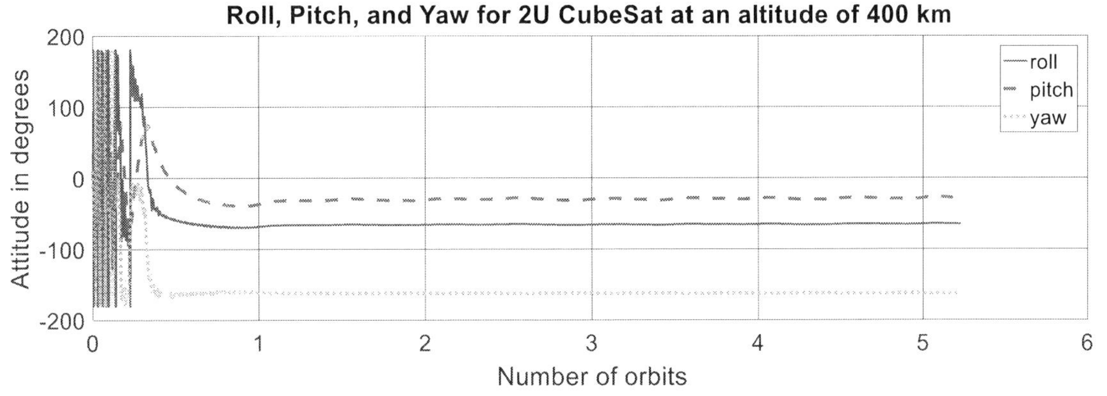

Figure 5. Plot of 2U CubeSat attitude (roll, pitch, and yaw) evaluated at altitude of 400 km, inlcuidng aerodynamic disturbance torque.

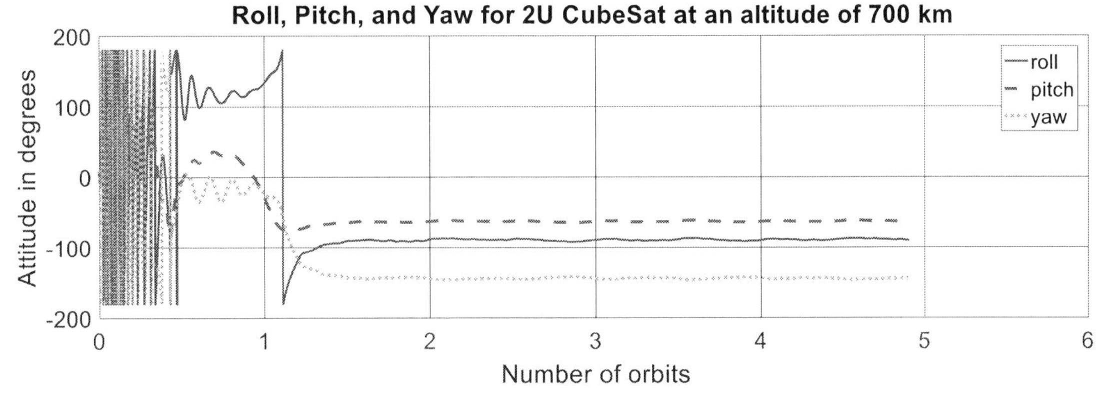

Figure 6. Plot of 2U CubeSat attitude (roll, pitch, and yaw) evaluated at altitude of 700 km, inlcuidng aerodynamic disturbance torque.

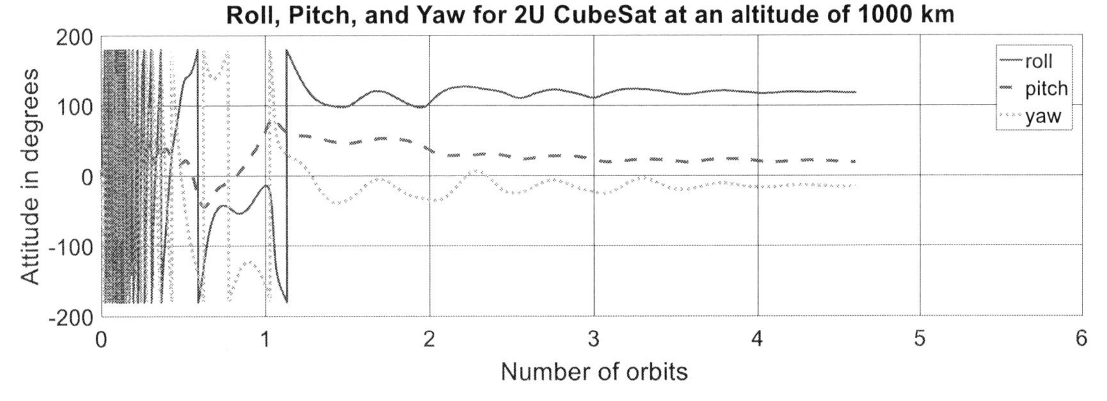

Figure 7. Plot of 2U CubeSat attitude (roll, pitch, and yaw) evaluated at altitude of 1000 km, inlcuidng aerodynamic disturbance torque.

In figure 5, it can be seen that the effect of aerodynamic torque is highly stabilizing. This is due to high atmospheric density at low altitude, contributing to larger aerodynamic torques. This observation is consistent in figures 6 and 7 where aerodynamic torque becomes less dominant with an increase in altitude. Therefore this may be used as a means of passive control alone for coarse pointing requirements for a 2U CubeSat.

4. Conclusion

Having considered the presence of magnetic torque for both analyses, the results establish a general idea of the effects of gravity-gradient and aerodynamic disturbances on the attitude of a 2U CubeSat at different altitudes in LEO. The effect of aerodynamic torque was seen to be more dominant. Consequently, passive aerodynamic stabilization may be achieved at lower altitudes, whereas gravity-gradient alone could not be used to stabilize attitude for this CubeSat configuration.

References

[1] D. Selva and D. Krejci, "A survey and assessment of the capabilities of Cubesats for Earth observation," *Acta Astronaut.*, vol. 74, pp. 50–68, 2012.
[2] P. C. Hughes, *Spacecraft Attitude Dynamics*. John Wiley & Sons, 1986.
[3] Z. Tudor, "Design and implementation of attitude control for 3-axes magnetic coil stabilization of a spacecraft," NTNU, 2011.
[4] F. Alvenes, "Satellite Attitude Control System," Norweigian University of Sceince & Tech, 2012.

IOP Conf. Series: Journal of Physics: Conf. Series **1152** (2019) 012025 doi:10.1088/1742-6596/1152/1/012025

Installation of UiTM's ground station – Connecting earth and space with amateur band communication

S N Mohamad Rahim[1], M H Jusoh[1,2,3] and A Taat[1,2]

[1]Faculty of Electrical Engineering, Universiti Teknologi Mara, Shah Alam, Selangor, Malaysia
[2]Applied Electromagnetic Research Group (AERG), Universiti Teknologi Mara, Shah Alam, Selangor, Malaysia
[3]Center for Satellite Communication (UiTMSAT), Universiti Teknologi Mara, Shah Alam, Selangor, Malaysia

E-mail: naddp1389@gmail.com

Abstract This paper portrays the installation of UiTM's Ground Station as being part of BIRDS Ground Station Networks; with its main task being to support BIRDS-2 nanosatellite's monitoring. UiTMSAT-1, UiTM's first nanosatellite, is one of the BIRDS-2 nanosatellites. It is expected to be deployed from International Space Station (ISS) by middle of 2018. The ground station is installed with two (2) Yagi antennas which operates in Amateur bands respectively at Very High Frequency (144 MHz -148 MHz) ranges and Ultra High Frequency (430 MHz - 438 MHz) ranges. Additionally, an APT antenna for 137 MHz-band polar-orbiting weather satellites is also installed at this ground station for earth weather monitoring. As for security and monitoring matters, lightning/surge protection system and surveillance system (CCTV) also has been installed. Conclusively, several tests had been done as confirmation on the equipment setup. The testing procedure includes receiving HORYU-4 and IO-86 satellite's signal, and evaluating NOAA satellites' processed data of South East Asia's coverage. The uplink communication test, with cooperation with satellite owner, is left for future planning.

1. Introduction

Installation work of UiTM's ground station started from September 2017. This ground station is built in anticipation of UiTMSAT-1's deployment by middle of 2018. This first Malaysian University's nanosatellite comes from Malaysia's collaboration with Japan, Bhutan and Philippines, in a program called Joint Global Multi-Nations Birds: BIRDS-2 CubeSat Project. BIRDS-2 satellites will be placed at Low Earth Orbit (LEO) area, approximately at altitude of 400 km, similar to the International Space Station (ISS). UiTM's ground station is a part of BIRDS Ground Station Network which consists of 10 different countries including BIRDS-1 participants [1][2]. It functions as the monitoring and communicating window between ground and BIRDS-2 satellite's constellation: BHUTAN-1 (Bhutan), MAYA-1 (Philippines) and UiTMSAT-1 (Malaysia). To note, in section 2, this paper will be updating on UiTM's ground station block diagram which includes the equipment's installation and safety concerns and in section 3, will present several testing results on ground station's equipment using chosen satellites.

2. Ground station's Setup

Generally, ground station is an important factor to ascertain the condition of the satellite's health and its mission's progress. The design for ground station must be compatible with the satellite communication's specification to ensure connectivity and must be securely maintained and controlled.

2.1. Block Diagram of UiTM's Ground station

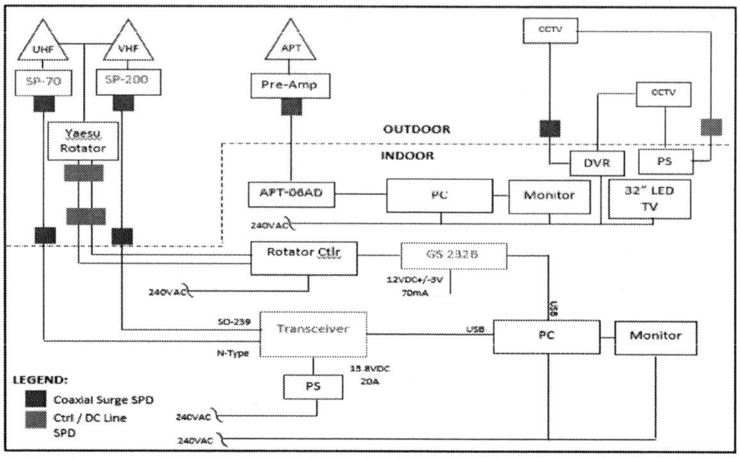

Figure 1. Block Diagram of UiTM's Ground Station

UiTM's Ground station is built at the rooftop of level 21, Menara 2, Faculty of Electrical Engineering, UiTM Shah Alam. Figure 1 shows the block diagram of UiTM's ground station. "INDOOR" refers to equipment inside control room which situated directly under the rooftop for ease of maintenance. "OUTDOOR" refers to equipment's installation includes Antennas such as UHF, VHF and APT; Yaesu Rotator that hold UHF and VHF antennas, and two CCTVs. Inside the control room, two computers were set up for separate tasks. One for VHF/UHF satellite monitoring and the other is for APT satellite receiving data. On the computer for VHF/UHF, GS-232B is connected via USB. This device is a USB-PC interface controller for rotator. It connects with a rotator controller which directly connected to the Yaesu rotator at the rooftop. It receives the satellite's elevation and azimuth data from SATPC32, a satellite tracking software which installed at control room's PC.

Figure 2. Yagi antennas on top of the antenna tower (Left-hand-side); KX-137 APT antenna (Right-hand-side)

Figure 2 shows the two installed Yagi antennas on a tower: one side for VHF band and the other side for UHF band. VHF antenna; model 2MCP22 [3] has frequency range of 144 MHz to 148 MHz.

On the other side, UHF antenna; model 436CP42UG [4] with frequency range of 430 MHz to 438 MHz. Both VHF and UHF Yagi antennas are circular polarized with gain of 14.39 dBic and 18.9 dBic respectively. Boom length of VHF antenna is 5.67 meter and boom length of UHF antenna is 5.75 meter. Current fixed height of the antenna tower is 3.5 meter. Added with height of antenna's rotator (20 cm), the whole installation conforms for a safe operation during satellite tracking condition.

Antenna rotator is an equipment that controls the antenna movement by azimuth and elevation. The rotator used in this ground station is Yaesu G-5500. This device controls the motor of its rotators. The azimuth rotator unit is securely mounted together with the elevation rotator unit at the top of antenna tower. A fiber glass cross boom (model FGCB60); for connecting both Yagi antennas; is installed at each side of the rotator. Yaesu G-5500 allows azimuth rotation of 360° in approximately 58 seconds and elevation up to 180° in 67 seconds [5]. This rotator helps to accurately point to the location of passing satellite for better connectivity.

For communication activities, ICOM IC-9100 Radio Transceiver is installed. The RF cables from both Yagi antennas installed at the rooftop are connected to the back of this radio transceiver. The radio covers High Frequency (HF) range, VHF and UHF ranges while supporting several modes of operation which includes SSB, CW and FM [6]. It connects to PC (installed with SATPC32) via USB and receives the calculated Doppler-compensated frequency of the tracked satellite. During tracking and communicating process, automatically, the radio will tune to the exact frequency calculated by the SATPC32 software.

For BIRDS-2 nanosatellites command and mission operation from ground station, a specific software is being developed using C# programming language by the BIRDS-2 members at Kyushu Institute of Technology, Japan. This software is a stand-alone application created with Microsoft Visual Studio. In the Main Tab, referring to Figure 3, there is serial port configuration setting which needs the ground station operator to input the correct ports (TNC and Radio) to establish connection with this software. Once the hardware has been confirmed and in compliance with the ground station operation software, satellite tracking can be started. In the event of any BIRDS Ground Station's operator received the BIRDS-2 nanosatellite's CW beacon, 20 characters (specific) can be inputted into the Morse Code Beacon Analysis. With that, ground station operator will have the tracked satellite's status for example battery condition and satellite's temperature. Additionally, the BIRDS-2 members also have been designing separate tabs for Missions operation and Terminal which act as Serial Monitor for downlink data/packet and acknowledgment from satellite. This design will allow ground operator to read data/packet from satellite in its decoded format.

Figure 3. BIRDS-2 Ground Station Operation Software - under development

IOP Conf. Series: Journal of Physics: Conf. Series **1152** (2019) 012025 doi:10.1088/1742-6596/1152/1/012025

Additionally, KX-137, a turnstile VHF antenna (APT) for polar-orbiting weather satellites [7] is also installed at the ground station rooftop. Its best feature includes wide angle reception with its omnidirectional and horizontally mounted cross-dipoles that makes circular polarized signal reception acceptable. At the altitude of approximately 71 meters from ground level, this antenna operates in the 137 MHz band and receives signal from several NOAA (National Oceanic and Atmospheric Administration) satellites such as NOAA 15, NOAA 18, and NOAA 19.

2.2. Safety Concerns

This section will describe the security measures taken which includes installation of lightning/surge protection system and surveillance system (CCTV).

2.2.1 Security camera. Two CCTVs were installed: one pointing outside of ground station's control room and another installed at the rooftop facing the Yagi antennas. The CCTVs are wired connected to a DVR which recording for whole 24 hours. The system is utilized for securing UiTM's control room and for monitoring the movement of Yagi antenna while in use and while in dormant state. The CCTV viewing can be remotely accessed via internet connection.

2.2.2 Lightning/Surge Protection. The ground station has the lightning protection system installed at the rooftop - a 7 meter-height of lightning arrester pole at the very end of the rooftop's perimeter. Attached with copper tape, it is grounded at the building grounding place at ground level. The antenna tower is situated straight 5-meters from this lightning pole. By using trigonometric functions and the 7-meter lightning pole's cone coverage at 60° angle, it is found that the maximum height covered at the tower's 5-meter distance is 4.1 meter. Hence, with the installed Yagi antenna's height of 3.7 meter, it is confirmed that the UiTM's ground station is covered.

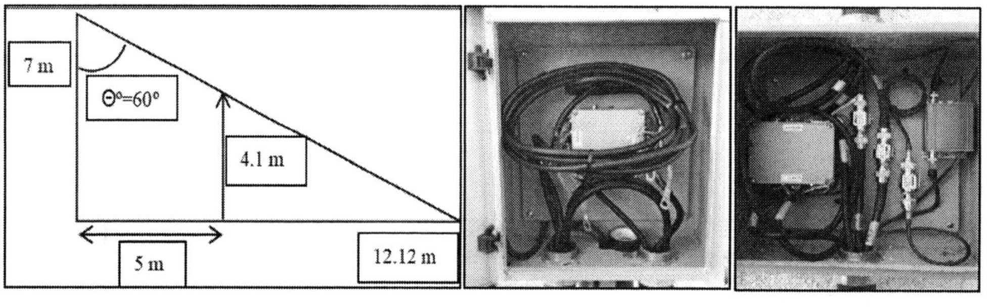

Figure 4. Left-hand-side: Calculation of Lightning pole's coverage; Middle: SPD box (antenna tower), Right-hand-side: SPD box (outside control room)

In addition to the lightning pole coverage, each equipment also protected by Surge Protection Device (SPD). As shown on Figure 4 Middle and Right-hand-side, SPD is installed at two places inside a weatherproof enclosure box: (1) at the antenna tower and (2) outside of control room. SPD at the antenna tower protects the rotator equipment while the SPD outside of control room protects all the outdoor equipment such as rotator's cable, Yagi and KX-137 antenna's RF cable and CCTV's cable. Both enclosure boxes' side are secured with copper tape as for grounding purpose. This security measure allows better protection of highly sensitive equipment outside and inside the control room such as antennas, rotator and radio transceiver.

3. Result and Discussion

Every equipment installed must be checked to confirm its function. In this matter, two satellites are chosen (HORYU-4 and IO-86) for UHF Yagi antenna test, and analyzing processed NOAA satellites' images from KX-137 antenna.

3.1. Receiving satellite's signal: HORYU-4 and IO-86 (LAPAN-A2)

IOP Conf. Series: Journal of Physics: Conf. Series **1152** (2019) 012025 doi:10.1088/1742-6596/1152/1/012025

HORYU-4 is a Japanese built nanosatellite (from Kyushu Institute of Technology) which launched into space at altitude of 575 km on February 17th, 2016. Its communication system includes VHF & UHF frequency ranges in addition to the S-band and L-band patch antenna. To track for HORYU-4's timing in the ground station's footprint, SATPC32 software is used. It calculates the expected time for satellite's arrival and loss of signal. With Continuous Wave (CW) mode transmitted from Horyu-4's satellite at centered frequency of 437.375 MHz [8], signal is successfully received by UiTM's ground station. To decode the signal received, a CW decoder is used: HORYU-4 CW Beacon Telemetry Decoder (downloaded from Mike Rupprecht, DK3WN webpage). This decoder will inform the satellite's health status as an example shown at Left-hand-side Figure 5 which was taken on November 18th, 2017.

IO-86 also known as LAPAN-A2 is an Indonesian microsatellite, developed by LAPAN (Lembaga Penerbangan dan Antariksa Nasional, or the "National Institute of Aeronautics and Space")[9]. The satellite was launched on September 28th, 2015 to circular near equatorial orbit at altitude of 650 km and it passes over Indonesia and other near equatorial region 14 times daily. IO-86 uses UHF frequency for TT&C (Telemetry Tracking and Command) functions. With arrival time estimated by SATPC32 software, the radio I9000 is set to received signal in Frequency Modulation (FM) mode centered at 437.425 MHz for telemetry beacon. The signal is captured as WAV file format and for decoding, FUNCube-1 Telemetry Decoder is used (downloaded from Mike Rupprecht, DK3WN webpage). Figure 5 on the Right-hand-side shows the IO-86's telemetry decoded data.

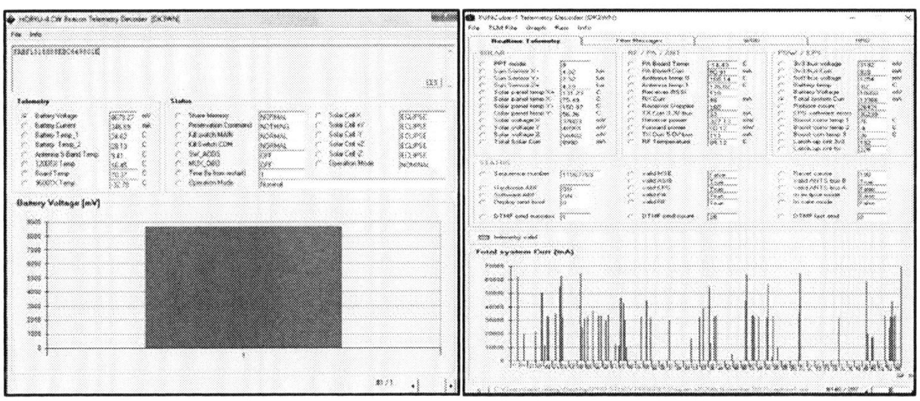

Figure 5. Left-hand-side: Decoded data of HORYU-4; Right-hand-side: Decoded data of IO-86

3.2. *Receiving satellite's signal: NOAA Satellites*

The KX-137 antenna has been fully operating from late October 2017 and received satellite APT (Automatic Picture Transmission) images in daily basis. The setup is completed with an APT-06 [10], a receiver device which handles the received signal from NOAA satellites and WXtoImg [11], a computer software that can process the satellite's data into readable images. On November 3rd, 2017, NASA published an article on its website [12], titled "Intensifying Typhoon Damrey Threatens Vietnam". The article stated that Typhoon Damrey originated in the South China Sea, west of the Philippines on November 1st, 2017, and has moved westward through the South China Sea. This statement matched the satellite images received by KX-137 antenna on UiTM's ground station. Table 1 shows part of the APT images that KX-137 antenna received dated from November 1st, 2017 until November 4th, 2017. The captured images show the typhoon Damrey's event time-by-time from accumulation to its dispersion.

Table 1. Typhoon's movement date-by-date

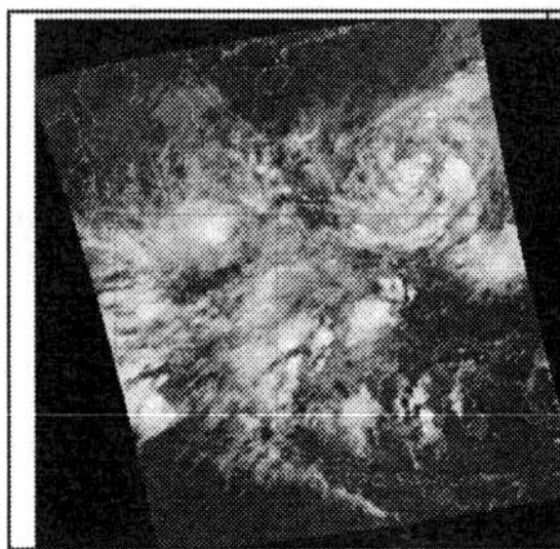	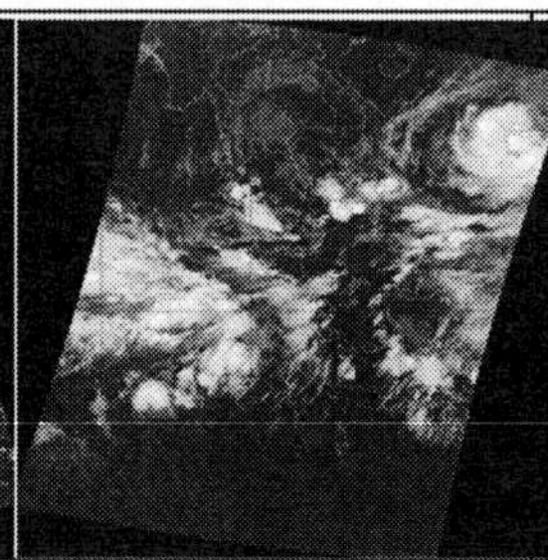	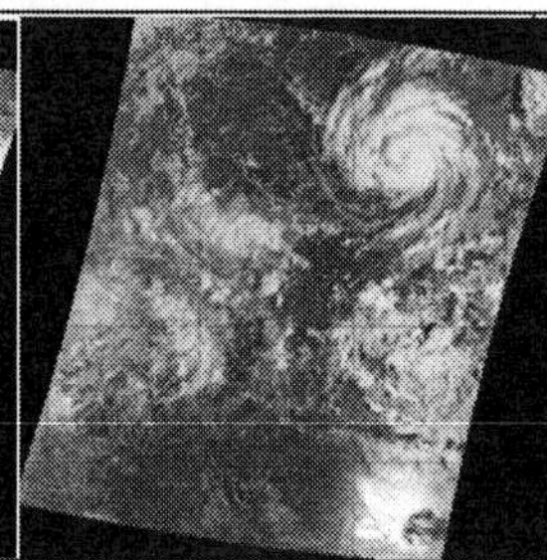	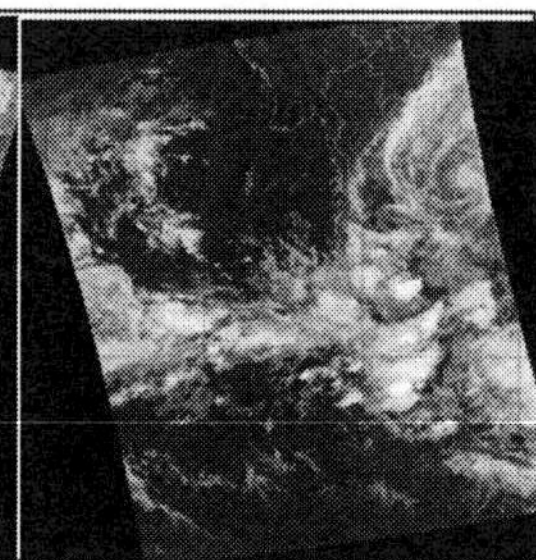
November 1st, 2017	November 2nd, 2017	November 3rd, 2017	November 4th, 2017

4. Conclusion

UiTM enhances Malaysia's development towards space-faring nation by participating the Joint Global Multi-Nation Birds Project (BIRDS-2). In preparation for the deployment of UiTMSAT-1, a ground station is constructed. UiTM's ground station can monitor and communicate with UiTMSAT-1 and track other satellites that use the VHF/UHF amateur frequency ranges as communication medium. This paper describes the ground station's block diagram, installation parts, and testing results. Future planning for uplink communication test is to be done with approval from satellites owners. In conclusion, UiTM's ground station is one move in supporting space research.

Acknowledgment

We would like to thank Assoc. Prof. Ir. Dr. Mohamad Huzaimy Jusoh as the project manager, and co-supervisor, Dr. Siti Amalina Enche Ab Rahim of the Faculty of Electrical Engineering, UiTM Shah Alam, the Electromagnetic Research Group (AERG) Faculty of Electrical Engineering, UiTM Shah Alam and the Center for Satellite Communication (UiTMSAT) Faculty of Electrical Engineering, UiTM Shah Alam for supporting this development. We also would like to extend our appreciation to BIRDS-2 team members and the company who have helped in the installation of UiTM's Ground Station.

References

[1] A. R. Khan, G. Maeda, H. Masui, J. project Members, and M. Cho, "Five-nations CubeSat Constellation; An Inexpensive Test Case for Learning and Capacity Building," in *3rd IAA Conference on University Satellite Missions and CubeSat Workshop*.
[2] G. Maeda, T. Taiwo, J. Javier, and M. Cho, "BIRDS Project Newsletter," no. 17, p. 81, 2017.
[3] I. M2 Antenna Systems, "Model No : 2MCP22 ASSEMBLY MANUAL." 2011.
[4] I. M2 Antenna Systems, "Model No : 436CP42UG ASSEMBLY MANUAL." 2010.
[5] Y. M. C. Ltd, "INSTRUCTION MANUAL G-5500." 2015.
[6] M. J. Wilson, "QST Product Review IC-9100," *American Radio Relay League (ARRL)*, 2012.
[7] W. Electronic, "WRAASE electronic - Hardware - KD-137 + AA-137." [Online]. Available: http://www.wraase.de/e_kd137.htm. [Accessed: 13-Jan-2018].
[8] M. Cho and H. Fukuda, "Flight results of new technology onboard a lean satellite HORYU-IV," in *IEEE Region 10 Conference (TENCON), International Proceedings*, 2017, pp. 3439–3442.
[9] R. H. Triharjanto, "Development of Micro-satellite Technology at the Indonesian National Institute of Aeronautics and Space (LAPAN)," in *17th CEReS International Symposium Proceedings*, 2012.
[10] W. Electronic, "Weather Satellite Receiver APT-06 / APT-06 AD Operating Manual." .
[11] Wxtoimg, "WXTOIMG (GUI)." Central North Publishing Limited. Available: http://wxtoimg.com/support/wxgui.pdf
[12] "Intensifying Typhoon Damrey Threatens Vietnam | Precipitation Measurement Missions." [Online]. Available: https://pmm.nasa.gov/extreme-weather/intensifying-typhoon-damrey-threatens-vietnam. [Accessed: 10-Nov-2017].

Enhanced satellite positioning methods using ultrawideband D2D-based localization for ultra-dense 5G wireless setting

Akeem Adebomehin[1,2] and Stuart Walker[2]
[1] Air Force Institute of Technology, Kaduna, Nigeria
[2] University of Essex, Colchester, United Kingdom

Emails: aaakeem2@gmail.com, stuwal@essex.ac.uk

Abstract. The drawbacks of legacy satellite positioning systems necessitate new procedures for high resolution location services. Moreover, location-aware devices, systems and networks are critical for 5G/future network setting that seeks to seamlessly connect global devices in an all-IP network. It is therefore believed that satellite positioning could be enhanced with impulse radio (IR) ultrawideband (UWB) localization approaches for accuracy-critical situations expected in 5G wireless. This paper presents results of research on enhanced satellite positioning methods for ultra-dense 5G setting. The methods combine information from satellites, fixed wireless networks, wireless LANs and cellular communication along with global cloud databases. This is conceptualized as an IR-UWB device-to-device (D2D) overlay network that seamlessly incorporates satellite data into the 5G terrestrial network environment. A link to at least one satellite on the global positioning constellation would suffice for accurate localization. Also presented is the IR UWB D2D-propagation-based combined localization and communication scheme (UD-CLOCS). The UD-CLOCS scheme transmits location and communication data in a single signal while performance evaluations have been impressive. It is believed that the solutions have potential for robust and cost-efficient satellite positioning as well as prospect to de-congest licensed spectrum in 5G wireless setting. The significance of the research outcomes is that satellite positioning signal and data could be seamlessly integrated into global 5G networks for enhanced positioning services as well as improved spectral efficiency.

1. Introduction

The 5G setting is envisaged as an all-IP seamless connection of global devices featuring gigabit data at less than 1ms end-to-end latency [1]. This would require high accuracy positioning schemes to perform better than legacy system in Figure 1; which requires at least three satellites on the GPS constellation for accurate position determination. Moreover, there are other drawbacks such as signal blocking effects in dense urban scenarios as well as harsh environments that affect legacy methods. There is thus the need for new schemes to enhance current satellite positioning systems in the 5G global network setting. Indeed, future wireless systems should possess seamless satellite positioning capabilities and IR UWB is considered to offer a promising alternative [2]. UWB technology capabilities such as robust communication in dense and multipath situations is considered suitable for enhanced satellite positioning applications in 5G wireless setting [3]. For location estimation purposes in this paper, the IEEE802.15.4a UWB localization enabled WPAN standards in [4] and [5] are adopted. The remaining sections highlight the UWB enhanced satellite positioning research along with performance evaluations.

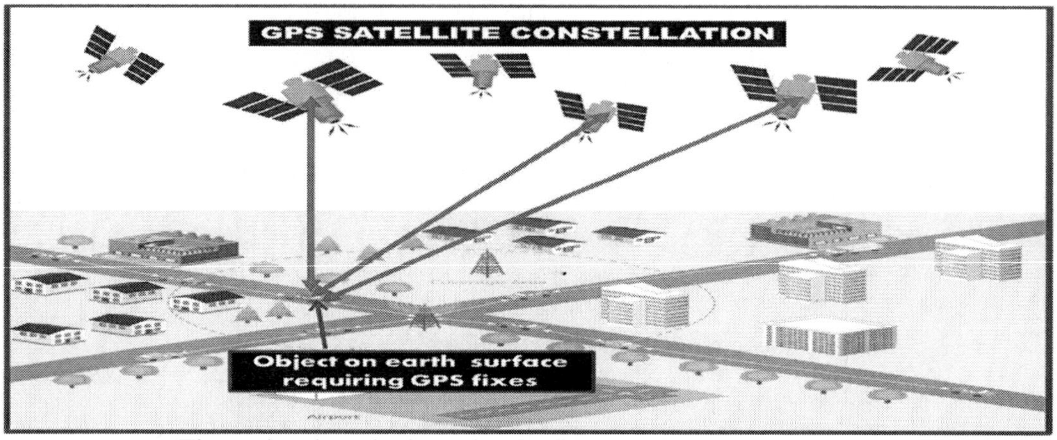

Figure 1. Adapted schematic of a GPS-based positioning system

2. IR UWB D2D-based enhanced satellite positioning

2.1. Prospects of IR UWB D2D enhanced satellite positioning

One of the challenges envisaged for the 5G environment is high accuracy satellite-based positioning due to expected ultra-dense setting. An IR-UWB network overlay is therefore conceptualized as highlighted in Figure 2 to address the need. The scheme localizes devices in a D2D fashion such that satellite is seamlessly incorporated with terrestrial communication systems while taking advantage of global positioning information on cloud databases. Also proposed as a complement enhancement for satellite positioning in ultra-dense 5G network setting is the IR UWB D2D-propagation-based combined localization and communication scheme (UD-CLOCS). The concept features wireless wide area network (WWAN) with piconet as its smallest component comprising of discrete micro-channel slots (DMCS). The method is a low system-load network and devices can join a cluster and update position estimate by executing a mathematical local cost-function. Additionally, the system also addresses key 5G prospects like higher device densities, smaller cells and globe-wide D2D cooperative positioning. Simulation results suggest D2D distribution with capacity to support a very high density of connected devices per square meter of urban environments as those expected in large-scale 5G/future networks.

3. UD-CLOCS IR-UWB WWAN for satellite positioning

3.1. Structure of the UD-CLOCS IR-UWB WWAN setting

The most important component of the UD-CLOCS IR-UWB WWAN piconet is the piconet coordinator (PNC). This comes with connected devices and data is shared in peer-to-peer (P2P) mode and is managed by the PNC in line with IEEE 802.15.3 standards in [6]. The PNC has basic timing as its major function as well as management of network resources. However, in the UD-CLOCS D2D WWAN, the piconet is IR-UWB-based which frees-up the licensed spectrum. It should be noted that each DMCS acts as separate channel used by any 2 devices in D2D connection. As shown in Figure 3, the DMCS is composed of three parts. These are: the channel time allocation period (CTAP), the beacon period (BP) and the channel access period (CAP). Simulation experiments suggest that the DMCSs possess enough UWB bandwidth. This therefore addresses the problem of in-band concurrent user interference as well as capacity to increase channel throughput. The observed effect on throughput is significant for satellite signal injection into ultra-dense 5G environment.

4. UD-CLOCS IR-UWB cluster for seamless satellite location services

4.1. Formation of cluster in a UD-CLOCS IR-UWB setup

The IR-UWB D2D communication overlay has the capacity to enhance satellite positioning methods along with improved spectral efficiency and reduced end-to-end latency in 5G setting. Earlier attempts at D2D communication in cellular network focused on distributed scheduling. Methods include [9] and

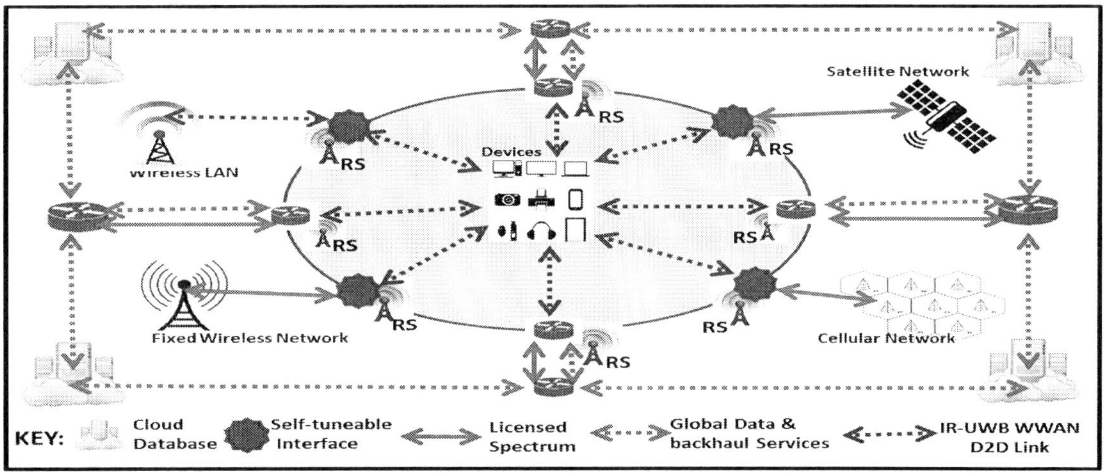

Figure 2. Architecture of IR-UWB WWAN D2D overlay network for enhanced satellite positioning

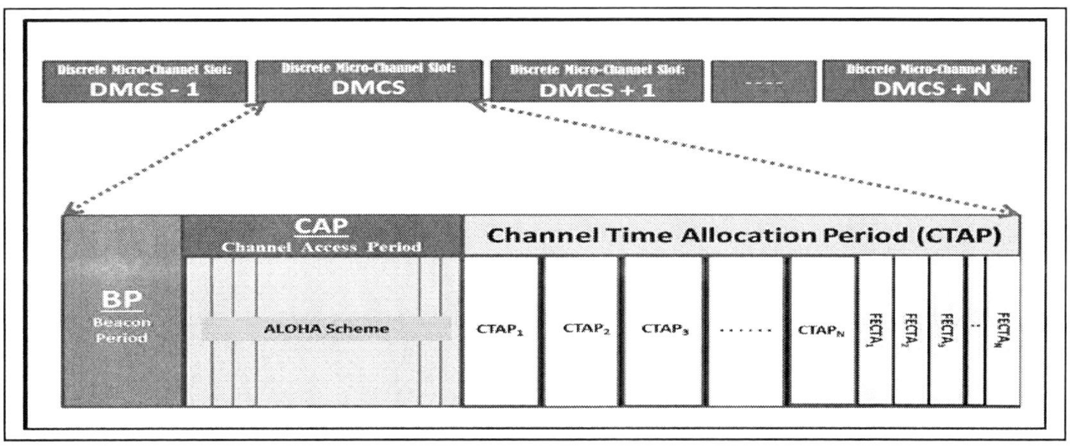

. ***Figure 3.*** *Schematic layout of UD-CLOCS DMCS*

many others like IEEE 802.11 protocol stack [10] as well as bluetooth and wifi-direct (Wifi-P2P). All were geared towards D2D connections that are efficient and effective in unlicensed bands such that also bypasses access points in a wireless environment [11]. The focus was to use radio technology for short range, low power as well as low cost wireless communication [12]. However, all these methods utilize different transmissions for the in-band and the out-band scenarios. Specifically, UD-CLOCS WWAN D2D features full–duplex IR-UWB communication within the same UWB band. This covers both communication and localization data transfers in the same signal band allocated for specific D2D link.

4.2. Satellite positioning cluster formation structure for UD-CLOCS IR-UWB scheme

Figure 4(a) highlights the UD-CLOCS D2D cluster formation scenario. A base station (BS) and a high density of relay stations (RS) deployments denoted as RSa, RS_b, RSc, ...RS_{n+1}. Also included are hypothetic clusters of many devices designated D_a, D_b, D_c...... D_n, D_{n+1}.nAll devices use IR-UWB interface for data transfer. Mobile devices have restricted battery power. Therefore, it is important to fashion a scheme that reduces power consumption while also agreeing with [13] regarding standard requirements for cluster, cluster head (CH) and cluster normal member (CNM). The considered devices are also able to determine their residual charge as well as signal-to-interference-plus-noise ratio (SINR). Moreover, the devices come with embedded respective WiFi, cellular and IR-UWB interfaces. Clusters

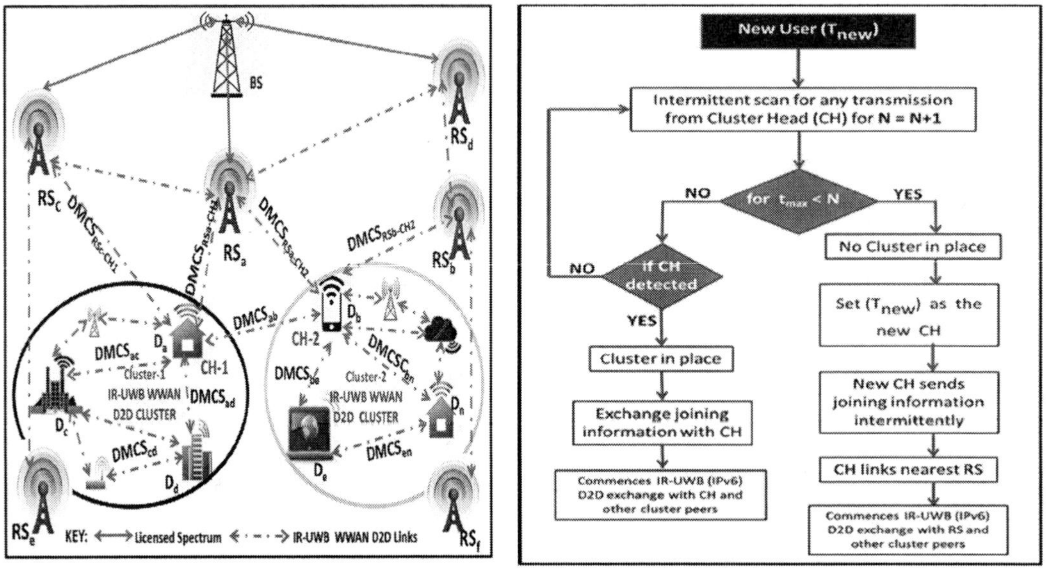

(a) **(b)**

Figure 4 Schematics of cluster formation method for UD-CLOCS IR-UWB WWAN D2D

additionally, have corresponding CH that is able to determine all of the CNMs' intent values (IV); while time is synchronized with the CH via the UD-CLOCS IR-UWB interface. Figure 4(b) details the cluster formation scheme for UD-CLOCS WWAN D2D IR-UWB. The procedure allows a new user/device (T_{new}) to initiate a D2D communication by a process of scan, detect and analysis of the joining instructions broadcasted by a CH periodically.

4.3. Satellite positioning cluster head transfer for UD-CLOCS scheme

A CH selection and transfer method is proposed bearing in mind limited cluster lifetime and system capacity. This combines the normalized residual charge of each CNM denoted as $R\text{-}CHG_{Nom}$ with normalized SINR denoted as S_{Nom} such that the following are defined:

- IV - CH updating rank variable.
- α - S_{Nom} weighted factor.
- $R\text{-}CHG_{Nom\text{-}1}$ - CNM residual charge when it becomes CH.
- $R\text{-}CHG_{Nom\text{-}2}$ - CH current residual charge.
- IVCH – current CH intent value such that Equation (1) is true.
- $T_{xpt\text{-}max}$ - UD-CLOCS IR-UWB WWAN D2D cluster maximum throughput.
- B - cumulative channel bandwidth of all DMCSs for cluster members D2D interchange.

$$IV = \alpha \times S_{Nom} + (1 - \alpha) \times R\text{-}CHG_{Nom} \quad (1)$$

A maximum SINR denoted as $SINR_{max}$ applies for cluster members. Therefore, $T_{x\text{-}max}$ could be expressed as shown in Equation (2) in accordance with Shannon's theorem. Hence, at specific given slots, each CNM sends its SINR and residual charge to the CH. The CH in-turn compares all received SINR with $SINR_{max}$ and selects α according to Equation (3) based on set parameters. Moreover, the CH calculates all the CNMs' IV to obtain the highest IV denoted IV_{max}. This enables selection of CNM with IV_{max} as the CH at the next step in accordance with Equation (4); else the CH continues to perform above actions which ensures extended cluster lifetime. Additionally, simulation results suggest that the process does not degrade UD-CLOCS WWAN throughput. Conversely, in a Wi-fi-based D2D, transmission between CH and CNMs are carried out with synchronized time reference via the WiFi-Direct interface using licensed spectrum [14]. Moreover, most of previous works on in-Band D2D communication focused mainly on resource allocation and interference from D2D users to the cellular networks [15].

However, UD-CLOCS IR-UWB D2D WWAN adjacent cluster cooperation contains all transmission within the allotted UWB band which is most significant. UD-CLOCS IR-UWB D2D WWAN therefore most probably offers a more efficient alternative in this circumstance.

$$T_{xpt\text{-}max} = B\log2(1+SINR_{max}) \quad (2)$$

$$\alpha \in [0, 1], S_{Nom} \in [0, 1],$$

$$R\text{-}CHG_{Nom} \in [0, 1] \quad (3)$$

$$IVCH \leq IV_{max} \quad (4)$$

5. UD-CLOCS Performance evaluation

5.1. UD-CLOCS cluster lifetime in comparison with number of active cluster members

This section presents evaluation of UD-CLOCS IR-UWB D2D WWAN cluster lifetime as against number of cluster members. It assumes that all cluster devices have the same charge and that the SINR follows a random distribution pattern. It further assumes that the SINR is constant for each CNM during the entire period of simulation. As results in Figure 5(a) indicate, for varying values of α (the weighted factor to the S_{Nom}); the mean lifetime of the cluster increases as the number of CNMs is increasing. Therefore, the cluster lifetime remains a constant value notwithstanding the growing number of CNMs. Nevertheless, UD-CLOCS IR-UWB D2D WWAN cluster scheme addresses low power scenario with the inequality expression in Equation (5). It very significant that increasing number of cluster members no matter the value of α results in increasing cluster lifetime for UD-CLOCS setting. This has potential to seamlessly accommodate satellite positioning signal and data into ultra-dense setting expected in 5G.

$$2\,(R\text{-}CHG_{Nom\text{-}2}) \leq (R\text{-}CHG_{Nom\text{-}1}) \quad (5)$$

5.2. Comparison of system throughput for UD-CLOCS, standard cellular and wifi

This section highlights comparison of system throughput for UD-CLOCS IR-UWB D2D WWAN, cellular and wifi. A working bandwidth of 20MHz is assumed for the cellular network. This is divided into 2 links and the cell can accommodate 2 cellular users at most for data transfer to BS. Also, one CH can connect with 7 CNMs at most. For the wifi simulation, this paper agrees with [16] and [17] that posit maximum throughput of wifi-direct network at 250 Mbps. Simulation results in Figure 5(b) show that the UD-CLOCS IR-UWB D2D WWAN cluster cooperation scheme is able to increases the system

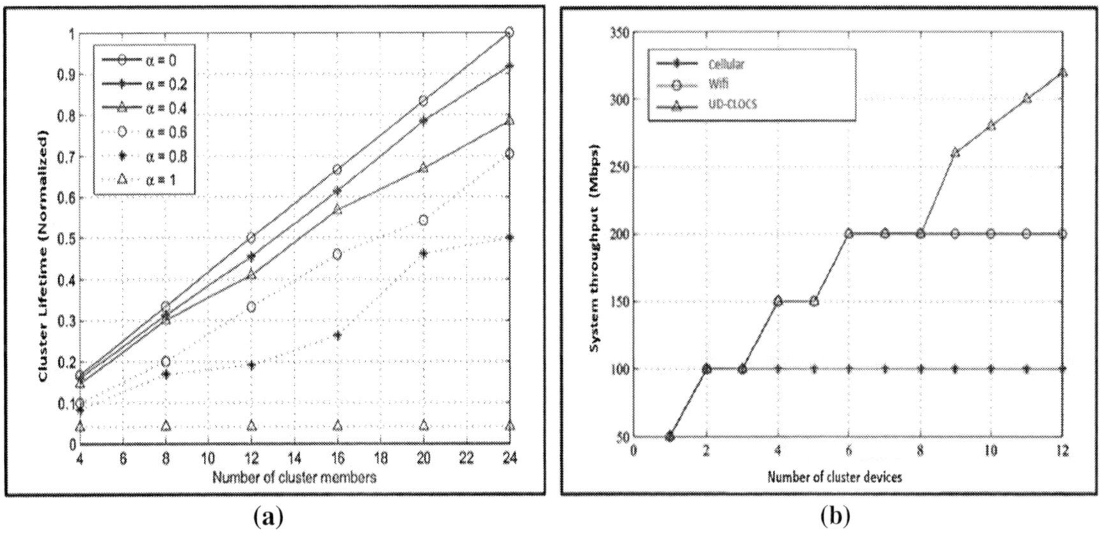

(a) **(b)**

Figure 5 Performance of UD-CLOCS cluster lifetime and comparison of its throughput capability.

throughput significantly by a factor to the magnitude of about 0.75 in concert with increasing number of cluster devices. This suggests suitability of UD-CLOCS to enable high data throughput which not only enhances satellite positioning but has additional potential for seamless, low latency and effective satellite data distribution over terrestrial networks in the 5G setting.

6. Conclusion

This paper presents an IR-UWB D2D overlay network; UD-CLOCS that seamlessly incorporates satellite positioning signal and data into the 5G all-IP terrestrial global network. Results of simulation experiments suggest that performance of UD-CLOCS cluster formation scheme, its cluster lifetime and system throughput would significantly enhance satellite positioning in 5G setting. Future work on this research will focus on further field work on incorporation of satellite signals and data in ultra-dense setting along with its effect on network performance. This is with a view to enhancing satellite positioning in 5G/future network.

7. References

[1] Frodigh, M., et al., "Future-generation wireless networks. Personal Communications", IEEE, 2001. 8(5): p. 10-17.
[2] Chih-Lin, I. et al. "Toward green and soft: a 5G perspective." IEEE Communications Magazine 52.2 (2014): 66-73.
[3] Roh, W. "5G mobile communications for 2020 and beyond-vision and key enabling technologies." Key note: at IEEE WCNC (2014).
[4] ISO/IEC_26907 (2007) High rate ultrawideband PHY and MAC standard. International Organization for Standardization.
[5] ISO/IEC_26908 (2007) MAC_PHY interface for ISO/IEC 26907.International Organization for Standardization.
[6] Li Y et al (2007) A frequency synchronization method for IRUWB system. In: IEEE WiCOM 2007
[7] Bonuccelli, Maurizio A. et al. "Tree slotted ALOHA: a new protocol for tag identification in RFID networks." Proceedings of the 2006 International Symposium on on World of Wireless, Mobile and Multimedia Networks. IEEE Computer Society, 2006.
[8] IEEE_802.15b (2005) Part 15.3b: wireless medium access control (MAC) and physical layer (PHY) specifications for high rate wireless personal area networks (WPAN): amendment to MAC sublayer. IEEE P802-15-3-DF8 draft Amendment, 2005.
[9] Xinzhou W, Tavildar S, Shakkottai S, Richardson T, Junyi L,Laroia R, Jovicic A (2010) FlashLinQ: A synchronous distributed scheduler for peer-to-peer ad hoc networks. In: 48th annual Allerton conference on communication, control, and computing (Allerton), Allerton. pp 514–521
[10] IEEE, Std 802.11-2012: Wireless LAN Medium Access Control (MAC) and Physical Layer (PHY) Specifications, (2012).
[11] Huang J, Qian F, Gerber A, Mao ZM, Sen S, Spatscheck O (2012) A close examination of performance and power characteristics of 4G LTE networks. pp 225–238
[12] McDermott-Wells P (2005) What is Bluetooth? Potentials. IEEE 23(5):33–35.
[13] Asadi A, Mancuso V (2013) WiFi Direct and LTE D2D in action Wireless Days (WD), 2013 IFIP, Valencia. pp 1–8.
[14] Asadi, Arash, and Vincenzo M. "WiFi Direct and LTE D2D in action." Wireless Days (WD), 2013 IFIP. IEEE, 2013.
[15] Tang, Huan, et al. "Enabling D2D communications through neighbor discovery in LTE cellular networks." IEEE Transactions on Signal Processing 62.19 (2014): 5157-5170.
[16] Conti, Marco, et al. "Experimenting opportunistic networks with WiFi Direct." Wireless Days (WD), 2013 IFIP. IEEE, 2013.
[17] Musaloiu-E, Razvan, and Andreas T. "Minimising the effect of wifi interference in 802.15. 4 wireless sensor networks." International Journal of Sensor Networks 3.1 (2008): pp 43-54.

IOP Conf. Series: Journal of Physics: Conf. Series **1152** (2019) 012027 doi:10.1088/1742-6596/1152/1/012027

Field-aligned currents (FACs) behaviour during the arrival of interplanetary magnetic shock

Farah Adilah Mohd Kasran[1,2], **Mohamad Huzaimy Jusoh**[1,2], **Binod Adhikari**[3], **Siti Amalina Enche Ab Rahim**[1,2]

[1] Faculty of Electrical Engineering, Universiti Teknologi MARA, 40450 Shah Alam, Malaysia
[2] Applied Electromagnetic Research Group, Universiti Teknologi MARA, 40450 Shah Alam, Malaysia
[3] Department of Physics, St. Xavier's College, Kathmandu, Nepal

farahadilah87@gmail.com

Abstract. In this paper, our study is focused on the FACs behavior during the SC event. SC consists of two events which are storm sudden commencement (SSC) and sudden impulse (SI) event. The difference between these two events is that the interplanetary shock due to SSC leads to geomagnetic storm while during SI event, there is no geomagnetic storm afterward. We performed the statistical analysis between the FACs value and the temporal variation of SYMH index (dSYMH/dt) during SC events from 2008 until 2015. The dSYMH/dt describe the interplanetary shock signature during the SC events. The results showed that the dSYMH/dt during SSC event has a better correlation than the dSYMH/dt during SI events. It suggested that during the SSC onset, there is the magnetic reconnection process between magnetosphere and ionosphere. So, the enhancement of ground magnetic field is contributed by both magnetospheric and ionospheric currents. However, during the SI onset, least connection process happen between magnetosphere and ionosphere thus, the ground magnetic disturbance was mainly associated with the magnetospheric current.

1. Introduction

Field-aligned currents (FACs) illustrates the coupling process between magnetosphere and ionosphere by which the energy associated with the solar events (solar flares, coronal mass ejection (CME) and coronal holes (CH)) are transferred and are converted into an ionospheric electric current. Generally, during the magnetic reconnection process, the value of FACs is enhanced [1]. Most of the study on FACs have only been carried out during the substorms [1]–[4], which are frequently occurred in auroral region. Plus, the substorm activity occurs when the magnetosphere-ionosphere is energized through the interplanetary convection electric field that governed by the negative value of interplanetary magnetic field (IMF) B_z component [5]. However, too little attention has been paid to compare the behavior of FACs during the SC events associated with the interplanetary shock waves. So, the present study thoroughly examined the magnetic reconnection process specifically during SC onset based on FACs value. SC is a clear onset in the magnetic field variation when the magnetospheric compression associated with interplanetary shock waves occurs. The SC event is categorized into two events which are storm sudden commencement (SSC) and sudden impulse (SI) event. The difference between these two events is that the interplanetary shock due to SSC leads to geomagnetic storm while during SI event, there is no geomagnetic storm afterward. The compression of Earth's magnetosphere due to interplanetary shocks yields to the enhancement of magnetopause current, ring current intensification, and equatorial electrojet, thus resulting in the important role of SC events to GIC activities in this region [6]–[9].

2. Data analysis

2.1. Solar wind an IMF data

Measurements of the solar wind and interplanetary magnetic field (IMF) parameters, including the solar wind velocity and orientation of IMF are made by Advanced Composition Explorer (ACE) satellite. ACE measurements are obtained from OMNI database that available online (http://nssdc.gsfc.nasa.gov/omniweb/). ACE satellite, which is located at LI point to 17 RE of the Earth, provides the measurement values for the solar wind and IMF near the outer boundary of Earth's magnetosphere.

2.2. Ground magnetic indices

In order to study the geomagnetic activity, different magnetic indices are analyzed. AU and AL indices are used to understand the influence of SI event impact to auroral electrojets. These indices are derived from magnetometer stations located in the auroral zone in the northern hemisphere. AU index is an indicator of the eastward auroral electrojet which flows on dayside and AL index represents the westward auroral electrojet which flows on nightside. Besides that, these indices are a good indicator to measure substorm activity in the auroral zone. Symmetric disturbance field in H (or known as SYM-H afterward) describes the variation of the ring current in the magnetosphere and tail current. The SYM-H index is made by averaging the horizontal magnetic field measured by several low-latitude and mid-latitude magnetometers in order to avoid the contamination of auroral and equatorial electrojets. Increase pattern of SYMH shows the compression phase of geomagnetic storm and it also shows the eastward magnetopause current. Westward ring current and main phase of the geomagnetic storm are represented by the decreasing pattern of SYMH. These indices (AU, AL, and SYMH) are obtained from World Data Center for Geomagnetism that operated by Kyoto University, Japan (http://wdc.kugi.kyoto-u.ac.jp/).

2.3. Calculation of field-aligned currents (FACs)

As the current requires a closed loop to circulate, the field-aligned currents (FACs) close the magnetospheric current loops to let it flow into and out the ionosphere. Regarding this process, it allows the solar energy transfer from the magnetosphere into the ionosphere. To calculate the FACs, we apply the formula (1) suggested by [1], [10].

$$FAC = 0.0328\left[np^{\frac{1}{2}}V_{sw}B_T sin\left(\frac{\theta}{2}\right)\right]^{\frac{1}{2}} + 1.4 \quad (1)$$

FAC is measured in μAm^{-2} based on the solar wind density in cm^{-3} (*np*), solar wind speed in km/s (V_{sw}), total of interplanetary magnetic field: $B_T = \sqrt{B_y{}^2 + B_z{}^2}$ and the angle of the magnetic field, (B_y/B_z):

- If $B_z > 0 \rightarrow \theta = tan^{-1}abs\left(\frac{By}{Bz}\right)$
- If $B_z < 0 \rightarrow \theta = 180 - tan^{-1}abs\left(\frac{By}{Bz}\right)$

3. Result and discussion

Figure 1 shows the different behavior of solar wind speed, V_{sw}, B_z component of interplanetary magnetic field (IMF) and ground magnetic SYMH index during SSC and SI events. The red-dashed boxes in both events represent the arrival of the interplanetary shock at 0914UT during SSC event and at 1214UT during SI event. The most obvious differences are the orientation of IMF B_z and the decreasing pattern of SYMH index. During SSC event, the IMF B_z turn southward after the short compression based on the positive IMF B_z at the moment of the interplanetary shock. The IMF magnitude during the southward pointing allows the solar wind energy transfer from the magnetosphere into the ionosphere and subsequently led to the magnetic field disturbances as seen in the negative value of SYMH index at around 2 hours after the SSC event. Meanwhile, during SI event, the IMF B_z remained at the positive value for about 8 hours after the interplanetary shock, which inhibits the magnetosphere-ionosphere coupling process happened. However, these sudden shocks have caused similar magnetospheric compression strength of 30nT/min based on dSYM/dt during both events. Since the reconnection between solar – earth magnetic field happened during SSC event, we can clearly see that higher FACs value during this event compared to SI event with 12.1μAm^{-2} and 8.2μAm^{-2} respectively.

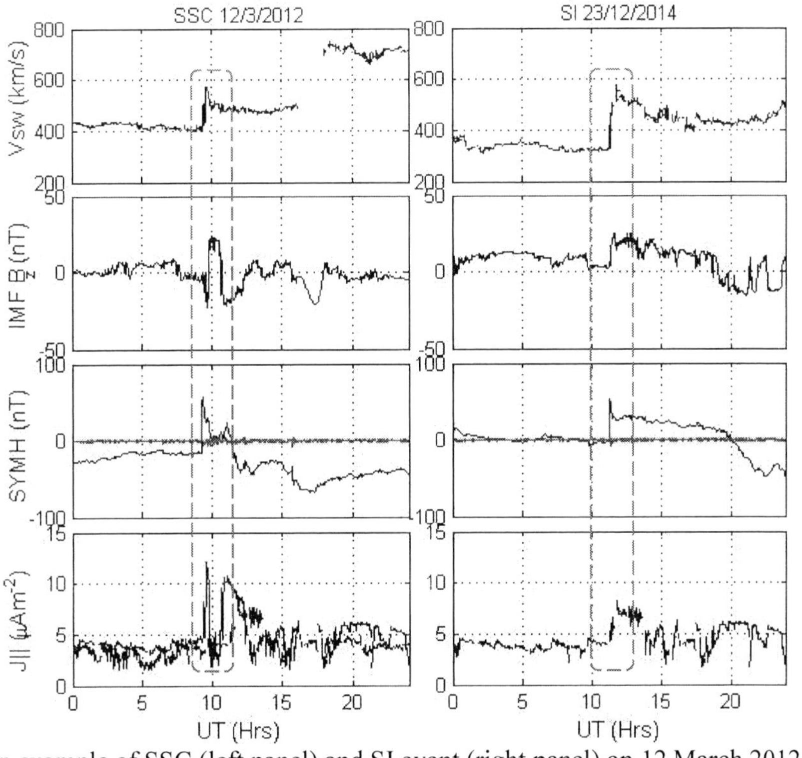

Figure 1. An example of SSC (left panel) and SI event (right panel) on 12 March 2012 and on 23 December 2014 respectively. From top panel to bottom: solar wind speed (V_{SW}), IMF B_z, ground magnetic index (SYMH) with its temporal variation (dSYMH/dt) in red line and field-aligned currents (FACs). Both squared red-dashed lines represent the arrival of interplanetary shock (SSC and SI)

The study of field-aligned currents (FACs) was very important to describe the solar energy transfer from magnetosphere into the ionosphere. The FAC value was controlled by the solar wind and the orientation of IMF B_z [1], [10]. In this study, we investigated the magnetic reconnection process specifically during SSC onset and SI onset by performing the correlation coefficient as shown in Figure 2. We calculated the correlation between the FACs value and the temporal variation of SYMH index (dSYMH/dt) during SC events from 2008 until 2015. Bear in mind that SC onset was associated with the interplanetary shocks events and the dSYMH/dt described the interplanetary shock signature mainly to the equatorial region [11]. The correlation was categorized into three parts: 1) overall SC onset (1st panel), 2) SSC onset (2nd panel) and 3) SI onset (3rd panel). In overall, there is a correlation between dSYMH/dt and FACs during all SC events. However, when we specifically classified the calculation into SSC and SI onsets, we have found a significant result where during SSC onset the correlation was higher than during SI onset. These correlation values explained that during the SSC onset, there was the magnetic connection between magnetosphere and ionosphere. So, the enhancement of ground magnetic field was contributed by both magnetospheric and ionospheric currents. However, during the SI onset, very low positive correlation was observed. Meaning that small reconnection process happen between these two layers and thus, the ground magnetic disturbance was mainly associated with the magnetospheric current. During the arrival of interplanetary shock, the generated magnetopause compression waves extend across the magnetosphere, thus enhancing the current connecting the magnetosphere and ionosphere [12].

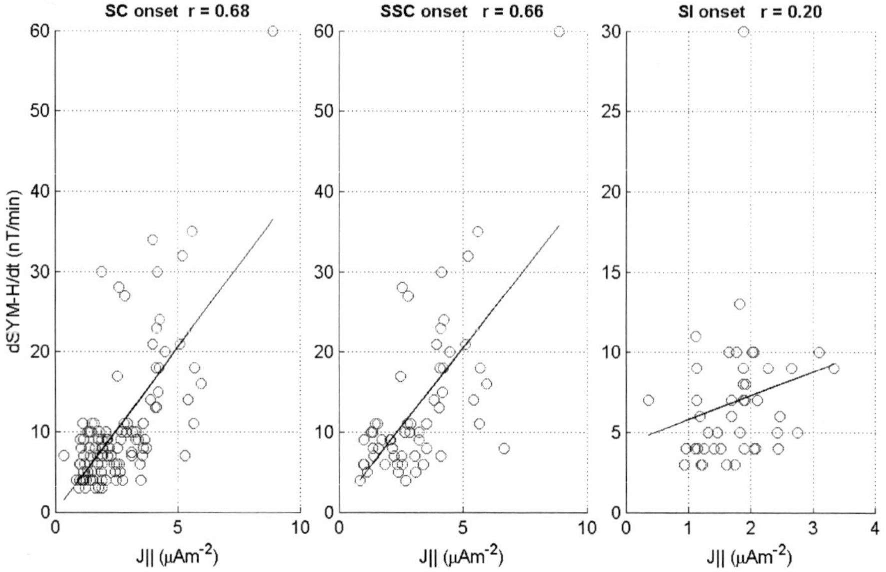

Figure 2. The correlation between the *dSYMH/dt* and FACs during SC (consisting of SSC and SI) events from 2008 until 2015 (in left panel). The second and third panels are the correlation during SSC and SI onset.

4. Conclusion and future recommendation

This study shows that stronger reconnection process between magnetosphere-ionosphere happened during SSC event compared to SI event. Recently, the study on the arrival of the interplanetary shock during SC event was related to the GIC risk in equatorial region [7], [11], [12] As the equatorial dB/dt level was similar despite SSC and SI events, these results depict that strong GIC activity may occur in equatorial region even with less contribution from the ionospheric current [11]. In other words, the ground technological system in equatorial region was exposed to the high risk of GIC even during quiet geomagnetic activity [11], [13]. So, the study on FACs will be more interesting if we extend the investigation to the GIC risk assessment in equatorial region.

5. References

[1] B. Adhikari, S. Dahal, and N. P. Chapagain, "Study of field-aligned current (FAC), interplanetary electric field component (Ey), interplanetary magnetic field component (Bz), and northward (x) and eastward (y) components of geomagnetic field during supersubstorm," *Earth Sp. Sci.*, vol. 4, pp. 257–274, 2017.

[2] A. Hasegawa and T. Sato, "Generation of field aligned current during substorm," in *A Series of Books on the Recent Developments of Space Science and of General Geophysics and Astrophysics Published in Connection with the Journal Space Science Reviews*, Akasofu SI., no. 78, Springer, Dordrecht, 1979, pp. 529–542.

[3] K. R. Murphy *et al.*, "The detailed spatial structure of field-aligned currents comprising the substorm current wedge," *J. Geophys. Res. Sp. Phys.*, vol. 118, pp. 7714–7727, 2013.

[4] Y. Ebihara and T. Tanaka, "Energy Flow Exciting Field-Aligned Current at Substorm Expansion Onset," *J. Geophys. Res. Sp. Phys.*, vol. 122, no. 12, pp. 12288–12309, 2017.

[5] W. D. Gonzalez, A. Joselyn, H. W. Kroehl, G. Ros, B. T. Tsuru, and V. M. Vasyliunas, "What is a geomagnetic storm ?," *J. Geophys. Res.*, vol. 99, no. A4, pp. 5771–5792, 1994.

[6] J. G. Kappenman, "An overview of the impulsive geomagnetic field disturbances and power grid impacts associated with the violent Sun-Earth connection events of 29--31 October 2003 and a

comparative evaluation with other contemporary storms," *Sp. Weather*, vol. 3, p. S08C01, 2005.

[7] B. A. Carter *et al.*, "Geomagnetically induced currents around the world during the 17 March 2015 storm," *J. Geophys. Res. A Sp. Phys.*, vol. 121, no. 10, p. 10,496-10,507, 2016.

[8] J. J. Zhang, C. Wang, T. R. Sun, and Y. D. Liu, "Risk assessment of the extreme interplanetary shock of 23 July 2012 on low-latitude power networks of 23 July 2012 on low-latitude power networks," *Sp. Weather*, vol. 14, 2016.

[9] J. G. Kappenman, "Storm sudden commencement events and the associated geomagnetically induced current risks to ground-based systems at low-latitude and midlatitude locations," *J. Sp. Weather*, vol. 1, no. 1016, 2003.

[10] T. Iijima and T. A. Potemra, "The relationship between interplanetary quatntities and Birkeland current densities," *Geophys. Res. Lett.*, vol. 9, no. 4, pp. 442–445, 1982.

[11] B. A. Carter, E. Yizengaw, R. Pradipta, A. J. Halford, R. Norman, and K. Zhang, "Interplanetary shocks and the resulting geomagnetically induced currents at the equator," *Geophys. Res. Lett.*, vol. 42, no. 16, pp. 6554–6559, 2015.

[12] B. O. Adebesin, A. Pulkkinen, and C. M. Ngwira, "The interplanetary and magnetospheric causes of extreme dB/dt at equatorial locations," *Geophys. Res. Lett.*, vol. 43, no. 22, p. 11,501-11,509, 2016.

[13] F. Adilah *et al.*, "Geomagnetic Disturbance During Quiet Period At Low-Latitude Stations," in *2015 International Conference on Space Science and Communication (IconSpace)*, 2015.

Acknowledgments
The authors would like to thank Universiti Teknologi MARA for management support throughout the project. We also gratefully acknowledge data providers. Solar wind and IMF parameters from ONMI database were provided by NASA's Goddard Space Flight Center (GSFC) Space Physics Data Analysis Web (CDAWeb). AU, AL, and SYM-H index values were provided by World Data Center for Geomagnetism, Kyoto University.

Assessment of the Geomagnetically Induced Current (GIC) at Low Latitude Region based on MAGDAS Data

[1]N Mohd Anuar, [1]F.A Mohd. Kasran, [3]M Abbas, [1,2]M.H Jusoh, [1]S.A Enche Ab Rahim, [4]N. Abdul Hadi, [5]A.Yoshikawa, [6]Z Mohd. Radzi, MAGDAS/CPMN

[1]Faculty of Electrical Engineering, Universiti Teknologi Mara, Shah Alam, Selangor, Malaysia.
[2]Applied Electromagnetic Research Group (AERG), Universiti Teknologi Mara, Shah Alam, Selangor, Malaysia.
[3]Radio Cosmology Laboratory, Faculty of Science, University of Malaya, 50603, Malaysia.
[4]Faculty of Mathematics, Science and Computer, Universiti Teknologi MARA, 40450 Shah Alam, Malaysia.
[5]International Center for Space Weather Science and Education, Kyushu University, Japan
[6]National Space Agency, 42700 Banting, Malaysia

E-mail: nabila199123@gmail.com

Abstract The geomagnetically induced current (GICs) impact is no longer concentrated at high latitude region strictly. The objective of this study was to extend the GICs observation conducted at equatorial region using geomagnetic data extracted from magnetometer installed at Langkawi National Observatory (LNO), Malaysia for the long-term analysis involving the solar cycle 24. The observation is now continued with the GICs comparison between LKW stations with five low latitude station which is BCL, PTN, MND, DAW, and TWV obtained from the magnetic data acquisition system (MAGDAS). The assessment of the GICs behavior revealed the possibility of LKW station experiences GICs is high, based on the time derivative of the horizontal component (dH/dt) value. Extend from that, the GICs estimation was successful done by applying the plane wave method as a model to calculate the North-South and the East-West component, Ex and Ey respectively. The equation of the GICs was provided by excluding the topology constant parameters, which is a and b constant and the value of the underground conductivity for low latitude station. Subsequently, the comparison was limited in years of 2008 – 2014 due to insufficient geomagnetic data. Indeed, the GICs calculation is able to derive the GICs estimation and showed a good agreement with the corresponding value of the time derivative of the horizontal component. However, the accurate current value (A) is able to gain by conducting a direct measurement of the high voltage power transmission system.

1. Introduction

The space weather perturbation is believed to affect the ground-based technology system in term of theoretical and experimental approached. The space weather perturbation mentioned above is the geomagnetically induced current (GIC). These GIC flow into the transformer architecture of the bulk power system through the winding where the current in the winding might be increased. More current

formed in the winding, more magnetic flux might be generated and the core might be received a partial saturation on the part of the transformer. The partial saturation is believed to damage the working principle of the transformer [1]. The AC waveform might be misinterpreted, the protective relays might be desperation and drives to trip out the power transmission line. More so, affected by that, the stability problem, the voltage drop, also the power blackout moment is possible to occur. Rapidly mentioned the famous event in 1989 which is the Hydro - Quebec power transmission blackout and the society suffered for nine hours without electricity [2]. The stimulation of the bulk power system enhances the consideration of awareness to identify to what extent the GIC can be impactful to the power transmission system [3]. The rate of the change of the total magnetic field (dB/dt) normally used to identify the GIC level at high latitude region [4]. The magnitude of dB/dt (nT/min) exceeding 30 nT was indicated the occurrence of the GIC. The high latitude regions are the one that achieves a high possibility that affected by the GIC event due to the auroral electrojet current as a driving mechanism. Previously, the dB/dt value during Hydro – Quebec power transmission is approximately 479 nT/min. Some of GIC cases, there was produced not only occurring due to the substorm, but the sudden impulse (SI) and the storm sudden commencement (SSC) also encourage the GIC to occur [5 and 6]. The authors [7] bring some information about the dB/dt value are not depending on the substorm as a single caused, but the SI event and the SSC had prepared the evidence that the value of dB/dt can still be high enough to penetrate in the power transmission system. This has been discussed by a great number of authors in literature, they proposed a model on how the ionospheric current penetrates in the lithosphere by applying the concept of Lorentz force. It has been reported that substations situated at the ends and at the corners of long transmission lines are prone to maximum GIC magnitudes [8, 9, and 10]. These are the substations which are only connected to one substation on one end and a transmission line on the other end. However, a number of the study was done in a region close to the auroral zone and may not necessarily agree with regions far from the auroral zone. An investigation was done on the Namibian network and reported that south northerly lines appeared to be more susceptible to GICs than the east-westerly lines [11]. This finding was also reported by [8] for the Japanese network. The method had been used previously is by determining the time derivative of the horizontal component as conducted by [12, 13, 2 and 14] to determine the GIC phenomena. Meanwhile, other studies are done by calculation approached [15, 16, and 17]. In addition, the on-site measurement that needs an equipment that must be conducted under the transmission line that operated with high voltage also had been used by [18, and 19].
However, the space weather perturbation that drives GIC during the descending, peak, and ascending phases of the solar cycle is still questionable due to the different mode of solar activity. Therefore, this study aims to observe and estimate the value of GIC by implementing the plane wave method as well as to investigate the dominant parameter of the space weather physic that contributes more to the higher time derivative of the horizontal component (dH/dt) at each low latitude station.

2. Technique and Procedure

2.1. Plane Wave Method

The plane wave method was successfully employed in this analysis. The details of the plane wave method fully described in the previous studies and the [20] are prepared as references. The primary source of GIC is geoelectric field caused by the ambient magnetic field changes on the Earth's surface. The input of geoelectric calculation consists of knowledge about the Earth's conductivity and about the magnetospheric-ionospheric currents or about the geomagnetic variations at the Earth's surface. The existing model such as simple plane wave model can be applied to calculate the geoelectric field of north-south, E_x, and east-west, E_y component:

$$E_{x,y}(t) = \frac{1}{\sqrt{\pi\mu_0\sigma}} \int_{-\alpha}^{t} \frac{g_{y,x}(t')}{\sqrt{t-t'}} dt' \qquad (1)$$

From the plane wave model, the geoelectric field value can be performed from the inverse Fourier transformation to the time domain and derivative of the magnetic component where $g_{x,y}(t') = {dB_{y,x}}/{dt'}$ g_y, μ_o is the vacuum permeability and σ is the Earth's conductivity. Derivatives of the ambient magnetic field in the north-south, B_x, and east-west, B_y is a good indicator for GIC activity and it can be considered as a major component of the geoelectric field.
On the other hand, the characterization of GIC will also be implemented using the rate of change of geomagnetic field data (*dH/dt*) from the available equatorial stations. The *dH/dt* measurement simultaneously with GIC records can provide the threshold level of derivative value that causing undesirable consequence in power grids. Thus, the significance of GIC can be characterized by measuring the variation of horizontal time derivatives of the geomagnetic field, at the identified threshold value.

Table 1: MAGDAS station used for this analysis.

MAGDAS Station	Longitude and latitude			
	GG Lat.	GG Lon.	GM Lat.	GM Lon.
BCL, Bac Lieu Vietnam	9.32	105.71	-0.36	178.36
LKW, Langkawi Malaysia	6.30	99.78	-3.30	172.44
PTN, Pontianak Indonesia	-0.07	109.31	-9.75	181.96
MND, Manado, Indonesia	1.44	124.84	-7.80	197.63
DAW, Darwin, Australia	12.41	130.92	-21.91	202.81
TWV, Townsville, Australia	-19.63	146.86	-28.73	220.30

The plane wave method required the *a* and *b* constant where it's referring to the resistance and voltage threshold value based on the topology of power transmission system of each station that definitely different. The underground conductivity also used in order to obtain the East-West and South-North electric field. In order to get the underground conductivity that should be set to be uniform value, the spherical harmonic analysis method has been implemented. The process flow and the description of the method were clearly explained in [21]. The equation of the underground conductivity is provided below.

$$\sigma_n^m = \frac{5.4 \ x \ 10^4}{m(\pi p)^2} \qquad (2)$$

3. Result and Discussion

Becoming a relevant question, what is the driven mechanism to enhance the dH/dt value at low latitude region in southern hemisphere whereas the northern hemisphere normally experiences a direct physical effect from the Sun activities. Basically, the Sun-Earth coupling system was produced an electric current around the Earth. The electric current generated the magnetic variation that can be measured from magnetometer installed at the Earth surface which is the ground level. Usually, the electric current can be measured experimentally but due to the lack of appropriate equipment, the variation of the magnetic field normally being used to estimate the electric current in the magnetosphere and ionosphere layer as well. The electric current formed at upper part might flow into the Earth atmosphere and induced into the underground level. A superfluity current that induced is harmful to the ground-based technologies system. This phenomenon so-called geomagnetically induced current. Based on [22], the GICs level is possible to high during the geomagnetic storm and the GICs level was enhanced by several factors of the magnetospheric source. The source of magnetospheric current including the magnetopause current; closely related with the sudden impulse (SI) event by observing the increase of solar wind dynamic pressure (SW Pdyn), the instability plasma in the magnetospheric current contribute to the magnetic pulsation also increased the GICs level. Second, the variation of ionospheric Sq current system by applying the dynamo process in such a way caused the high level of GICs for the moment.

Before the analysis of the GICs behavior, a brief overview of the significant event is presented in Table 2. The nine GICs event was initiated or driven by the solar activities and it was classified under storm commencement which is sudden impulse (SI) when there is no further disturbance or sudden storm commencement (SSC) due to the subsequence of the magnetic storm. From the listed nine significant GICs event experiences at LKW station, the comparing step between LKW and other low latitude station to investigate the higher dH/dt managed to cover only a period of time which is from 2008 – 2014 due to insufficient geomagnetic data can be obtained at low latitude station. The analysis carried out in this studies have been placed in accordingly in a few sub-section to ensure the flow of work is manageable as well as the contribution was clearly defined. The sub-section is pointed below.

3.1 Characteristics of Sudden Commencements (SC)

A brief discussion on the characteristic of the sudden commencement that highlighted before continued at this session. The disturbance originated from the Sun was created the geomagnetic storm that carries a huge plasma moving downward to the Earth. The energetic plasma carried by the geomagnetic storm with the high-speed solar wind leads a coronal mass ejection (CME) or coronal holes (CH). The CH event was formed by the process namely co-rotating interaction regions (CIR). The driven mechanism of the geomagnetic storm is CME event during the active year, while the CH able to drive geomagnetic storm during inactive years. The impactful CME or CH event leads to high solar wind dynamic pressure. The high pressure of the solar wind pushing the magnetosphere and increased the magnetopause current. This moment can be observed by the signature of the horizontal geomagnetic variation that increased sharply so-called SSC or SI event depending on the further signature after the main phase of the SI or SSC. Normally, the SIs or SSCs is closely related to the high time derivative of the magnetic component dB/dt (nT/min) due to the energetic compression to the magnetosphere layer. Author [2] point out the condition classified as SI if there is no further disturbance while the classification of SSC is depending on the consequence of the magnetic disturbances. Based on Table 2, a single SI event was observed on 23rd November 2014 while the rest was classified as SSC event. The observation of H-component is due to the variation of the magnetic field. The magnetic field can be measured by using the ground-based magnetometer installed at the Earth surface to estimate the physical performance of the Sun-Earth system. Normally, the SI and SSC event always associated with the enhancement of the magnetopause current.

Table 2: The nine (9) significant date with the maximum dH/dt recorded in Universal Time (UT) and Local Time (LT). The classification of the geomagnetic disturbance also added. The dash (-) represent two meaning, the data is not available or there is no GICs event on the date at that stations.

Date	UT (Hr)	LT (Hr)	dH/dt (nT/min)						SC Class
			LKW	BCL	PTN	MND	DAW	TWV	
27-Mar 08	06:18	14:18	-33.66	-	-	-	-	-	SSC
24-Jan 12	15:07	23:07	34.83	-	-	-	-	-	SSC
15-Jul 12	01:52	09:52	30.3	-	-	-	-	-	SSC
17-Mar13	06:00	14:00	41.2	46.52	38.01	-	-	-	SSC
2-Oct 13	01:58	09:58	88.28	92.98	33.70	-	-	-	SSC
8-Feb 14	07:15	15:15	30.8	32.79	-	-	-	-	SSC
16-Feb 14	04:50	12:50	35.11	33.23	-	-	-	-	SSC
12-Sep 14	15:54	23:54	31.46	36.24	-	-	-	-	SSC
23-Dec 14	11:25	19:25	40.44	40.12	-	-	-	-	SI

Other than the enhancement of the magnetopause current, the geomagnetic activity associated with the severe disturbance leads to the significant effect such as Aurora at high latitude region normally. Nevertheless, low latitude region also experiences a significant effect from the geomagnetic disturbance for the sometimes. Based on the period conducted in this studies, the SSC event is frequently occurred compared to SI event. Previously, the researchers revealed the primary cause of the SSC event is the magnetospheric shock that initiated by the extreme solar activities. The intense disturbance from the Sun able to affect the until low latitude region, not to mention high latitude region. According to [x], the output on the studies revealed the low latitude region is more affected due to the SSC event and that the reason only a single SI event can be observed in Table 2. The source of the SSC event comes from several factors such as coronal mass ejection (CME), coronal holes (CH), high-speed solar wind, (HSSW). Due to this factors, the magnetospheric shock could be occurred expressly and leave a significant effect on low latitude region. We analyze the SSC and SI event to observe the effect which is the GICs purposely. The intense geomagnetic storm or disturbance from the Sun indirectly leads to a serious harm to the underground based technology system.

IOP Conf. Series: Journal of Physics: Conf. Series **1152** (2019) 012028 doi:10.1088/1742-6596/1152/1/012028

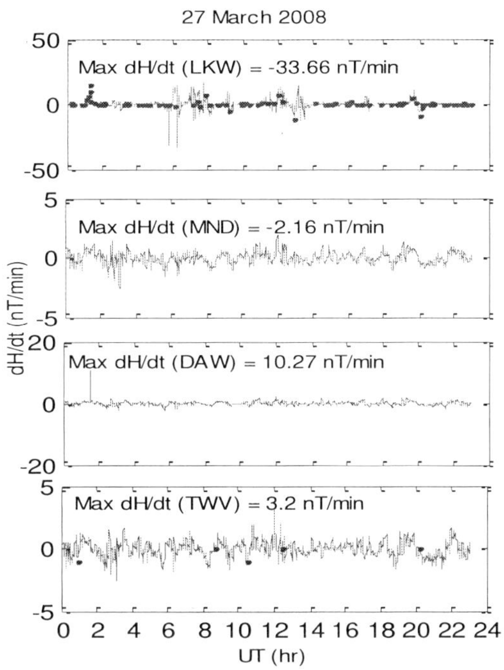

Figure 1: The time derivative of the horizontal component, dH/dt of LKW, MND, DAW, and TWV station. It shows the only LKW station experience the GIC on 27 March 2008 magnetic disturbance.

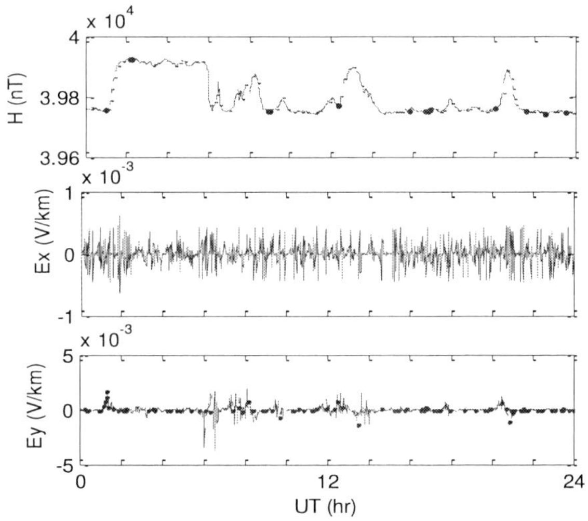

Figure 2: The variation of H-component at LKW station with the variation of North-South and the East-West component, Ex and Ey for LKW station on 27 March 2008.

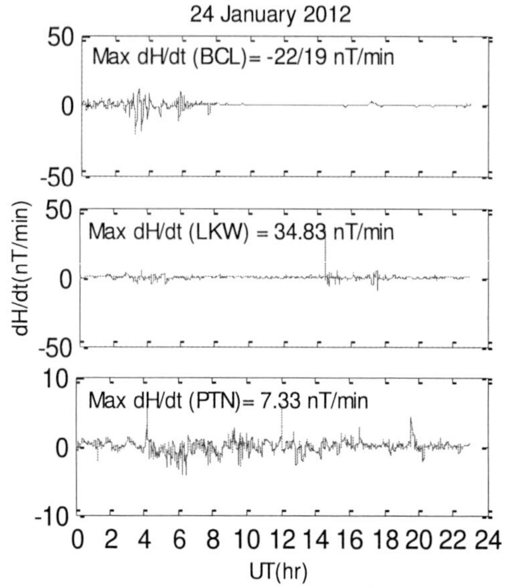

Figure 3: The time derivative of the horizontal component, dH/dt of BCL, LKW, and PTN station. It shows the only LKW station experience the GIC on 24 January 2012 magnetic disturbance.

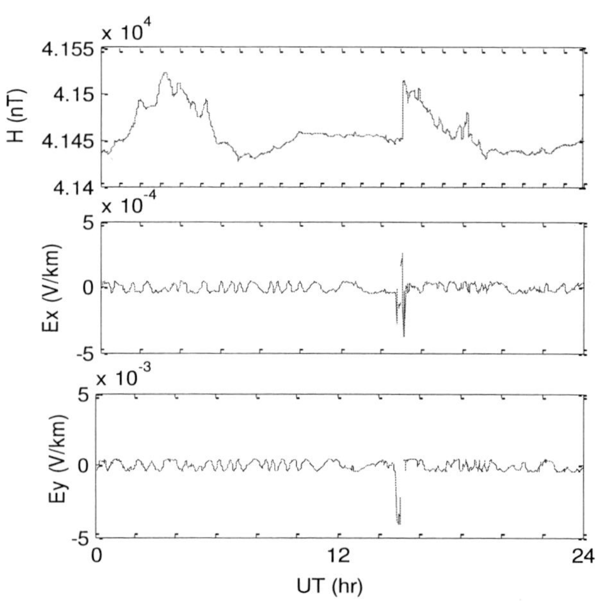

Figure 4: The variation of H-component at LKW station with the variation of North-South and the East-West component, Ex and Ey for LKW station on 24 January 2012.

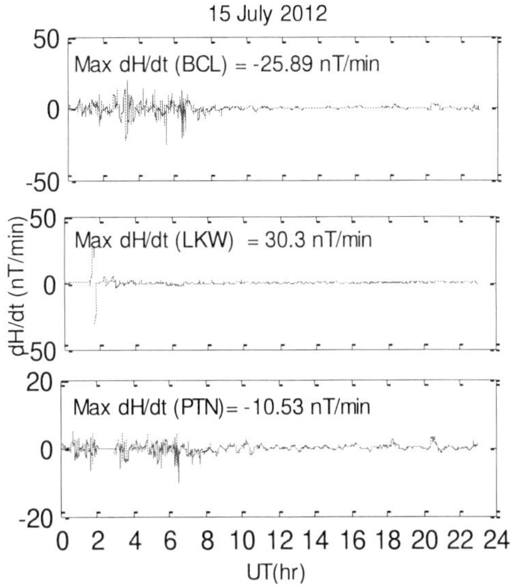

Figure 5: The time derivative of the horizontal component, dH/dt of BCL, LKW, and PTN station. It shows the only LKW station experience the GIC on 15 July 2012 magnetic disturbance.

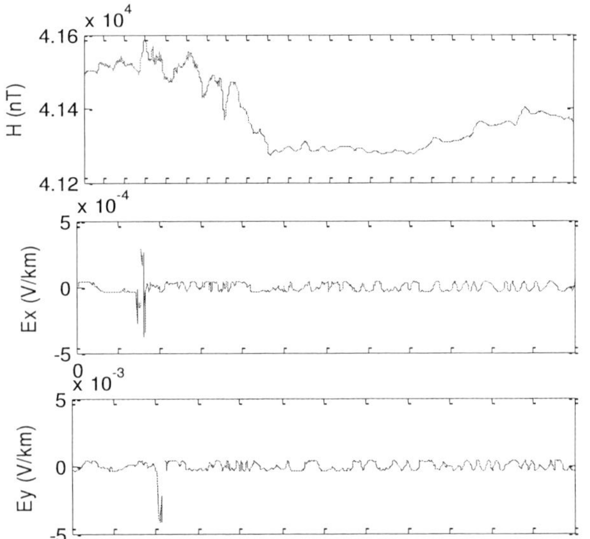

Figure 6: The variation of H-component at LKW station with the variation of North-South and the East-West component, Ex and Ey for LKW station on 15 July 2012.

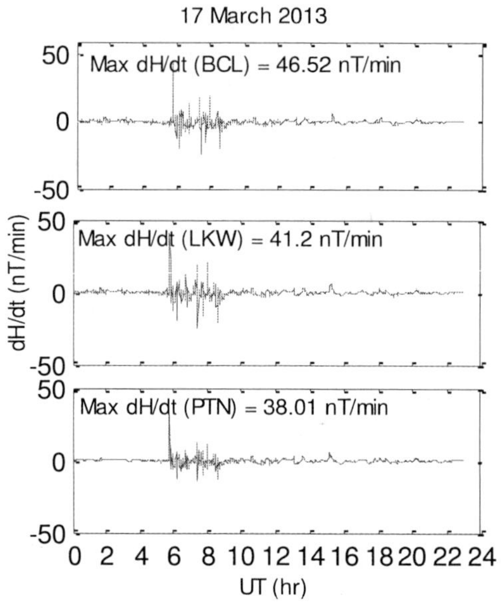

Figure 7: The time derivative of the horizontal component, dH/dt of BCL, LKW, and PTN station. It shows the BCL, LKW, and PTN station was experiences high dH/dt leading by BCL station with 46.52 nT/min on 17 March 2013.

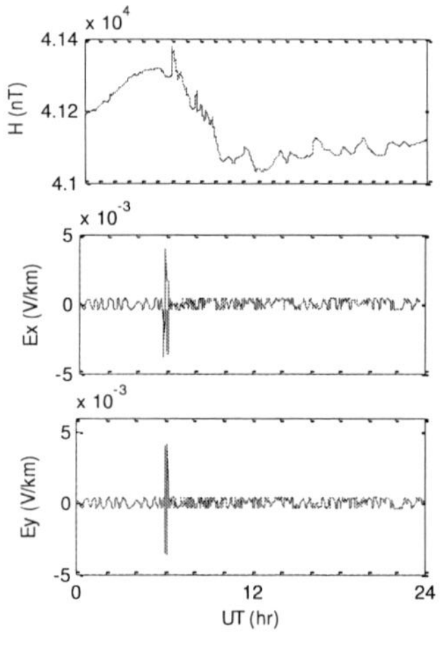

Figure 8: The variation of H-component at BCL station with the variation of North-South and the East-West component, Ex and Ey for LKW station on 17 March 2013.

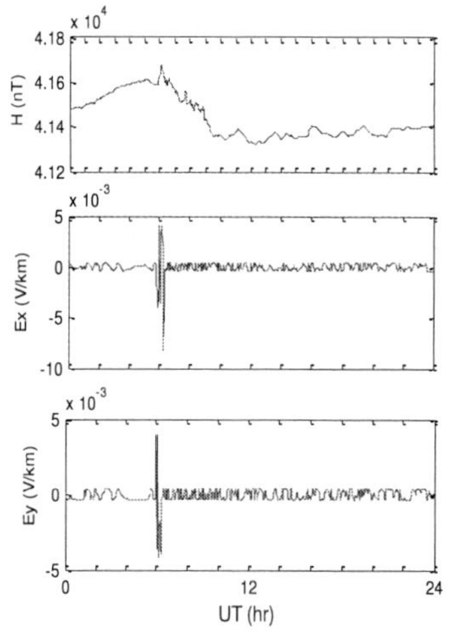

Figure 9: The variation of H-component at LKW station with the variation of North-South and the East-West component, Ex and Ey for LKW station on 17 March 2013.

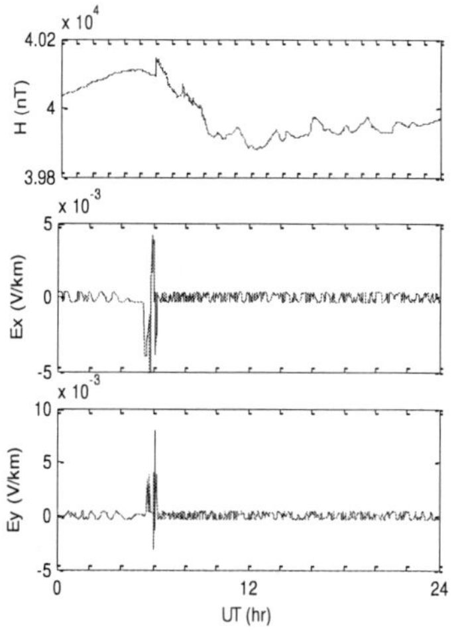

Figure 10: The variation of H-component at PTN station with the variation of North-South and the East-West component, Ex and Ey for LKW station on 17 March 2013.

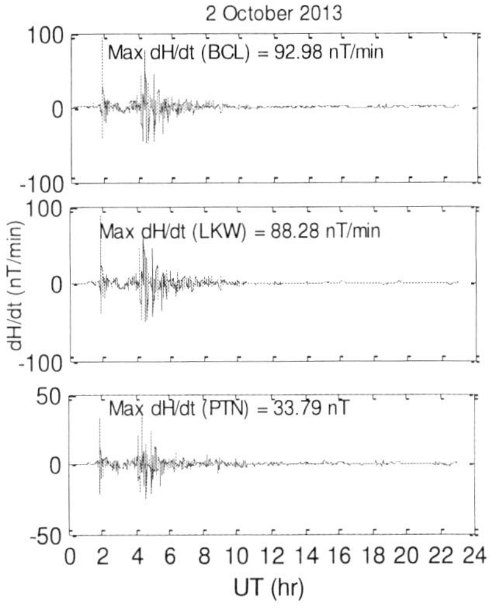

Figure 11: The time derivative of the horizontal component, dH/dt of BCL, LKW, and PTN station. It shows the BCL, LKW, and PTN station was experiences high dH/dt leading by BCL station with 92.98 nT/min on 2 October 2013.

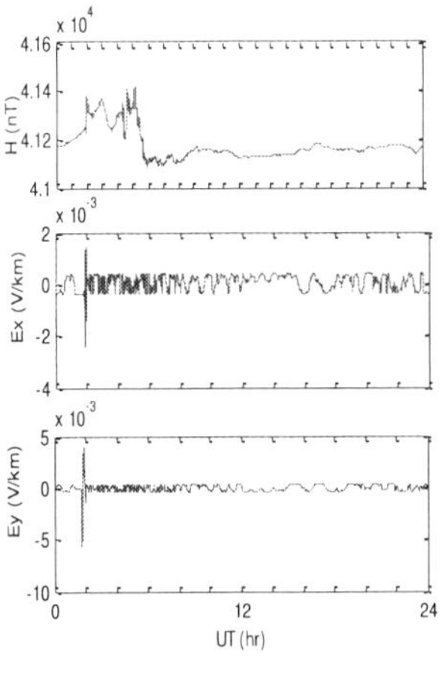

Figure 12: The variation of H-component at BCL station with the variation of North-South and the East-West component, Ex and Ey for LKW station on 2 October 2013.

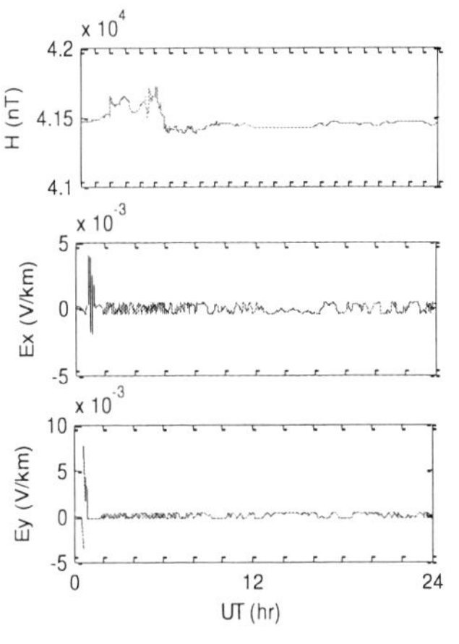

Figure 13: The variation of H-component at LKW station with the variation of North-South and the East-West component, Ex and Ey for LKW station on 2 October 2013.

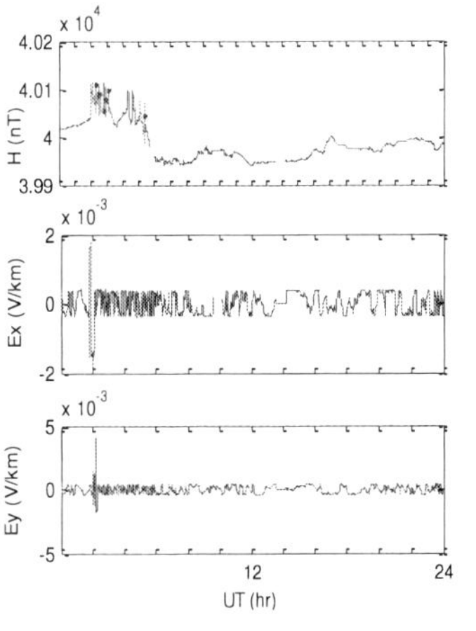

Figure 14: The variation of H-component at PTN station with the variation of North-South and the East-West component, Ex and Ey for LKW station on 2 October 2013.

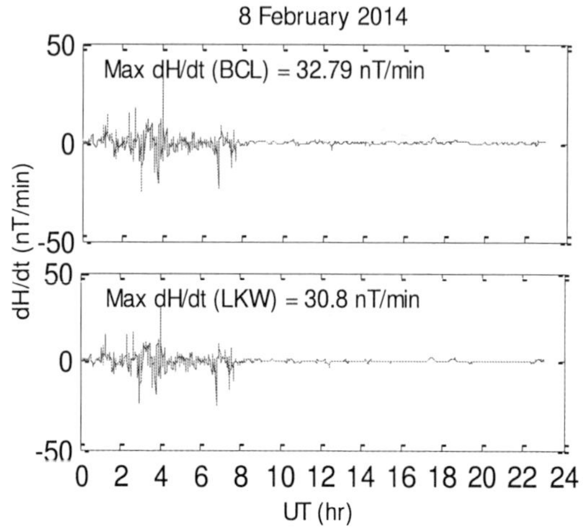

Figure 15: The time derivative of the horizontal component, dH/dt of BCL and LKW station. It shows the BCL and LKW, the station was experiencing high dH/dt leading by BCL station with 32.79 nT/min on 8 February 2014.

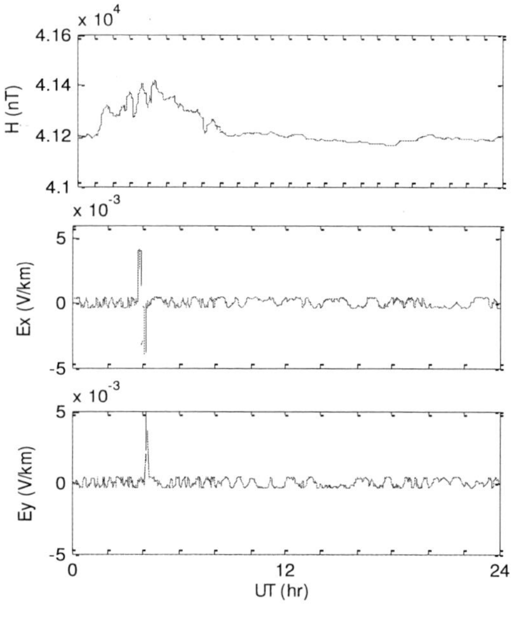

Figure 16: The variation of H-component at BCL station with the variation of North-South and the East-West component, Ex and Ey for LKW station on 8 February 2014.

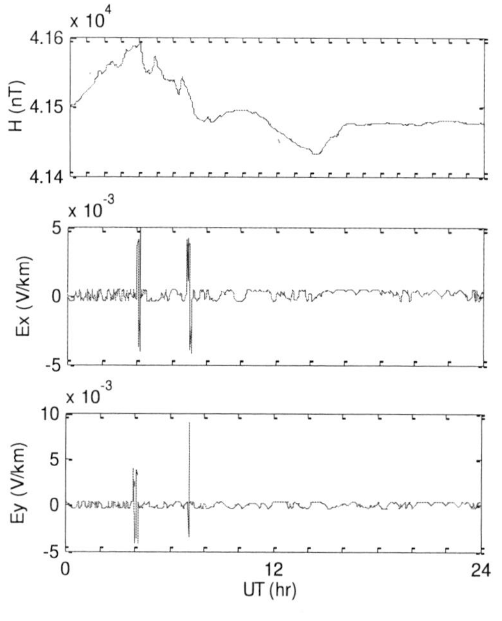

Figure 17: The variation of H-component at LKW station with the variation of North-South and the East-West component, Ex and Ey for LKW station on 8 February 2014.

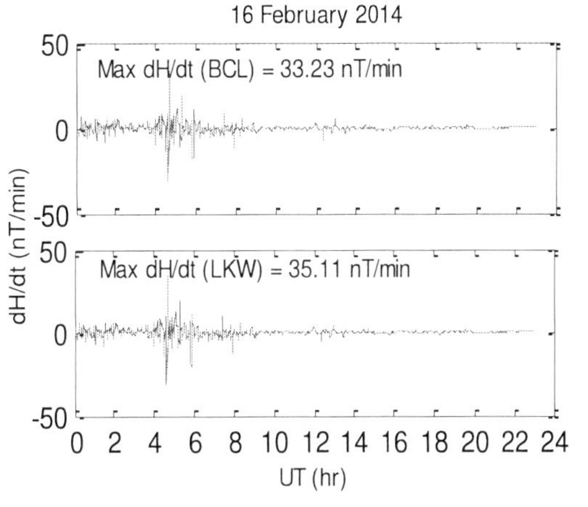

Figure 18: The time derivative of the horizontal component, dH/dt of BCL and LKW station. It shows the BCL and LKW, the station was experiencing high dH/dt leading by LKW station with 35.11 nT/min on 16 February 2014.

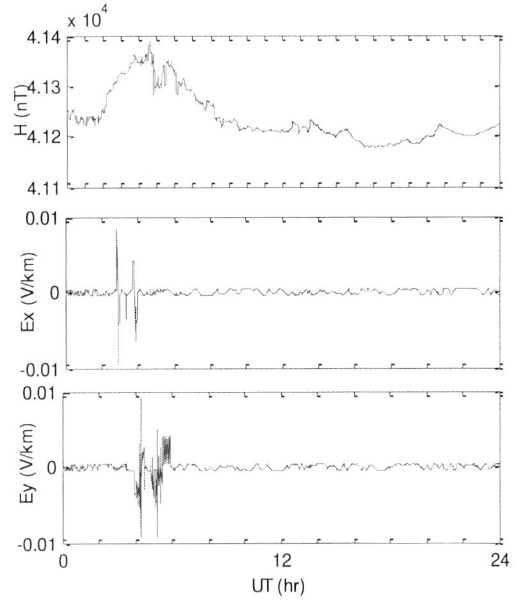

Figure 19: The variation of H-component at BCL station with the variation of North-South and the East-West component, Ex and Ey for LKW station on 16 February 2014.

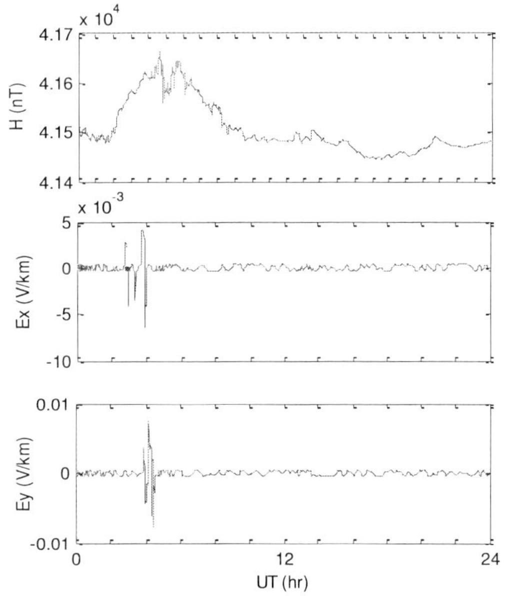

Figure 20: The variation of H-component at LKW station with the variation of North-South and the East-West component, Ex and Ey for LKW station on 16 February 2014.

IOP Conf. Series: Journal of Physics: Conf. Series **1152** (2019) 012028 doi:10.1088/1742-6596/1152/1/012028

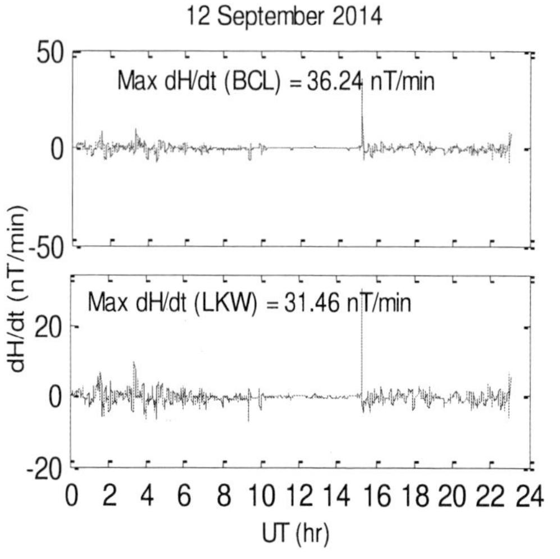

Figure 21: The time derivative of the horizontal component, dH/dt of BCL and LKW station. It shows the BCL and LKW station was experiences high dH/dt leading by BCL station with 36.24 nT/min on 12 September 2014.

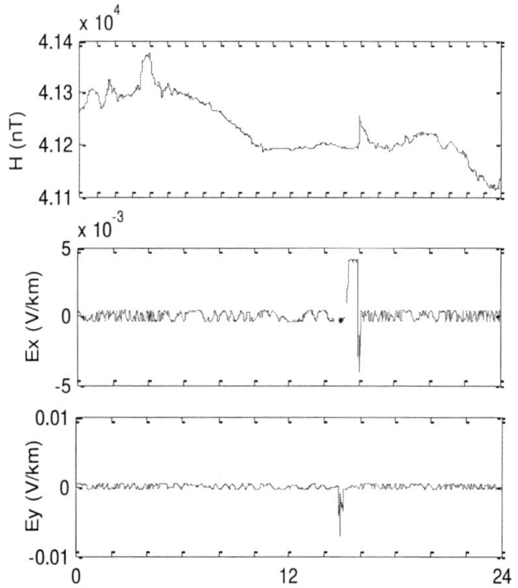

Figure 22: The variation of H-component at BCL station with the variation of North-South and the East-West component, Ex and Ey for BCL station on 12 September 2014.

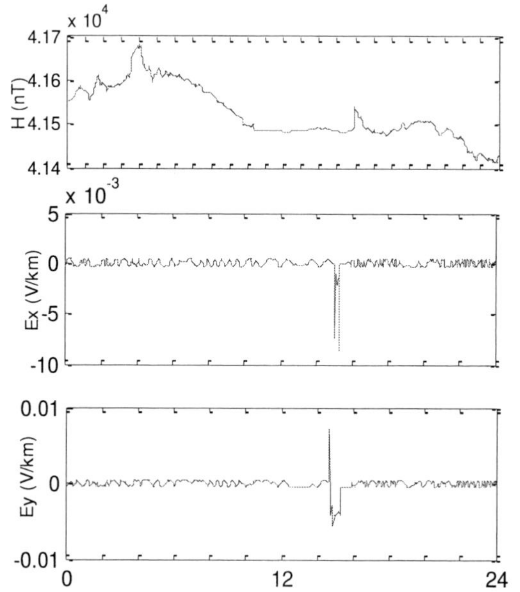

Figure 23: The variation of H-component at LKW station with the variation of North-South and the East-West component, Ex and Ey for LKW station on 12 September 2014.

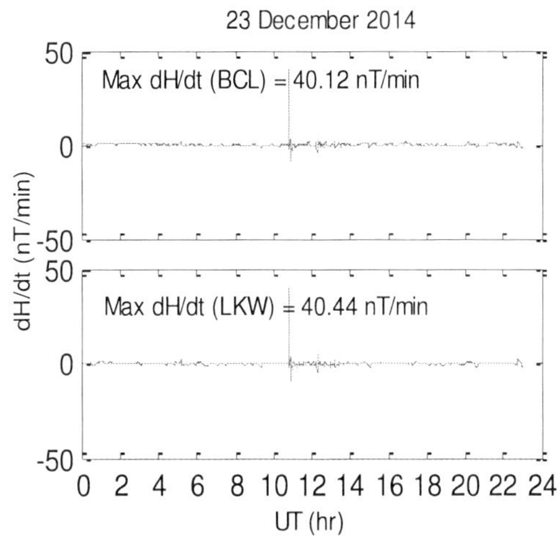

Figure 24: the time derivative of the horizontal component, dH/dt of BCL and LKW station. It shows the BCL and LKW station was experiences high dH/dt leading by LKW station with 40.44 nT/min on 23 December 2014.

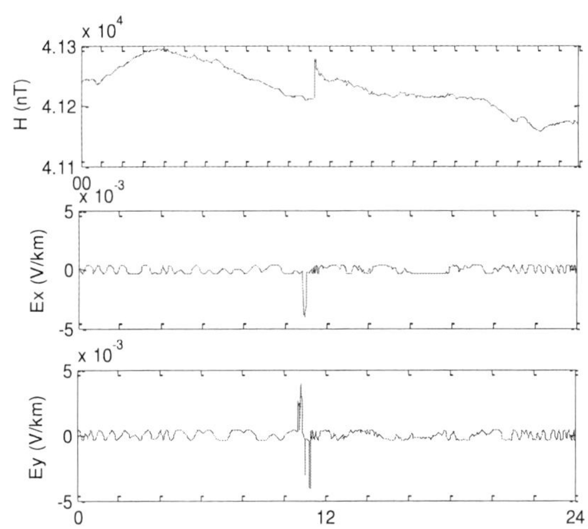

Figure 25: The variation of H-component at BCL station with the variation of North-South and the East-West component, Ex and Ey for BCL station on 23 December 2014.

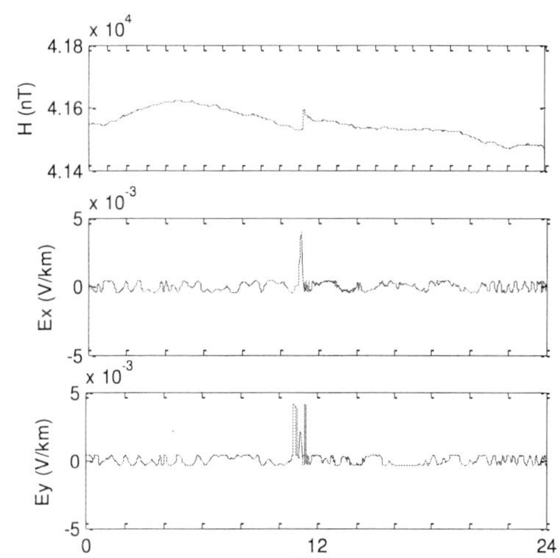

Figure 26: The variation of H-component at LKW station with the variation of North-South and the East-West component, Ex and Ey for LKW station on 23 December 2014.

3.2 The North-South and the East-West electric field component, Ex and Ey to the GICs cases.

The dH/dt illustration presents the comparison between LKW stations and other low latitude stations where a comparison was made based on the availability of data. The value of dH/dt exceeding 30 nT/min is taken apart further more detailed by plotting the variation of the North-South and East – West electric field, Ex, and Ey to estimate the GIC value of each station.

By working the plane wave method, a few parameters need to be prepared and substituted in the formula. There is a constraint to get the constant *a* and *b* but the author comes out with a mathematical equation with the constant was being excluded in the equation to give some idea on how the GICs equations look likes for each of low latitude station. The constant *a* and *b* is the parameter on the power grid topology and the value is depending on the power transmission system. The *a* and b parameters realized to be dissimilar to one and another power grid transmission. Plus, the power grid system also different depending on the location of the station itself. This equation also required the value of underground conductivity, where it needs a method to calculate the underground conductivity. So far, the spherical harmonic analysis (SHA) has been used for a long time ago to calculate the underground conductivity. Based on the SHA approached, the maximum value of the underground conductivity is around 0. 1975 s/m for the depth of 973 km conducted well in Malaysia region in the year of 2016 and 2017 [22]. In this study, the conductivity value is very likely can be used for LKW station but it not suitable for other stations. However, the plane wave method decided the conductivity value is assumed to be uniform, not directly proportional to the depth. However, the value of the underground conductivity might be change depending on the Sun activities. Based on that reason, the conductivity value was excluded in the GICs calculation due to the analysis made is in different years. A further experiment to continue this study was properly planned to conduct a direct measurement under high voltage power transmission line in nearest time. The value of *a* and *b* can be acquired from a long-term measurement located at LKW station to get the accurate value. The GIC calculation is important because the currents are able to trigger and disrupting the transformer on the power grid system. The transmission structure easily can distract from the unusual current flow in the winding of the transformer. Other than that, the undesired current might heating the structure of the transformer, reduced the lifespan of the transformer, and the road to the worst case is the whole transmission system is expected can deliver instability voltage to lead a serious harm which is power blackout in particular time.

By applying SHA concept, we manage to obtain the value of the underground conductivity with the downward increased pattern, which is directly proportional to the depth. The SHA approached presented in this analysis is based on the 2-D electrical conductivity – depth estimation because it uses the vertical and horizontal geomagnetic component as the primary data to extract the value of underground conductivity. While, the underground conductivity required to implement the plane wave method is 1-D conductivity, where it contradicts with the SHA method. However, the 1-D conductivity can be obtained by using only vertical geomagnetic component or the second approach is uses the ratio of the vertical and horizontal component as the primary data to compute the conductivity value. We also found the method to modify the 2-D SHA method to obtain a uniform underground conductivity is by applying the C-responds approached according to [24]. Outweighed by the reason, we found [25] was conducted the GICs calculation at mid-low-latitude regions by implementing the plane wave method. The value of the underground conductivity used in their analysis was obtained experimentally using the magneto-telluric equipment where the equipment having a limited capability to deeply penetrated. Hence, the value of underground conductivity used is around 0.001 s/m with respect to the depth below than 100 km. The below table shows the typical model of the underground conductivity using SHA concept.

Table 3: Typical the underground conductivity – depth estimation Earth model.

Depth (km)	Conductivity (s/m)
100 km	$\sigma = 0.017\ s/m$
350 km	$\sigma = 0.0455\ s/m$
650 km	$\sigma = 0.1294 s/m$
950 km	$\sigma = 0.1983\ s/m$

The underground conductivity model in Table 3 shows the downward increased patterns to the depth. It explained that, the increasing of depth (km), the more conductive that area. However, this analysis was decided to perform the GICs equations for each of GICs event without substitute the underground conductivity value where the value lets as the constant parameter. The value of underground conductivity change will cause the Ex and Ey value also changed. Due to that reason, the underground conductivity was excluded from the equation.

The GICs equation provided below is the estimation of current itself. Against, there are three constant parameters were excluded from the GICs equation. The constant *a*, *b*, and underground conductivity, σ value were unsubstituted in the equation due to explanation in sub-section ii and sub-section iii. However, the GICs equation represents each event are still continued. Only the stations experience GICs that provide the equation. The Ex and Ey value were taken on the same time with the maximum amplitude of dH/dt. The variation of Ex and Ey seemed smaller if compared with the previous work. We believed that issues raised up due to unsubstituted underground conductivity value plane wave equation. The equation prepared as follows;

IOP Conf. Series: Journal of Physics: Conf. Series **1152** (2019) 012028 doi:10.1088/1742-6596/1152/1/012028

Table 4: The equation formed using the plane wave method, with a constant parameter for LKW station, as the majority of GICs event experiences on.

Date	GICs equation for LKW station
27-Mar 08	$GIC = a(\frac{0.0004}{\sqrt{\sigma}}) - b(\frac{0.00331}{\sqrt{\sigma}})$
24-Jan 12	$GIC = a(\frac{0.0033}{\sqrt{\sigma}}) - b(\frac{0.00411}{\sqrt{\sigma}})$
15-Jul 12	$GIC = a(\frac{0.0047}{\sqrt{\sigma}}) - b(\frac{0.0029}{\sqrt{\sigma}})$
17-Mar 13	$GIC = a\left(\frac{0.0004}{\sqrt{\sigma}}\right) + b(\frac{0.00331}{\sqrt{\sigma}})$
2-Oct 13	$GIC = a\left(\frac{0.0047}{\sqrt{\sigma}}\right) + b(\frac{0.0071}{\sqrt{\sigma}})$
8-Feb 14	$GIC = a(\frac{0.0022}{\sqrt{\sigma}}) - b(\frac{0.0049}{\sqrt{\sigma}})$
16-Feb 14	$GIC = a(\frac{0.0027}{\sqrt{\sigma}}) - b(\frac{0.00331}{\sqrt{\sigma}})$
12-Sep 14	$GIC = a(\frac{-0.0082}{\sqrt{\sigma}}) - b(\frac{0.0047}{\sqrt{\sigma}})$
23-Dec 14	$GIC = a(\frac{0.003.11}{\sqrt{\sigma}}) - b(\frac{0.0009}{\sqrt{\sigma}})$

Table 5: The equation formed using the plane wave method, with a constant parameter for BCL station, as the station also experiences the GICs event in this period of studies.

Date	GICs equation for BCL station
17-Mar 13	$\mathrm{GIC} = \mathrm{a}(\frac{0.0034}{\sqrt{\sigma}}) + \mathrm{b}(\frac{0.0019}{\sqrt{\sigma}})$
2-Oct 13	$GIC = a(\frac{0.0010}{\sqrt{\sigma}}) - b(\frac{0.0062}{\sigma})$
8-Feb 14	$GIC = a\left(\frac{0.0039}{\sqrt{\sigma}}\right) + b(\frac{0.0024}{\sqrt{\sigma}})$
16-Feb 14	$GIC = a(\frac{0.0082}{\sqrt{\sigma}}) - b(\frac{0.0071}{\sqrt{\sigma}})$
12-Sep 14	$GIC = a(\frac{0.0037}{\sqrt{\sigma}}) - b(\frac{0.00471}{\sqrt{\sigma}})$
23-Dec 14	$GIC = a(\frac{0.0026}{\sqrt{\sigma}}) - b(\frac{0.0037}{\sqrt{\sigma}})$

Table 6: The equation formed using the plane wave method, with a constant parameter for PTN station, as the low latitude station rarely experiences the GICs event.

Date	GICs equation for PTN station
17-Mar13	$GIC = a(\frac{0.0043}{\sqrt{\sigma}}) - b(\frac{0.0010}{\sqrt{\sigma}})$
2-Oct 13	$GIC = a\left(\frac{0.0019}{\sqrt{\sigma}}\right) + b(\frac{0.004}{\sqrt{\sigma}})$

Briefly discussion on the variation of the Ex and Ey that becoming an important parameter to plug in into the GIC equation. For the four years analysis based on different phases of SC, the LKW station is the most experiences GIC event, followed by BCL and PTN station. While, the MND, DAW, and TWV do not experience on the GIC event. By thoroughly looking, the value of Ex and Ey responded with the positive and negative number. Most of the Ey variation at LKW station is negative except on 2 October 2013. On the other hand, The Ey variation at BCL station was responded with the negative value except for 17 March 2013. While the PTN station a contradict version based on two events. On the same time, we manage to observe the variation of Ex that usually demonstrates a positive value but it shows the opposite sign for two events on 12 September 2014 and 23 December 2014 for LKW and BCL station respectively. The matter that needs to be exposed is the positive and the negative value of Ex and Ey is represent the power transmission system connected trend. The trend of the power transmission structure has been marked as an issue that needs to be proposed for further investigation. Need to be reminded that, the topology of the power transmission structure is not to be similar to other where is based on region. The maximum amplitude of dH/dt is mainly related to the variation of Ex and Ey as the peak point of electric field component can be observed in the variation. Consequently, the GIC equation provided in the table is verifiable by conducting a direct measurement under high voltage power transmission line for a long-term analysis. Previously, the studies on the GICs values were used both approaches namely calculation and experimentally measurement.

3.3 GIC penetration based on skin depth.

On the whole, becoming another relevant question, how far the GICs can penetrate into the Earth Lithosphere and what is approached to estimate the depth of penetration. More so, we carried out this issues by explaining the electromagnetic waves propagate in Ultra-Low frequencies (ULF) range in between 300 Hertz - 3 kiloHertz. Previously, the ULF propagation used for earthquake prediction using three different model so-called ULF emission, micro-fracturing, and the ULF polarization. In this context, the ULF propagate landward experiences a basic process of wave propagation where the wave incident at the Earth surface and reflect windward. The wave then propagates to transmit the magnetic field depending on the Earth conductivity. Loses of the magnetic field intensity is should have in the middle of the incident and reflect wave process. However, the Earth underground conductivity plays a significant role in deciding the ration of the reflected waves. A direct hypothesis can be made. If the underground conductivity is infinite value, the incident and the reflected magnetic field is approximately similar. While, if the underground conductivity is a finite number, which is not uniform, the reflected magnetic field intensity can be properly observed by depending on the skin depth. How far (km) the electric current can flow around the surface so-called skin depth. The less skin depth, the far current able to flow. The frequency and the resistivity of the layer of Lithosphere is dependence parameter to the skin depth. The skin depth is directly proportional to the resistivity but inversely proportional to the frequency. It brings an explanation the short period variation, closely related to the frequency value might able to penetrate to the shallow depth and vice versa.

Table 7: The skin depth calculation based on the conductivity and the period of the wave propagation. This table was fully copied and rewrite from [26] without changing any parameter, values, or formula.

Skin depth $\boldsymbol{\delta}$ (km)			Conductivity (s/m)		
			10^{-1}	10^{-2}	10^{-3}
T (period)	Sec	10	5.03	15.91	50.32
		45	10.68	33.76	106.76
		150	19.40	61.64	194.92
	Min	50	87.18	275.66	871.73
		150	150.99	477.46	1509.88

$$\delta = \left(\frac{T}{\pi\mu\sigma}\right)^{1/2} \tag{3}$$

Table 7 shows the significant parameter to calculate the skin depth in km using equation 3. This table was properly constructed by [26]. The underground conductivity value was divided into three range while the period, T obtained by considering the ULF value is below than 10 Hertz. Extracted information from this established table, the more conductive region might produce a thin skin depth while the less conductive region performs a high skin depth value. The way to mitigate the skin depth is the period must be short, refer the table 7. Other than that, the permeability also plays a significant role to reduce skin effect but for ULF wave propagation, the permeability used is the physical constant of the free space ($\mu = 4\pi\, x\, 10^{-7}$). The changing of permeability value can be only for material for design concept but not appropriate with this topic. Based on the summary provided before, the GICs current able to penetrated unlimited depth only depending on the underground conductivity. Merely, the SHA method only can provide the maximum value of the underground conductivity around 950 km down to Earth Lithosphere.

Conclusion

A long-term analysis has been done well and the GIC estimation is managed to conduct even the equation consist of the constant parameter. The issue easily can be solved by conducting a direct measurement on that particular area. Remaining investigation on the variation of Ex and Ey with the variation of solar wind parameter and the solar activities was planned to conduct in order to strengthen the understanding of the relationship between an upper part that contribute to the geomagnetically induced current event. Several conclusions can be underlined, namely;

a) The value of Ex and Ey is varied and peak at the point of the maximum amplitude of the time derivative of the horizontal component (nT/min).
b) The changing of conductivity may allow the Ex and Ey also changed. The underground conductivity was excluded in the equation due to uncertainties issues.
c) The estimation of the GIC value is depending on the Ex and Ey value also the underground conductivity.
d) The more conductive region, the more current may deeply penetrate.

e) The GIC value calculated in this study applies only to that particular station provided due to the calculation is based on the magnetic field on the same date.
f) The Ex and Ey value obtained in this study represent the value of the time derivative of the horizontal component of each station.
g) The GICs estimation can be verified by conducting the long-term direct measurement under high voltage power transmission at each of station.

Acknowledgment

This project is financially supported by MITRA Perdana (600-IRMI/PERDANA 5/3/MITRA (005/2018)-1 provided by Universiti Teknologi MARA for the management support and consideration throughout the project. The work of A.Y was supported by MEXT/JSPS KAKENHI Grant 15H05815. The authors also would like to acknowledge OMNIWeb data, ACE satellite, magnetic data acquisition system (MAGDAS) for contributing to data collection and analysis for the research.

References

[1] Fiori, R. A. D., Boteler, D. H., & Gillies, D. M. (2014). Assessment of GIC risk due to geomagnetic sudden commencements and identification of the current systems responsible. Space Weather, 12(1), 76-91.

[2] Kasran, F. A. M., Jusoh, M. H., Yoshikawa, A., & Radzi, Z. M. (2018). The Time Derivative of the Horizontal Geomagnetic Field for the Low Latitude MAGDAS Langkawi Station for the Estimation of Geomagnetically Induced Current. In Space Science and Communication for Sustainability (pp. 57-71). Springer, Singapore.

[3] Ngwira, C. M., Pulkkinen, A., Wilder, F. D., & Crowley, G. (2013). Extended study of extreme geoelectric field event scenarios for geomagnetically induced current applications. Space Weather, 11(3), 121-131.

[4] E. O. Falayi and N. Beloff (2012) Modelling of geomagnetically induced currents during geomagnetic storms using geoelectric fields and auroral electrojet indices. Indian J. Phys 86:423–429

[5] Marshall, R. A., Gorniak, H., Van Der Walt, T., Waters, C. L., Sciffer, M. D., Miller, M., & Wilkinson, P. (2013). Observations of geomagnetically induced currents in the Australian power network. Space Weather, 11(1), 6-16.

[6] Marshall, R. A., Dalzell, M., Waters, C. L., Goldthorpe, P., & Smith, E. A. (2012). Geomagnetically induced currents in the New Zealand power network. Space Weather, 10(8).

[7] Kappenman, J. G. (2006). Great geomagnetic storms and extreme impulsive geomagnetic field disturbance events–an analysis of observational evidence including the great storm of May 1921. Advances in Space Research, 38(2), 188-199.

[8] Magnetic Components. (2015). Retrieved from http://www.geomag.nrcan.gc.ca/mag_fld/comp-en.php.

[9] Campbell, W. C. (2007). Introduction to Geomagnetic Fields. Cambridge University Press. 4-5.

IOP Conf. Series: Journal of Physics: Conf. Series **1152** (2019) 012028 doi:10.1088/1742-6596/1152/1/012028

[10]Yumoto, K. (2005). MAGDAS Project. Retrieved from http://www.serc.kyushuu.ac.jp/magdas/MAGDAS_Project.htm

[11] K. Yumoto, S. Ikemoto, M. G. Cardinal, M. Hayakawa, K. Hattori, J. Y. Liu, S. Saroso, M. Ruhimat, M. Husni, D. Widarto, E. Ramos, D. McNamara, R. E. Otadoy, G. Yumul, R. Ebora, and N. Servando, "A new ULF wave analysis for Seismo-Electromagnetics using CPMN/MAGDAS data," Phys. Chem. Earth, Parts A/B/C, vol. 34, no. 6–7, pp. 360–366, 2009

[12] R. Pirjola, A. Viljanen, A. Pulkkinen, S. Kilpua, and O. Amm (2004) Chapter 13 Ground Effects of Space Weather pipelines in Effects of Space Weather on Technology Infrastructure, I. A. Daglis, Ed. Kluwer Academic Publisher, p 235–256.

[13] E. H. Bernhardi, T. A. Tjimbandi, P. J. Cilliers, and C. T. Gaunt (2008) Improved calculation of geomagnetically induced currents in power networks in low- latitude regions. PSSC Proceedings 677.

[14] A. Pulkkinen, R. Pirjola, and A. Viljanen (2007) Determination of ground conductivity and system parameters for optimal modeling of geomagnetically induced current flow in technological systems. Earth, Planets Sp 59:999–1006.

[16]W. A. Radasky and J. G. Kappenman (2010) Impacts of Geomagnetic Storms on EHV and UHV Power Grids. Paper presented at 2010 Asia-Pacific International Symposium on Electromagnetic Compatibility, Beijing, China, 12 - 16 April 2010/.

[17]E. H. Bernhardi, T. A. Tjimbandi, P. J. Cilliers, and C. T. Gaunt (2008) Improved calculation of geomagnetically induced currents in power networks in low- latitude regions. PSSC Proceedings 677.

[18]C. Liu, L. Liu, R. Pirjola, and Z. Wang (2009) Calculation of geomagnetically induced currents in mid- to low-latitude power grids based on the plane wave method : A preliminary case study. Sp. Weather 7:1–9.

[19]R. Pirjola (2012) Geomagnetically Induced Currents as Ground Effects of Space Weather. InTech, p 27 - 44.

[20]W. A. Radasky and J. G. Kappenman (2010) Impacts of Geomagnetic Storms on EHV and UHV Power Grids. Paper presented at 2010 Asia-Pacific International Symposium on Electromagnetic Compatibility, Beijing, China, and 12 - 16 April 2010.

[21] Abidin, Z. Z., Jusoh, M. H., Abbas, M., & Yoshikawa, A. (2018). Application of solar quiet (Sq) current in determining mantle conductivity-depth structure in Malaysia. Journal of Atmospheric and Solar-Terrestrial Physics.

[22] Barbosa, C., Alves, L., Caraballo, R., Hartmann, G. A., Papa, A. R., & Pirjola, R. J. (2015). Analysis of geomagnetically induced currents at a low-latitude region over the solar cycles 23 and 24: comparison between measurements and calculations. Journal of Space Weather and Space Climate, 5, A35.

[23] F. A. Mohd Kasran, M. H. Jusoh, S. N. A. Ahmad, and S. N. Ibrahim, "Geomagnetic Disturbance During Quiet Period At Low-Latitude Stations," in 2015 International Conference on Space

Science and Communication (IconSpace), 2015, no. August, pp. 273–276.

[24] Püthe, C., & Kuvshinov, A. (2013). Determination of the 1-D distribution of electrical conductivity in Earth's mantle from Swarm satellite data. Earth, Planets and Space, 65(11), 4.

[25] Liu, C. M., Liu, L. G., Pirjola, R., & Wang, Z. Z. (2009). Calculation of geomagnetically induced currents in mid - to low - latitude power grids based on the plane wave method: A preliminary case study. Space Weather, 7(4).

[26] Yumoto, K., Ikemoto, S., Cardinal, M. G., Hayakawa, M., Hattori, K., Liu, J. Y., & Ramos, E. (2009). A new ULF wave analysis for Seismo-Electromagnetics using CPMN/MAGDAS data. Physics and Chemistry of the Earth, Parts A/B/C, 34(6-7), 360-366.

Characteristic of the Disturbed Days Ionospheric Current System in the 180-Degree Magnetic Meridian

Siti Nurbaiti Ibrahim[1*], Mohammad Huzaimy Jusoh[1,2], Ahmad Asari Sulaiman[1,2], Akimasa Yoshikawa[3] and Zahira Mohd. Radzi[4]

[1]Faculty of Electrical Engineering, Universiti Teknologi MARA Malaysia, Selangor, Malaysia.

[2]Applied Electromagnetic Research Group, Advance Computing and Communication Communities of Research, Universiti Teknologi MARA Malaysia, Selangor, Malaysia.

[3]International Center for Space Weather Science and Education, Kyushu University, Japan.

[4]National Space Agency, Banting, Malaysia.

sitinurbaiti.ibrahim@gmail.com*

Abstract. The behavior of the ionospheric current during disturbed days using 10 geomagnetic observatories along 180-degree magnetic meridian was analysed. Separation of external and internal current during geomagnetic storm on 12 April until 15 April 2012 data was carried out using spherical harmonic analysis method. The contour of day to day variation of total current, external current and internal current were distinguished based on geomagnetic storm phase. These current then were correlate with the solar wind parameter, IMF Bz, and geomagnetic indices in order to identify their relationship. The results reveal that these current intensities were found higher during the initial phase. This due to the enhancement of the magnetic field during the initial phase caused by the increment of solar wind dynamic pressure and southward IMF Bz. Hence, the induced current identified in the internal current. This study demonstrated that the electric current induction can be obtained from the weak geomagnetic storm.

1. Introduction

A magnetic field recorded in ground-based magnetometer consists of a summation of various sources of signals. The geomagnetic variation field is comprised of contribution in external currents in the ionosphere and magnetosphere and internal (earth's dynamo core, magnetization field of rocks in the crust and induced current generated by external field) currents in the Earth. The most significant contribution is from the current generated from fluid dynamo process in the Earth's core that produce magnetic field up to ~ 30000-60000 nT on the surface called as main-field. The others source only up to few percents contributes in the geomagnetic field [1]. One of the sources from the external field is electric current flowing in the ionosphere that referred as solar quiet(Sq) current. Sq is a dynamo current driven by thermo-tidal action in the ionospheric E-region between 90 and 130km. The term Sq is due to the solar regular variation in sunlit direction during a quiet day. This resultant current would be controlled by ionospheric conductivity and the electric field [2]. During geomagnetically disturbed condition, the magnetic field from the external source is mainly contaminated from ionosphere and magnetosphere current (solar wind variation depends on the direction of interplanetary magnetic field) and the increasing of internal current (telluric current) induced from the external source. Thus, by separating these field, the magnetic variation can be inferred in the purely original source.

A spherical harmonic analysis is a mathematical modeling technique for separating external/internal field in Sq analysis has been adequately demonstrated in [3][4][5][6]. However, these studies have examined only separating current only during quiet days. Although there is some research has been carried out on separation external and internal current during the disturbed period[7][8][9], however, the effect of day to day geomagnetic storm variation on the longitudinal variation on these current has not been closely examined.[9] studies external/internal field associated with Dst index (disturbance time index) during a geomagnetic storm. They found that the large contribution on the internal field is during the main phase of the geomagnetic storm. However, this analysis only limited a few stations in latitudinal variation. A geomagnetic storm is a temporary disturbance of the Earth's magnetosphere caused by intense energy from solar wind that interacts with the Earth's magnetic field. One of the factors that cause the magnetic storm is solar Coronal Mass Ejection(CME).

In this study, the aim was to examine the day to day characteristic of ionospheric current during the disturbed period of the geomagnetic storm due to CME on 13 April 2012. The geomagnetic data used from 12 April until 15 April 2012 were obtained from 10 geomagnetic stations located longitudinally along the 180° magnetic meridian. In addition, these regions actually were not have been studied yet particularly in external/internal current analysis. Separation of external current and the internal current was carried out by employing the spherical harmonic analysis. These results were then distinguished based on the geomagnetic phase such as initial phase, main phase and recovery phase to differentiate their physical mechanism. In order to analyze the characteristic of these current and to identify any possible factors, the data of solar wind parameters, interplanetary magnetic field, and geomagnetic indices were provided.

2. Data and Methods

2.1. Geomagnetic Data

The minute geomagnetic data were extracted from ten geomagnetic observatories along 180° Magnetic Meridian obtained from Magnetic Data Acquisition System of Circum-pan Pacific Magnetometer Network (MAGDAS/CPMN) and International Real-Time Magnetic Observatory Network (INTERMAGNET). The map and details of the stations are shown in Figure 1 and Table 1 respectively. In this study, the geomagnetic coordinate is deployed due to the most natural phenomenon in the magnetic field. Geomagnetic field components of horizontal(H), declination (D) and vertical (Z) were carried out in this analysis.

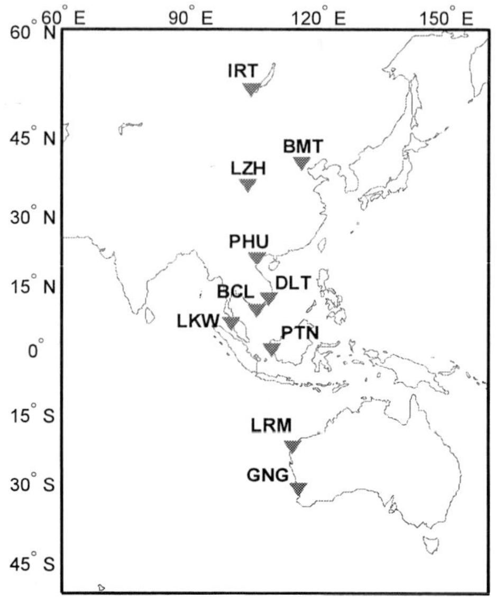

Figure 1: Map of geomagnetic observatories used.

Table 1: Details of stations used.

Station Name	Abbrev.	Array	Geographic		Geomagnetic		Time Zone (GMT)
			Lat.(°)	Long.(°)	Lat.(°)	Long.(°)	
Irkutsk	IRT	Intermagnet	52.27	104.45	42.37	177.3	+8
Beijing Ming Tombs	BMT	Intermagnet	40.3	116.2	30.51	187.59	+8
Lanzhou	LZH	Intermagnet	36.1	103.84	26.17	176.51	+8
Phuthuy	PHU	Intermagnet	21.03	105.95	11.06	178.33	+7
Dalat	DLT	Intermagnet	11.94	108.48	2.13	181.0	+7
Bac Liue	BCL	MAGDAS	9.32	105.71	-0.6	178.11	+7
Langkawi	LKW	MAGDAS	6.3	99.78	-3.42	172.35	+8
Pontianak	PTN	MAGDAS	-0.07	109.31	-10.0	181.71	+7
Learmonth	LRM	Intermagnet	-22.22	114.1	-32.06	187.07	+8
Gingin	GNG	Intermagnet	-31.36	115.71	-41.14	189.16	+8

2.2. IMF, Solar Wind Parameters, and Geomagnetic Indices Data

To investigate the possible mechanism that influences the behavior of current during the geomagnetic storm; the solar wind parameter, interplanetary magnetic field, and geomagnetic indices were applied. The data of solar wind parameters: solar wind speed (Vsw) and solar wind pressure (Pdyn), interplanetary magnetic field (IMF Bz) and geomagnetic indices (Kp*10 index, Ap index, Dst index, AL, and AU index) were obtained from OMNIWeb Plus website at https://omniweb.gsfc.nasa.gov/. The data of solar wind parameters and IMF Bz were present in a minute to show the real-time of disturbance. For estimating the rate of solar wind input energy transfer into the magnetosphere, the Akasofu's epsilon [10] was used from the equation:

$$Input\ Energy\ (\varepsilon) = V_{sw} \times B^2 \times F(\theta) {I_0}^2 \quad (1)$$

where V_{sw} is a solar wind speed, B is an interplanetary magnetic field, I_0 is 7 Earth radii, Re and F(θ) is a function IMF clock angle θ (By/Bz).

2.3. Spherical Harmonic Analysis

In order to separate current, the mathematical model called as spherical harmonic analysis is applied. As devised by Gauss,1839 was ascribed by [11], the magnetic potential function, *V* at any location on the earth can be expressed as in converging series of terms:

$$V = a\sum_{n=1}^{\infty}\left[\left(\frac{r}{a}\right)^{n} S_n^e + \left(\frac{a}{r}\right)^{n+1} S_n^i\right] \tag{2}$$

These series allow a separation of the external and internal current source over a sphere (radial distance, r) by two series of terms. First term is r^n that has increasing power of r, that means as r direction increases, the field would increase; approaching into an external current source. The other series of the term is $(1/r)^n$, that increase with reciprocal power of r. As the r increase, the term becomes smaller as if the field approaching the internal current source in reduced r direction. Thus, such series represent the magnetic potential function, *V* in the ionospheric current analysis is a combination of external (ionospheric source) and internal (induced) source of the geomagnetic field.

The magnetic potential that computed from the daily mean values in UT time that represent both external and internal current in spherical harmonic can be expressed by:

$$V_n^m = C + a\sum_{n=1}^{\infty}\sum_{m=0}^{n}\left\{\left(a_n^{me}\left(\frac{r}{a}\right)^{n} + a_n^{mi}\left(\frac{a}{r}\right)^{n+1}\right)cos(m\phi) + \left(b_n^{me}\left(\frac{r}{a}\right)^{n} + b_n^{mi}\left(\frac{a}{r}\right)^{n+1}\right)sin(m\phi)\right\}P_n^m(\theta) \tag{3}$$

where *C*, *a*, *r*, *θ* and *ϕ* denote as constant of integration, the geocentric radius, the earth radius, the geomagnetic colatitude and the local time of the geomagnetic station. The a_n^{me}, a_n^{mi}, b_n^{me} and b_n^{mi} represent as Legendre polynomial coefficients where e and i representing as the external and internal values respectively. $P_n^m(\theta)$ is Legendre polynomial function of colatitude θ only. The integers n and m are called as degree and order respectively; where n has a value of 1 or greater, and m is always less than or equal to n ($m \leq n$). In the present analysis, the magnetic potential computed in order m=4 and degree n=12.

The equivalent current function, $J(\phi)$ for an hour of a day can be obtained from:

$$J(\phi) = \sum_{m=1}^{4}\sum_{n=m}^{12}(U_n^m\, cos(m\theta) + V_n^m\, sin(m\theta))P_n^m \tag{4}$$

For the representation of external current;

$$U_n^m = -\left(\frac{5R}{2\pi}\right)\left(\frac{2n+1}{n+1}\right)a_n^{me}\left(\frac{a}{R}\right)^n \tag{5}$$

$$V_n^m = -\left(\frac{5R}{2\pi}\right)\left(\frac{2n+1}{n+1}\right) b_n^{me} \left(\frac{a}{R}\right)^n \tag{6}$$

and also for the internal current;

$$U_n^m = -\left(\frac{5R}{2\pi}\right)\left(\frac{2n+1}{n}\right) a_n^{mi} \left(\frac{R}{a}\right)^n \tag{7}$$

$$V_n^m = -\left(\frac{5R}{2\pi}\right)\left(\frac{2n+1}{n}\right) b_n^{mi} \left(\frac{R}{a}\right)^{n+1} \tag{8}$$

where R is the Earth radius in kilometers and a is the radius of a sphere at which current could flow. Since the value of a is approximately equal to R, thus ratio a/R is 1.

3. Results and Discussion

The first set of analysis was to examine the variation of external source based on solar wind parameters, interplanetary magnetic field, and geomagnetic indices. Figure 2 shows the variation of interplanetary magnetic field (IMF Bz), solar wind input energy (ε), solar wind speed (Vsw), solar wind dynamic pressure (Pdyn), and geomagnetic indices (AL, AU, Kp, Ap, and Dst index) during the minor geomagnetic storm that been analysed on 12 April 2014 until 15 April 2014. The graph of variations was distinguished based on geomagnetic storm phase as demonstrated in Dst index plot. The geomagnetic indices of Kp, Ap, and Dst index were applied to measure the rate of disturbance in geomagnetic field in middle and low latitude. The AL and AU indices used to be proxy of auroral disturbance in high-latitude. AU is proxy of eastward auroral electrojet that flow in dayside and AL is proxy of westward auroral electrojet in nightside. These indices were included due to the possibility of the auroral electrojet that can travel equatorward during high magnetic activity[12].

On 12 April 2014, the sudden commencement (SC) occurs at 04UT followed by prolonged initial phase. During the SC, at a prior time, the Dst index have a small increase, then as the IMF Bz is turned to southward; there is a small drop in Dst index and the geomagnetic indices of Kp and Ap are increases with the value -4 and 22nT. Simultaneously, the input energy abruptly increases up to $1.5\text{x}10^{19}$ erg/s and solar wind dynamic pressure enhance up to 8.77nPa. The increment is due to pressure from the shock hits the magnetosphere and cause the magnetopause current (Chapman Ferarro current) increase. Moreover, it also causes the horizontal magnetic field component in ground-based having a positive sudden increase[13]. This SC, however, was not so related to the solar wind speed due to slow solar speed range (~300km/s). As soon as the Bz is positive and gradually varying, all the parameters have been in lower variation and having a prolonged initial phase for about few hours.

Later, at 15UT on the same day, the IMF Bz changes to direction into southward and all the parameters varies abruptly indicating the main phase process is started. During the main phase, as the Bz is negative, the magnetic reconnection occurred and magnetosphere convection enhanced. The ring current growth and cause the negative perturbation in the Earth's magnetic field[14]. As shown in Figure 2, the magnitude of input energy transfers into magnetosphere increases and peak about $1.75\text{x}10^{19}$ erg/s. The solar wind speed experiences the high-speed solar wind that increases up to ~600 km/s. In contrast, the value of solar wind dynamic pressure was enhanced but not so high compared during SC time. Moreover, the value of AL and AU index was found increased at 19UT. However, the AL index shows a higher variation index (large decrease) compared to the AU index. This due to the increment of westward auroral electrojet current in nightside. Note that at 19UT, all selected stations having nighttime. Based on the Dst index, the value of the index was gradually decreased until -60nT at 04UT on 13 April 2014. This decrease is caused by the growth of ring current. At the same time, the Kp*10 index and Ap index increased with the value of 50 and 48nT respectively, representing the geomagnetic disturbance is weak storm type G1. Further, as the Bz is northward the magnetic field begins to recover over within three days. At this phase, the energy transfer becomes weaken.

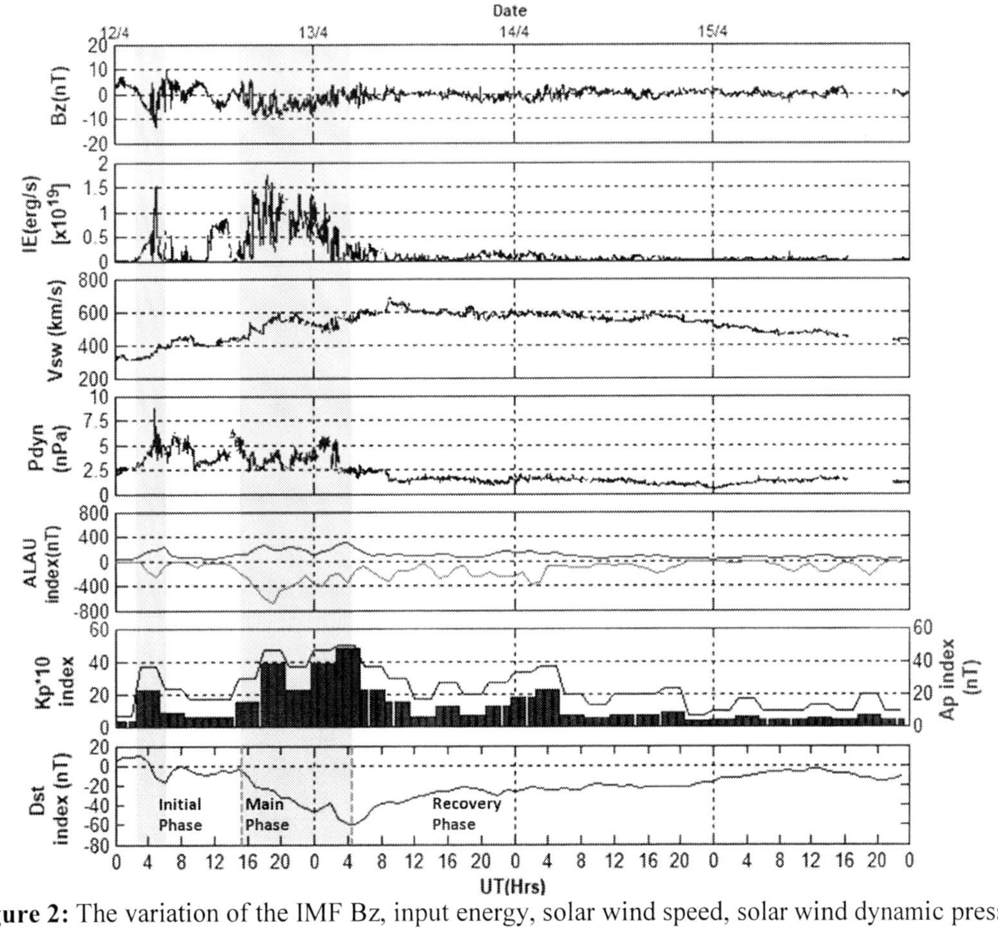

Figure 2: The variation of the IMF Bz, input energy, solar wind speed, solar wind dynamic pressure including geomagnetic indices including AL, AU, Kp, Ap and Dst index during 12 April 2014 until 15 April 2014 in UT hours.

Figure 3 shows the contour map of day to day current intensity of equivalent current, external current and internal current over four days of geomagnetic storm phases. The Dst index is included in the graph is used to illustrate the phase of the geomagnetic storm. The purple lines indicating the starting process of initial phase and recovery phase and the red line represents the prior of the main phase of geomagnetic storms happen. Two-current vortices form that are oppositely directed at 00UT to 10UT, demonstrating the equivalent current of the daytime ionosphere. The positive peaks of the current indicating the current flows in the counter-clockwise direction in the northern hemisphere. And also negative peaks in southern hemisphere represent the current flows in a clockwise direction. Near to equatorial station (BCL and LKW), the current lines are close together denoted as equatorial electrojet current. However as shown in the external plot, the value was very low due to computation in spherical harmonics regarding the highest order[6].

In the view of the current pattern and their intensity based on geomagnetic phase, the initial phase shows highest current intensity and greater peak compared during the main phase and recovery phase Referring to total current result (upper panel), the value of this current is higher compared to external and internal current mainly during the initial phase. The intensity of the total current is representing the summation of external and internal current. During the initial phase, the intensity of all currents shows the highest value. The maximum of the total current in the northern hemisphere is 18.3 x10^6 A and in the southern hemisphere is -13.8 x10^6 A at peak of current vortices. For the the external current, the value in the northern is peak at value 12 x10^6 A and the southern peak value is -11 x10^6 A. The lowest current is at internal current where the northern peak is 9 x10^6 A and southern peak is -9 x10^6 A. It can be seen contribution of initial phase physical process that causes the enhancement of magnetic field. Note that, the data are computed from daily average, thus it can reduce the effects of the magnetospheric source. The values of calculated current also reached up to $\sim 10^7$ A using equations (5) until (7). These measured values were generally agreement with results obtained by [15] that studied the equivalent current during the quiet days using the same equations.

Eventually, as the main phase started (red line), the value of all current gradually decreases. It is apparent that the current reduced due to negative perturbation on the geomagnetic field due to the enhancement of ring current. When the Bz is changing to negative value started at 15UT, the storm started to occur during nighttime. As can be seen in Figure 3 in total current and external current graph, there is a small current vortex occurred in the mid-latitude station in the northern hemisphere. This current might be attributed to movement of auroral electrojet to the equatorward. But in this case, it reaches only in mid-latitude. The main phase process ends at 04UT on 13 April 2014 where it is noon time. At the present time, the peak of total current decrease with the value of 14 x10^6 A in the northern hemisphere and -8 x10^6 A in the southern hemisphere. The peak of external current value for northern and southern hemisphere are 7 x10^6 A and -5 x10^6 A respectively. The value in internal current also reduced with about ± 1x10^6 in both hemispheres compared to the external current. Later, as the recovery phase started, the total current and external current changes gradual increase until 15 April that represents regular daily variation. This is due to the dissipation of transfer energy. However, in contrast with total and external current, the internal current changes gradually to a lower value. Hence, for this reason, the increment during initial phase showing that there is a telluric current induced by the external source. Moreover, the behavior of internal current during the last day of recovery phase show that internal source with the small value of induced current.

Clearly, as expected, the enhancement during the initial phase in the external current is comprised of the magnetospheric current that depends on the IMF direction and the ionospheric current that have varied in their ionospheric conductivity. Current flowing in the magnetosphere and ionosphere also can be induced in the ground. The changing of the magnetic field can cause current induction in any conductive place. During nighttime, the currents can receive a contribution from the magnetosphere. The smaller current vortices in nightside can be attributed to the westward auroral electrojet indicate by AL indices in Figure 2. This auroral electrojet driven magnetospheric field-aligned current can be propagated to lower latitude in the conducting E-region [16]. The value of northern hemisphere also was found higher compared to southern due to April month used in this study where the spring equinox in the northern hemisphere and autumn equinox in the southern hemisphere. In addition, the low current intensity at the equatorial station in southern hemisphere shown in the external current plot could be attributed to the interhemispheric current that cancels out the external current in that particular hemisphere [17]. Moreover, the concentration of current is less in the southern hemisphere due to only four stations used.

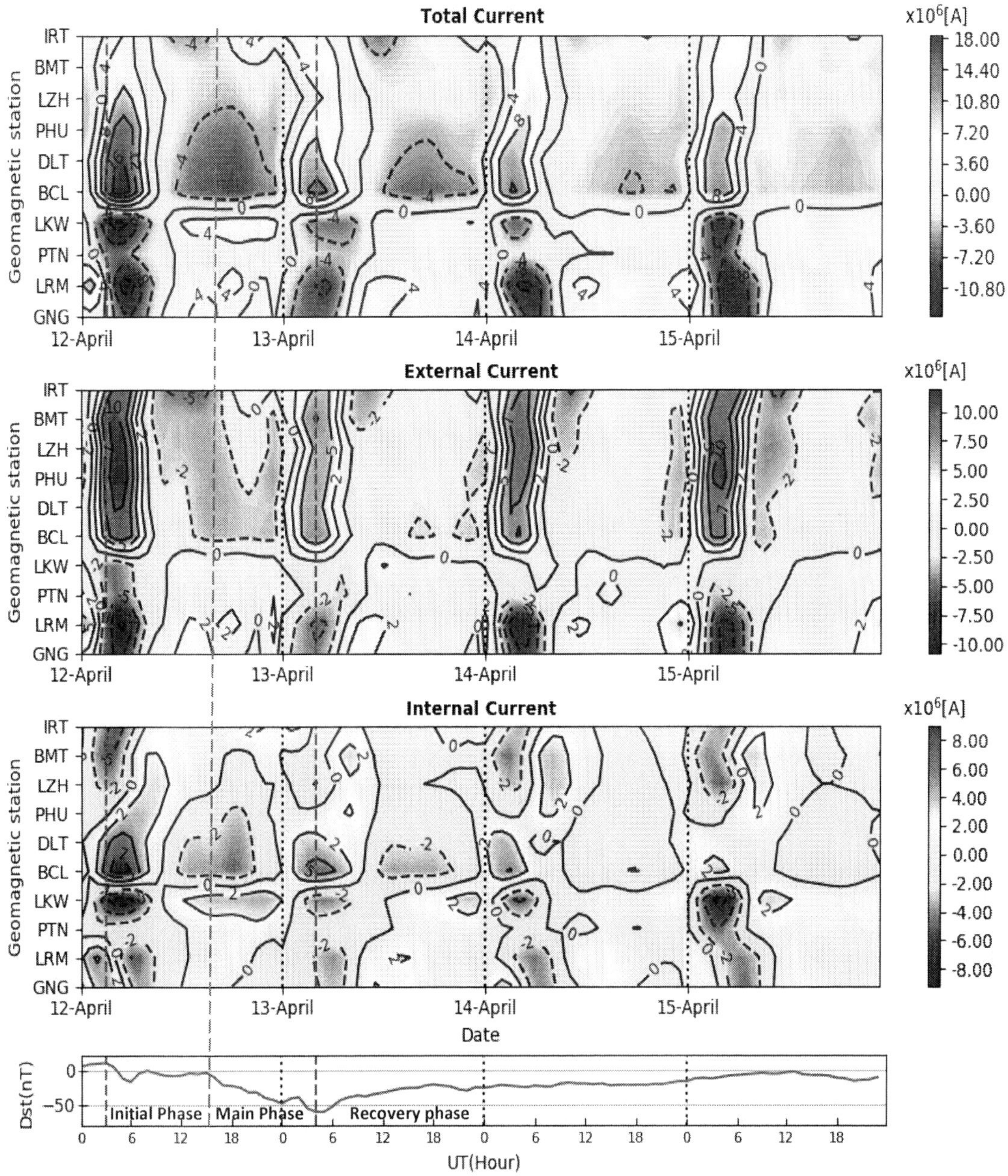

Figure 3: The contour of day to day variation of total current(upper), external current (middle) and internal current(bottom) during 12-15 April 2012. The Dst index is used to represent the flow of geomagnetic storm occurred.

4. Conclusion

The present study was set out to determine external/internal current separation during the disturbed period and correlate with the solar wind parameter, IMF, and geomagnetic indices in order to identify their relationship. The result has shown that the variation of the total, external and internal current based on geomagnetic storm phase are associated with these parameters. The representation of current computed from the longitudinal stations also shows the real situation of current vortices during a geomagnetic storm. The results of current separation have been indicating that external current is from magnetospheric current and ionospheric current source while the internal current from the Earth's main field and also induced current by an external source. During the initial phase, these current enhanced due to reconnection process and generating the induced current in the internal current. Even though the data was obtained from daily means baseline that reduced the magnetospheric source, the results however also show the induction of currents in the external and internal currents. Overall, the presented results show that the SHA method can be used to determine the current induction mainly during initial phase due to external source obtained from a weak geomagnetic storm event. The findings of the study can extend our knowledge of the behavior purely current. The study was limited to only one case of a geomagnetic storm. More cases of the geomagnetic storm are required to better understand the characteristic of separated current associated with the external source. The fewer stations used in southern hemisphere also cause the concentration of current was reduced. Thus, it is preferable using more stations to study the high resolution of current features. Further work in determining the conductivity needs to be carried out in order to identify the characteristic of an internal source. It would be interesting also to further research on understanding the effect of geomagnetically induced current (telluric current) in the internal field.

Acknowledgments

The authors have appreciated the MAGDAS/CPMN Group for providing data. This work by Akimasa Yoshikawa has been supported by MEXT/JSPS KAKENHI Grant 15H05815. The author also wants to thanks to national institute that provides geomagnetic observatory data for the INTERMAGNET's network.

References

[1] C. Amory Mazaudier, "Electric Current Systems in the Earth 's Environment," Nigerian Journal of Space Research, vol. 8, pp. 178–255, 2009.

[2] H. Maeda, "Variation in geomagnetic field," Space Science Reviews, vol. 8, no. 4, pp. 555–590, 1968.

[3] R. J. Stening and D. E. Winch, "The ionospheric Sq current system obtained by spherical harmonic analysis," Journal of Geophysical Research: Space Physics, vol. 118, no. 3, pp. 1288–1297, 2013.

[4] N. Olsen and C. Stolle, "Magnetic Signatures of Ionospheric and Magnetospheric Current Systems During Geomagnetic Quiet Conditions—An Overview," Space Science Reviews, vol. 206, no. 1–4, pp. 5–25, 2017.

[5] W. H. Campbell and E. R. Schiffmacher, "Quiet ionospheric currents of the northern hemisphere derived from geomagnetic field records," Journal of Geophysical Research, vol. 90, no. A7, p. 6475, 1985.

[6] S. Matsushita and H. Maeda, "On the geomagnetic solar quiet daily variation field during the IGY," Journal of Geophysical Research, vol. 70, no. 11, pp. 2535–2558, 1965.

[7] A. Suzuki, "UT and Day-to-Day Variations in Equivalent Current Systems for World Geomagnetic Variations," Journal of Geomagnetism and Geoelectricity, vol. 31, no. 1948, pp. 21–46, 1979.

[8] U. Schmucker, "4 Magnetic fields the earth," in Subvolume B, vol. 27, no. 14, Berlin/Heidelberg: Springer-Verlag, 1980, pp. 31–73.

[9] L. V. T. Häkkinen, T. I. Pulkkinen, H. Nevanlinna, R. J. Pirjola, and E. I. Tanskanen, "Effects

of induced currents on Dst and on magnetic variations at midlatitude stations," Journal of Geophysical Research: Space Physics, vol. 107, no. A1, 2002.
[10] T. Aoki, "On the validity of Akasofu's e parameter and of the Vasyliunas et al. general formula for the rate of solar wind-magnetosphere energy input," Earth, Planets and Space, vol. 57, no. 2. pp. 131–137, 2005.
[11] W. H. Campbell, Introduction to geomagnetic fields. Cambridge University Press, 2003.
[12] T. N. Davis and M. Sugiura, "Auroral electrojet activity index AE and its universal time variations," Journal of Geophysical Research, vol. 71, no. 3, pp. 785–801, Feb. 1966.
[13] S.-I. Akasofu, "The solar wind-magnetosphere energy coupling and magnetospheric disturbances," Planetary and Space Science, vol. 28, no. 5, pp. 495–509, May 1980.
[14] W. D. Gonzalez et al., "What is a Geomagnetic Storm," Journal of Geophysical Research-Space Physics, vol. 99, no. A4, pp. 5771–5792, 1994.
[15] T. N. Obiekezie and F. N.Okeke, "Solar quiet day ionospheric source current in the West African region," Journal of Advanced Research, vol. 4, no. 3, pp. 303–305, 2013.
[16] W. H. Campbell, "The upper mantle conductivity analysis method using observatory records of the geomagnetic field," Pure and Applied Geophysics, vol. 125, no. 2–3, pp. 427–457, 1987.
[17] Z. Z. Abidin, M. H. Jusoh, M. Abbas, O. S. Bolaji, and A. Yoshikawa, "Features of the inter-hemispheric field-aligned current system over Malaysia ionosphere," Journal of Atmospheric and Solar-Terrestrial Physics, no. October 2017, pp. 1–7, 2018.

IOP Conf. Series: Journal of Physics: Conf. Series **1152** (2019) 012030 doi:10.1088/1742-6596/1152/1/012030

The first solar-powered Magdas-9 installation and possible geomagnetically induced currents study at Johor, Malaysia

Zatul Iffah Abd Latiff[1,2], Mohamad Huzaimy Jusoh[1], Farah Adilah Mohd Kasran[1], Siti Amalina Enche Ab Rahim[1], Munirah Onn[4], Aznilinda Zainuddin[2], Muhammad Asraf Hairuddin[2], Nurhani Kasuan[2], Muhammad Muzamil Mustam[2], Suhana Mohamed[3] and Yoshikawa Akimasa[5]
[1]Faculty of Electrical Engineering, Universiti Teknologi MARA, Shah Alam 40450, Malaysia
[2]Faculty of Electrical Engineering, Universiti Teknologi MARA Johor, Masai 81750, Malaysia
[3]Faculty of Business Mangement, Universiti Teknologi MARA, Johor, Masai 81750, Malaysia
[4]Faculty of Applied Science, Universiti Teknologi MARA, Johor, Masai 81750, Malaysia
[5]International Center for Space Weather Science and Education, Kyushu University, Fukuoka, Japan

Email: zatuliffah87@gmail.com

Abstract. The Magnetic Data Acquisition System (MAGDAS-9) was successfully installed at Universiti Teknologi MARA Johor, Pasir Gudang Campus, Malaysia (JOH station), under the joint collaboration between International Center for Space Weather Science and Education (ICSWSE), National Space Agency (ANGKASA) and UiTM Johor. This is the first solar-powered magnetometer under the Magnetic Data Acquisition System of Circum-pan Pacific Magnetometer Network (MAGDAS/CPMN). In this paper, construction process, installation process and analysis of first geomagnetic observations at MAGDAS JOH station are presented. In addition, analysis on time derivative of horizontal component of geomagnetic field (dH/dt) to indicate the possibility of Geomagnetically Induced Currents (GIC) occurrences is also presented in this paper. The obtained geomagnetic components: H, D, Z and Total F are approximately similar to World Magnetic Model 2015 (WMM2015) while the diurnal variation from this geomagnetic observation showing a good variation pattern. The analysis of dH/dt has indicated 2 significant events of GIC.

1. Introduction

MAGDAS is a real-time Magnetic Data Acquisition System introduced by International Center for Space Weather Science and Education (ICSWSE) which was deployed for the International Heliophysical Year programme (IHY; 2007-2009) [1][2][3]. By using MAGDAS system, a few scientific objectives can be conducted which are a real time monitoring and modelling of (1) the global 3-dimensional current system, (2) the ambient plasma density and (3) electric field in order to understand Sun-Earth Coupling System[4]. In present, MAGDAS/CPMN was constructed along the 3 unique chains of magnetic observatories; (1) 210° magnetic meridian, (2) African Longitude Sector and (3) Magnetic Equator as shown in Figure 1[5]. Currently 73 stations are actively operating.

1.1 Installation of MAGDAS in Malaysia

Until December 2017, there are five (5) installed MAGDAS's magnetometers in Malaysia. Malaysia is a strategic location for the magnetometer installation due to its geographic location that is near to the geomagnetic equatorial. On 8th September 2006, the first MAGDAS magnetometer was installed in Malaysia which is located at the Langkawi National Observatory (LNO), Langkawi district (LKW) at Kedah state. The geographic and geomagnetic latitudes coordinates of LKW station are (6.30°, 99.78°) and (-3.30°, 172.44°) respectively. Next on 30th March 2013, the second installation of MAGDAS-9 magnetometer was installed in Universiti Malaysia Sabah (UMS), Sabah (SBH). The geographic and geomagnetic latitudes coordinates of SBH station are (6.02°, 116.07°) and (-3.56°, 188.66°)

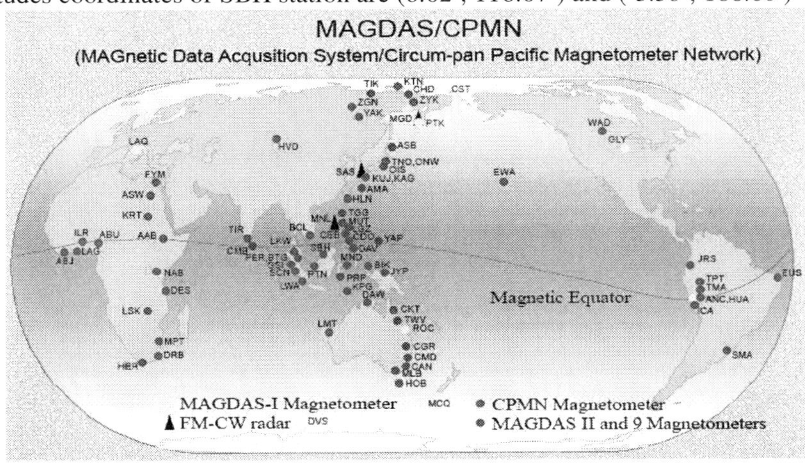

Figure 1. MAGDAS/CPMN (MAGnetic Data Acquisition System/Circum-pan Pacific Magnetometer Network (reference : http://magdas2.serc.kyushu-u.ac.jp/station/index.html)

respectively. The third installation of MAGDAS-9 magnetometer was installed on 25th October 2016 at University Perguruan Sultan Idris (UPSI), Perak state (PER) station. The geographic and geomagnetic latitudes coordinates of PER station are (3.72°, 101.53°) and (-5.92°, 174.14°) respectively. Next, on 19th November 2016, installation of fourth MAGDAS II magnetometer in Malaysia is located in Agensi Angkasa Negara (ANGKASA) at Banting, Selangor state, (BTG) station. The geographic and geomagnetic latitudes coordinates of BTG station are (2.78°, 101.53°) and (-6.86°, 174.10°) respectively. The latest installation was at UiTM Johor Kampus Pasir Gudang, (JOH) station on 17th February 2017. JOH station uses the latest ICWSE magnetometer called MAGDAS-9. MAGDAS-9 is more sensitive and provides accurate geomagnetic field data. The geographic and geomagnetic latitudes coordinates of JOH station are (1.53°, 103.87°) and (-7.99°, 176.79°) respectively. This station was selected for MAGDAS-9 magnetometer installation due to a few requirements which are to fill in observation gap between Davao (DAV) station in Philippines and Manado (MND) station in Indonesia for precise geomagnetic measurement of 210° magnetic Meridian, and it is located in equatorial regions and near to geomagnetic equator. The next MAGDAS installation will be followed by "TER" station located in Universiti Sultan Zainal Abidin (UnisZa), Terengganu, Malaysia. The installation locations are selected based on the requirement for the magnetometer installation which is having low disturbance from any human activities. As shown in figure 2, it can be observed that there are a few degrees of latitude differences between all installed magnetometers covering upper, middle and lower part of peninsula of Malaysia. Table 1 shows the geographic and geomagnetic coordinates for five installed magnetometers in Malaysia.

1.2 World's Magnetic Model

In order to validate the geomagnetic components, the results obtained from the magnetometer installed in JOH station were compared with the World Magnetic Model (WMM). The WMM 2015 is the latest world magnetic model used as standard or reference to provide information on magnetic declination at any desired location on the globe. It also provides the complete geometry of the field from 1 km below the earth's surface to 850 km above the surface [6]. In this paper, WMM2015 is used to compare with the geomagnetic field measurement taken at JOH station for duration of 3 months. The WMM2015 valid from 1st January of 2015 to 31st December of 2019.

Figure 2. Location of MAGDAS installation in Malaysia (reference : http://go2travelmalaysia.com/

Table 1. The Coordinates location of MAGDAS station in Malaysia

Stations Location	Station Name	Abbreviation	Geographic Coordinate	Geomagnetic Coordinate
Langkawi National Observatory (LNO),	Langkawi	LKW	(6.30°, 99.78°)	(-3.30°, 172.44°)
Universiti Malaysia Sabah (UMS)	Sabah	SBH	(6.02°, 116.07°)	(-3.56°, 188.66°)
University Perguruan Sultan Idris (UPSI),	Perak	PER	(3.72°, 101.53°)	(-5.92°, 174.14°)
Agensi Angkasa Negara ANGKASA, Selangor	Banting	BTG	(2.78°, 101.53°)	(-6.86°, 174.10°)
Universiti Teknologi MARA UITM, Johor	Johor	JOH	(1.53°, 103.87°)	(-7.99°, 176.79°)

1.3 Geomagnetically Induced Currents (GIC)

Geomagnetically induced current (GIC) is the ground level manifestation of solar activity which potentially can cause detrimental effects to earth surface technologies and buried equipment such as power network, electrical equipment, pipelines, railways, telecommunication technology and oil and gas technology [16-18]. During geomagnetic disturbances, the rise of solar wind input energy contributes to rapid changing of magnetic fields and subsequently the GIC will be generated through a process called electromagnetic induction. The phenomena of GIC can be linked with the Faraday's Law of induction where the induced electric field on the earth's surface can be computed from the rate of change of magnetic field as presented in eq. 1.

$$\nabla \times E = \frac{-\partial B}{\partial t} \quad (1)$$

GIC are also considered as quasi-dc current because of their low frequency characteristic and it flows along electric transmission lines and electrically-conducting infrastructure.

2. Methodology

2.1 Overview of Magnetometer System in MAGDAS

As mentioned in [7], MAGDAS system is a real time monitoring system where the measured data is sent to ICSWSE via the internet. The MAGDAS-9 is a second generation of magnetometer system after the first generation of magnetometer system called MAGDAS-I. As presented in figure 3, MAGDAS-9 unit consists of 3-component ring-core fluxgate type magnetic sensor (magnetometer) with 7 meter cable, pre-amplifier (preamp), GPS (Global Positioning System) antenna with cable, data logger for data control and 70 meter cable. The type of geomagnetic sensor used is 3-axial components ring-core flux gate with ±70,000nT/32bits dynamic range and a 0.01nT resolution. The magnetic field data are obtained with the sampling rate of 16 Hz and then the averaged data are transferred to ICSWSE at real time. The ambient magnetic fields expressed by H (geomagnetic northward), D (geomagnetic eastward) and Z (vertical downward) are digitized by using field-cancelling for the dynamic range of ±70,000 nT/32 bits [7][8].

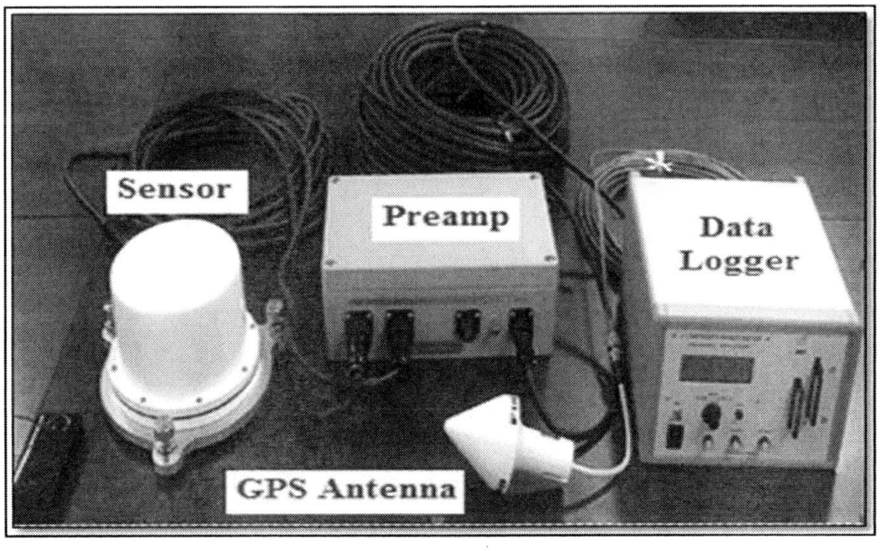

Figure 3. The component of MAGDAS-9 magnetometer system

2.2 *Installation of MAGDAS-9 at JOH station*

The first job that need to be implemented before installing the magnetometer in JOH station was surveying the potential location for instrument placement. For the isolation of noise or interferences contributed by external factor, the criteria of placement that need to be considered is (1) the distance from the sea must be greater than 1000 meter to minimize the noise and avoid the effect from the sea [9] and (2) distance from road or building must be more than 100 meters to minimize the noise contributed by human activities [10]. For the installation of MAGDAS-9 in JOH station, the magnetometer sensor is placed around 10 kilometres from the sea, 700 metres from nearest building and 3 kilometres from main road. The location of MAGDAS-9 installation at JOH station is as illustrated in Figure 4.

The architecture layout of MAGDAS-9 is as illustrated in figure 5. The installation of MAGDAS-9 can be divided into 3 stages which are (1) Construction of data logger cabin, pre-amplifier hut and sensor hut, (2) Installation of solar panel and (3) Installation of MAGDAS-9 unit. The cabin is where the data logger and GPS are placed. They are connected to the Amplifier Hut and Sensor Hut where the pre-amplifier and the magnetometer sensor are set up. The distance between data logger cabin and pre-amplifier hut is 40 meters while the distance from pre-amplifier hut to sensor hut is about 5 meters.

2.3 *Horizontal component of geomagnetic field*

The horizontal component of geomagnetic field is extracted from the Johor station (JOH), one of the observatories under the MAGDAS network and it can be obtained from Eq. 2. In this study, in order to identify the possibility of GIC's occurrence, time derivative of horizontal component of geomagnetic field (Eq. 3) was used. Previous studies [17, 19-24] discovered that the time derivative of horizontal component exceeding 30 nT/min (dH/dt > 30 nT/min), indicates the possibility of GIC presences at the ground surface thus can cause an adverse effect to power network. Therefore in this study, H-component data ranged from March 2017 to October 2017 were observed and the derivative of H-component was performed.

Figure 4. Location of MAGDAS-9 Installation at JOH station

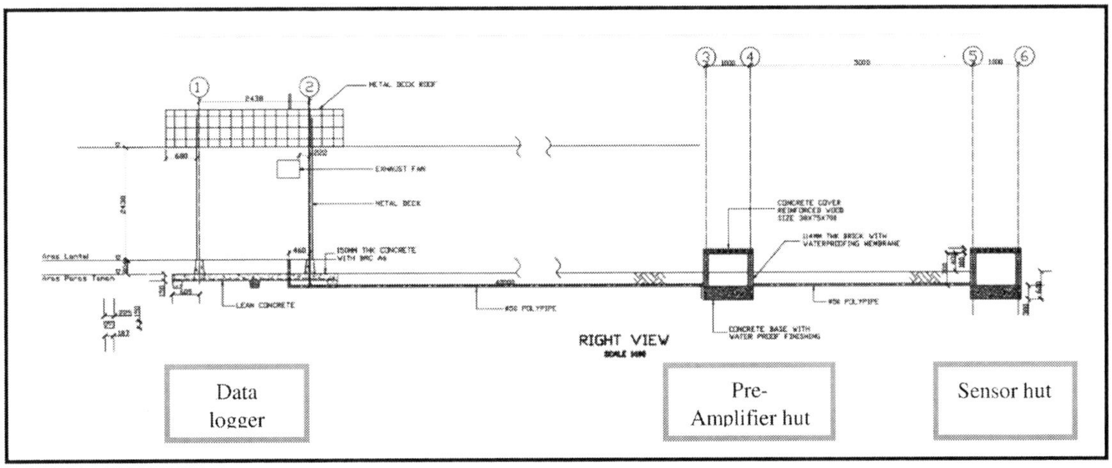

Figure 5. The Architecture layout of MAGDAS-9 Installation at JOH Station

$$H = \sqrt{x^2 + y^2} \tag{2}$$

$$\frac{dH}{dt} = \sqrt{dx/dt^2 + dy/dt^2} \tag{3}$$

3. Result and Discussion

3.1 First observation at JOH station

In this paper, the preliminary results of geomagnetic components; H, D, Z and Total F obtained from the first MAGDAS-9 measurement at JOH station are presented. Figure 6 shows the geomagnetic observations obtained by using MAGDAS-9's magnetometer from 13 June 2017 – 17 June 2017. As seen on Figure 6, the variation of H, D, Z and total F for above mentioned date shows a similar pattern and consistent for each component. The variation of obtained geomagnetic component values are tabulated in table 2. The value of geomagnetic components; H, D, Z and Total F are approximately similar to the WMM2015 (figure 7). Based on this model, Malaysia is located in ranges of between ± 40000 nT for H and F elements. As seen on table 2, the acquired value of H and Total F elements are almost identical. These results can be related to the main field of inclination (I) of WMM2015 as shown in figure 8 and the formula of inclination, *I* by [11] as presented in Eq. 4 :

$$\cos I = \frac{H}{F} \tag{4}$$

where the degree of inclination (I) is approaching 0 for Johor area since the obtained value for H and F elements from the geomagnetic measurement in JOH station is almost similar. Next analysis is on the diurnal variation where the analysis is performed based on the data collected on 15 June 2017. As indicated in figure 9, diurnal variation of geomagnetic components H, D, Z and F are observed in 24 hours local time (LT). The amplitude of each geometric component element indicates increment during daytime from 0800 LT to 1400 LT. From the observation, the highest amplitude of geomagnetic component is during noon time. During night-time, low variation of geomagnetic component is observed from 1900 LT to 0700 LT. For H component, the highest peak value is 4.081×10^4 nT during noontime (1200LT to 1300LT). The amplitude of declination (D) component during peak hour (1100LT) is approaching to 0 nT. The highest amplitude for the Z component is

IOP Conf. Series: Journal of Physics: Conf. Series **1152** (2019) 012030 doi:10.1088/1742-6596/1152/1/012030

-9760 nT at 1400LT. As for F element, the amplitude and peak hour is almost identical with H element. During night-time, the variation of geomagnetic components is maintained from 1900LT to 0700LT.

Table 2. The variation of geomagnetic components based on the first observation at JOH station

Geomagnetic Component	Geomagnetic Variation (nT)
H component	4.078×10^4 nT until 4.085×10^4 nT
D component	-38 nT until 10.24 nT
Z component	-9778 nT until -9753 nT
Total F	4.192×10^4 nT until 4.196×10^4 nT

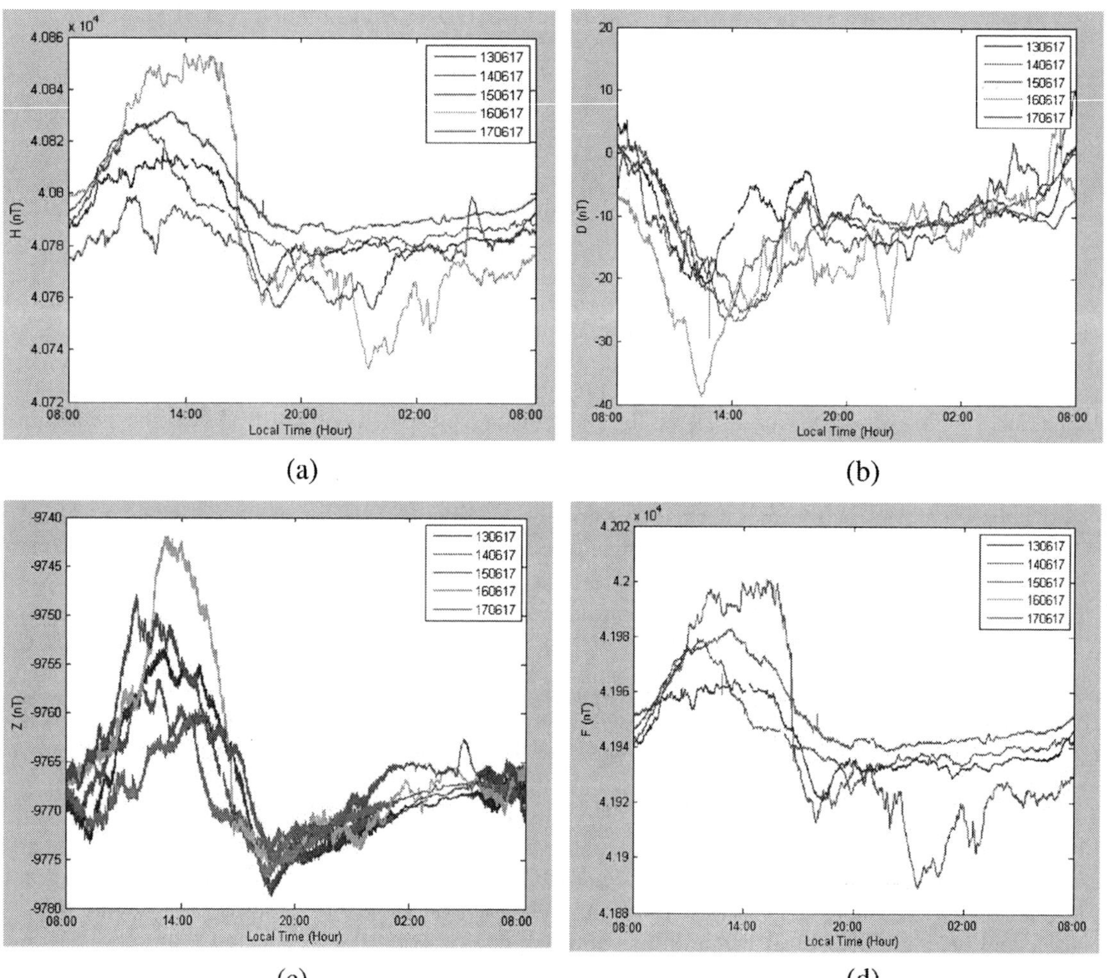

(a) (b) (c) (d)

Figure 6. Geomagnetic data obtained from MAGDAS-9's magnetometer from 13 June 2017 till 17 June 2017; (a) Horizontal (H) component, (b) Declination (D) component, (c) Vertical Downward (Z) component, (d) Total Field (F) component.

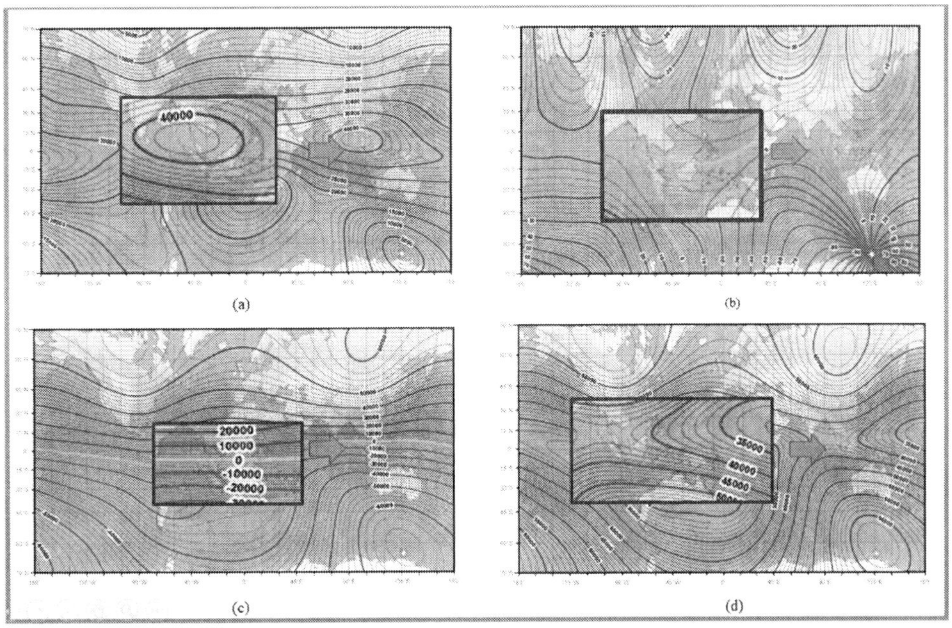

Figure 7. The Mercator projection of (a) Main Field Horizontal Intesity (H). Malaysia is located in range of 40 000 nT (b) Main Field Declination (D). Malaysia lies near green line 0° (c) Main Field Down Component (Z) Malaysia lies between -20 000 to 10 000 nT (d) Main Field Total Intensity (F). Malaysia is located in range of ±40 000 nT.

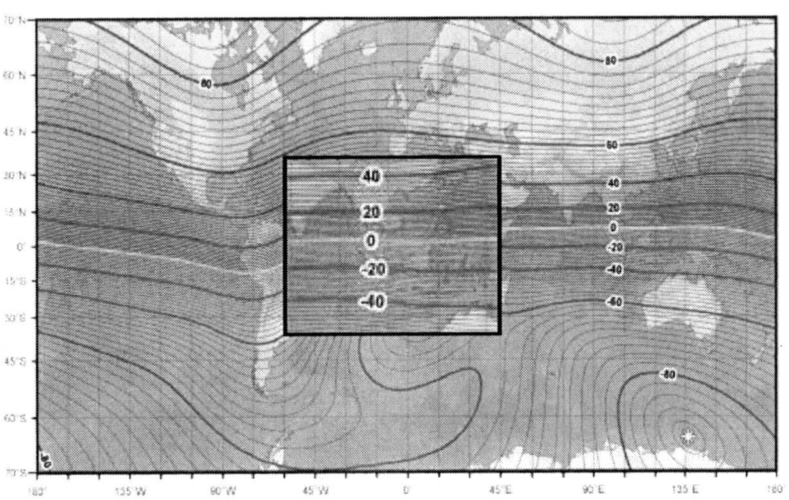

Figure 8. The Mercator projection of Main Field Horizontal Inclination (I). Contour interval is 2°. The colour contour represents as red positive (down); blue negative (up) and green is zero line. Malaysia is near to 0°

IOP Conf. Series: Journal of Physics: Conf. Series **1152** (2019) 012030 doi:10.1088/1742-6596/1152/1/012030

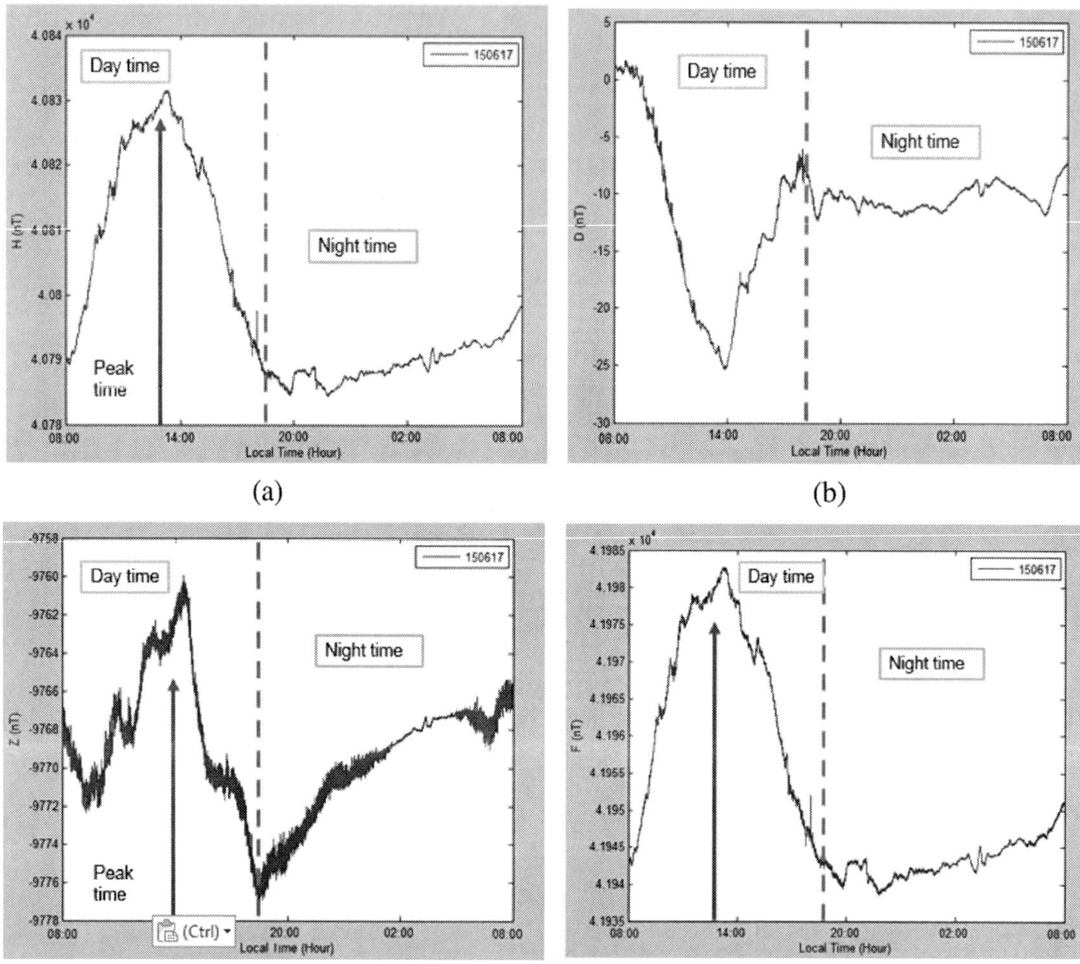

(a) (b)

(c) (d)

Figure. 9. Diurnal variation of geomagnetic components on 15 June 2017; (a) Horizontal (H) component, (b) Declination (D) component, (c) Vertical Downward (Z) component, (d) Total Field (F) component.

3.2 Analysis on possible occurrence of geomagnetically induced currents (GIC) at JOH stations

The possibility of the occurrence of GIC at surface of the ground can be obtained by performing time derivative of horizontal geomagnetic field (dH/dt). The value of derivative, dH/dt > 30 nT indicates the presence of GIC. Based on analysis, 2 significant events with dH/dt > 30 nT and corresponding dates are illustrated in figure 10. Table 2 summarise the maximum value, dH/dt and local time for these significant events.

Table 3. Maximum values of dH/dt from the observed event during March17-October17 with corresponding time (UT & LT)

Date	Maximum dH/dt (nT/min)	Time (UT)	Time (LT)
8 June 2017	32.9 nT	03:30	11:00
4 July 2017	73.54 nT	09:00	17:00

(a)
(b)
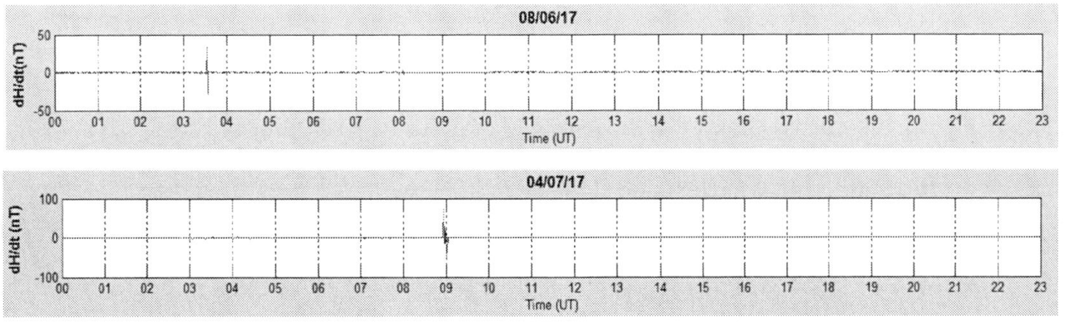

Fig. 10. Time derivative of Horizontal Component of gemagnetic field (dH/dt) exceeding 30 nT/min; (a) dH/dt on 8 June 2017, (b) dH/dt on 4 July 2017.

3.3 Discussion
The first observation of the geomagnetic variation obtained at JOH station shows a good representation of geomagnetic elements and approximately similar to WMM2015. The daily and diurnal variation of geomagnetic field as observed in figure 6 and figure 9 is attributed by the ionospheric dynamo [12]. Ionospheric dynamo describes the electric field and current that result from the motion of neutral atmosphere current where the electrodynamic effects arise from collisions between neutral and plasma particles contributed to the geomagnetic field variation [13]. The ionospheric dynamo usually refers to the dynamo process at E-region height. The diurnal pattern is significantly increased during day-time and almost constant during night-time. These results are consistent with the previous studies of [14] and [15]. The electron density is known to be maximum close to noon time since the E-region drift velocities are maximum due to rising of the sun at E-layer height.

The analysis on the possibility of GIC occurrence at JOH station is a preliminary study to indicate or prove the existence of GIC study at equatorial region and low latitude region. Next analysis that can be carried out is investigating the significance of space weather parameter to the GIC occurrence specifically focused on low latitude and equatorial region. By performing this kind of analysis, an understanding on physical mechanism between the space weather parameter and GIC occurrence can be developed.

4. Conclusion
A MAGDAS-9 magnetometer is successfully installed at JOH station on 17th February 2017. Based on the first observations conducted at JOH station, it indicates reliable pattern of geomagnetic components. The amplitude variation of each geomagnetic component indicates an approximation to WMM2015. Therefore the observed data at JOH station are reliable and acceptable to be analysed for any related geomagnetic research puposes. The analysis on possibility of GIC occurrence have shown the existence of Geomagnetically Induced Currents at JOH station, therefore analysis can be extended

into low latitude and equatorial region and also understanding on the physical mechanism beyond the generation of GIC.

Acknowledgements
The authors would like to thank International Center for Space Weather Science and Education (ICSWSE) for providing MAGDAS's instrumentation and installation training. The authors also want to acknowledge the National Space Agency of Malaysia (ANGKASA-MOSTI) for the guidance and assistance provided during the installations and UiTM Johor (Pasir Gudang Campus) for providing the location for the installation of MAGDAS magnetometer.

6. References

[1] K. Yumoto and C. Group, "A Review of MAGDAS / CPMN Project During IHY," Niger. J. p.
Res., vol. 8, no. March 30, pp. 349–390, 2010.
[2] Yumoto K. The STEP 210 degrees magnetic meridian network project. J Geomagn Geoelectr. 1996;48(11):1297–309.
[3] Maeda G, Yumoto K. Progress report on the deployment of MAGDAS. In: Earth, Moon and Planets. 2009. p. 271–6.
[4] K. Yumoto and S. Sub-committee, "International Heliophysical Year Activities in Japan," Data
Sci.J., vol. 8, no. March, pp. 14–23, 2009.
[5] Campbell WH. Introduction to Geomagnetic Fields: Second Edition [Internet]. Introduction to Geomagnetic Fields. 2003. 337 p. Available from: http://adsabs.harvard.edu/abs/2003igf..book.....C
[6] Chulliat, A., S. Macmillan, P. Alken, C. Beggan, M. Nair, B. Hamilton, A. Woods, V. Ridley, S. Maus and A. Thomson, 2015, The US/UK World Magnetic Model for 2015-2020:Technical Report, National Geophysical Data Center, NOAA. doi: 10.7289/V5TB14V7
[7] Abe S, Yumoto K, Ikeda A, Uozumi T, Maeda G. Data and information activities of icswse, Kyushu university, Japan. Data Sci J [Internet]. 2013;12. Available from: http://www.scopus.com/inward/record.url?eid=2-s2.0-84875394922&partnerID=40&md5=96f4d5e18888f693c88efb08581fa41e
[8] Ibrahim SN, Jusoh MH, Sulaiman AA, Ahmad SNA, Makmud MZH, Musta B, et al. First geomagnetic observation at sabah, malaysia by using MAGDAS array. Int J Simul Syst Sci Technol. 2017;17(41):30.1-30.8.
[9] Parkinson WD, Jones FW. The geomagnetic coast effect. Vol. 17, Reviews of Geophysics. 1979. p. 1999–2015.
[10] 1. Jankowski J, Sucksdorff C. Guide for Magnetic Measurements and Observatory Practice [Internet]. Iaga. 1996. 238 p. Available from: http://iugg.org/IAGA/iaga_pages/pdf/IAGA-Guide-Observatories.pdf
[11] N. Basavaiah, Geomagnetism: Solid Earth and Upper Atmosphere Perspectives. Springer Netherlands, 2012.
[12] 1. Chapman S, Bartels J. Geomagnetism. Vol. 1, The International Series of Monographs on Physics. 1940. 1-543 p.
[13] Yamazaki Y, Maute A. Sq and EEJ—A Review on the Daily Variation of the Geomagnetic Field Caused by Ionospheric Dynamo Currents. Vol. 206, Space Science Reviews. 2017. p. 299–405.
[14] R. G. Rastogi, "Quiet day variation of geomagnetic H-field at low latitudes," J.Geomag. Geoelectr., vol. 28, pp. 461–479, 1976.
[15] E. López, G. Maeda, K. Vicente, K. Yumoto, N. Vasquez, H.Matsushita, A. Shishime, and C.

Vásconez, "First Magdas Equipment in Ecuador," Sun Geosph., vol. 9, no. 1–2, pp. 31–34, 2014.

[16] Pirjola R, Viljanen A, Pulkkinen A et al (2004) Ground effects of space weather pipelines in effects of space weather on technology infrastructure. In: Daglis IA (ed). Kluwer Academic Publisher, p 235–256

[17] Pirjola R (2012) Geomagnetically induced currents as ground effects of space weather. InTech, p 27–44

[18] Pirjola R (2002) Review on the calculation of surface electric and magnetic fields and of geomagnetically induced currents in ground-based technological systems. Surv Geophys 23:71–90

[19] Viljanen A, Pulkkinen A, Amm O et al (2004) Fast computation of the geoelectric field using the method of elementary current systems and planar Earth models. Ann Geophys 22:101–11323:71–90

[20] Falayi EO, Beloff N (2012) Modelling of geomagnetically induced currents during geomagnetic storms using geoelectric fields and auroral electrojet indices. Indian J Phys 86:423–429.

[21] Bernhardi EH, Tjimbandi TA, Cilliers PJ et al (2008) Improved calculation of geomagnetically induced currents in power networks in low–latitude regions. In: PSSC Proceedings 677

[22] Trivedi NB, Vitorello I, Kabata W et al (2007) Geomagnetically induced currents in an electric power transmission system at low latitudes in Brazil: a case study. Sp Weather 5:1–10

[23] Falayi EO, Beloff N (2010) Estimating geomagnetically induced currents at subauroral and low latitudes to assess their effects on power systems. Can J Pure Appl Sci 4:1187–1197 [24] Ngwira CM, McKinnell L-A, Cilliers PJ (2011) Geomagnetic activity indicators for geomagnetically induced current studies in South Africa. Adv Sp Res 48:529–534

The Response of Geomagnetic H and Z Components During X-Class Solar Flare

A Taat[1,2], M Abbas[3,4] and M H Jusoh[1,2]

[1]Faculty of Electrical Engineering, Universiti Teknologi Mara, Shah Alam, Selangor, Malaysia
[2]Applied Electromagnetic Research Group, Universiti Teknologi Mara, Shah Alam, Selangor, Malaysia
[3]Department of Physics, Kebbi State University of Science and Technology, Aliero, Nigeria
[4]Radio Cosmology Laboratory, Department of Physics, Faculty of Science, University of Malaya, Malaysia

afifahtaat.uitm@gmail.com

Abstract. This paper presents the study of solar flare effects (SFE) on H and Z components of the geomagnetic field. The raw data of H and Z components are extracted from Magnetic Data Acquisition System (MAGDAS) at Langkawi station (LKW:6.3°, 99.78°) and INTERMAGNET website for Lanzhou (LZH: 36.1°,103.84°) and Yakutsk (YAK: 61.96°,129.66°) stations. There are eight (8) events of X-class solar flares from 2011 to 2014 were selected to shows the changes and enhancement of H and Z components with different latitudes. The regression analysis is used to prove the correlation between the enhancement of geomagnetic field and the x-ray flux. The statistical results show the ratios of $\Delta H_{SFE}/ \Delta H_o$ and $\Delta Z_{SFE}/ \Delta Z_o$ is greater than zero for all stations and proven the SFE has affected the H and Z components and not just an augmentation of pre-flare ionospheric currents.

1. Introduction

Solar Flare Effect (SFE) or a magnetic crochet arises from the increased ionization in the D and E layers of the ionosphere caused by the massive increase in X-ray radiation generated by the solar flare. This ionization changes the properties such as conductivity of these ionospheric layers allowing electric currents to flow more easily. It is the magnetic effect of these currents which produce the jump in the Earth's magnetic field. As the flare declines, the ionospheric layers quickly return to their previous state, the electric currents in the layers return to normal, and the change in the magnetic field ends. A solar flare is contained of x-ray flares and the enhancements are classified as based on their strengths in terms of peak burst intensity (I_B) measured by satellite in the wavelength band of 1-8A. It is classified as A, B, C, M, and X according to the peak of x-ray flux and expressed in the unit of Wm^{-2}.In 1961, [1] suggested that the solar flare (magnetic crochet) related current was regarded as an augmentation of the Sq current system. Other researchers revealed the significant differences between the crochet current system and Sq. The studied was explained by solar flare effect on the geomagnetic field globalized and suggested the level of ionospheric current is a slightly lower than the layer for the Sq current system [2][3]. Besides, [4] made a study of 15 typical SFE during the IGY and found a perfectly simultaneous relationship between optical and radio solar flares, sudden ionospheric disturbances, and geomagnetic SFE. The effect of solar flare during the normal electrojet period produced a positive change in H, a negative change in Y and Z-components [5]. Taking advantages of that, this paper presents the correlation coefficient between the changes of H and Z components (ΔH_{SFE}, ΔZ_{SFE}) with values of pre-flare amplitude (ΔH_O, ΔZ_O) variation and the intensity of x-ray flux.

2. Methodology

The present study has been carried out using geomagnetic data from Magnetic Data Acquisition System (MAGDAS) at Langkawi, Malaysia (LKW) station and INTERMAGNET website for Lanzhou, China (LZH) and Yakutsk, Russia (YAK) stations as shown in Table 1. The eight (8) events of intense solar flares and the raw data of x-ray flux were extracted from National Centers for Environmental Information (NOAA), Boulder, USA while the quiet days data were accessed from World Data Center for Geomagnetism, Kyoto, Japan.

Table 1. Geographic coordinates of geomagnetic observatories.

Station	IAGA	Country	Geo.Lat.(N°)	Geo.Long.(E°)
Langkawi	LKW	Malaysia	6.30	99.68
Lanzhou	LZH	China	36.10	103.84
Yakutsk	YAK	Russia	61.96	129.66

2.1. Theoretical and Estimation Calculation

This study was used estimated calculation by [6] and [7].The Pre-flare and SFE amplitude variations in H and Z components defined from 0000h to start time of solar flare occurred. Therefore, the value of ΔH_O and ΔZ_O is expressed as follows:

$$\Delta H_O = H_{min} - H_{00} \quad (1)$$

$$\Delta Z_O = Z_{min} - Z_{00} \quad (2)$$

The SFE on H and Z components can be written as:

$$\Delta H_{SFE} = H_{max} - H_{min} \quad (3)$$

$$\Delta Z_{SFE} = Z_{max} - Z_{min} \quad (4)$$

Where H_{max} and Z_{max} are the amplitude value of geomagnetic components during the peak time of solar flare occurrence as shown in Figure 1.

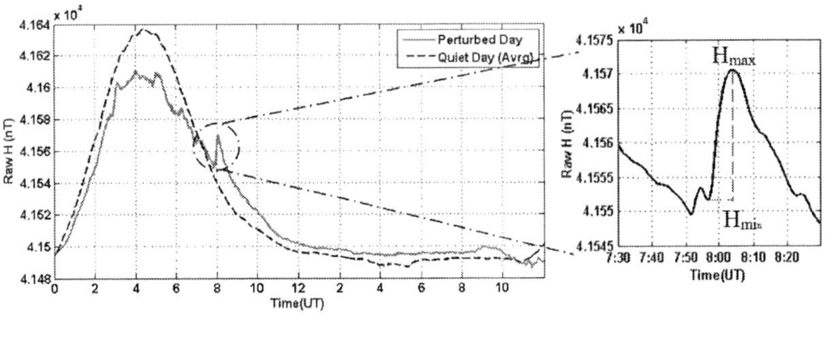

Figure 1. Plots of H component during the perturbed day (X-class solar flare) and quiet day and the illustration of ΔH_{SFE}.

As refer to [8], the portions of SFE current flowing in the E-layer was expressed as follows:

$$i_o = i_{oE} + i_{oR\ (Sq\ current)} \tag{5}$$

Where i_o is a total Sq current. The i_{oE} flows in the maximum level of E-Region and remaining portion i_{oR} to flowing in other regions. The additional current i of SFE is made up of i_E in the E-region maximum and a portion i_D in other regions. Therefore, the equation defined as follows:

$$i_o = i_E + i_{D\ (SFE\ current)} \tag{6}$$

Hence, the total current flowing during SFE:

$$i_o + i = i_{oE} + i_E + i_{oR} + i_D \tag{7}$$

The magnetic horizontal intensity is proportional to the current and the equation can be simplified as follows:

$$\frac{i}{i_o} = \frac{\Delta H_{SFE}}{H_O} = \frac{i_E}{i_{oE}} \tag{8}$$

Hence,

$$\Delta H_{SFE} = \frac{\Delta H_O}{i_{oE}} \tag{9}$$

In this study, the equation (6) can be generally defined as,

$$\Delta H_{SFE} = S\Delta H_O + \varphi \tag{10}$$

Where S = $\frac{\Delta H_{SFE}}{\Delta H_O}$ is slope and φ is intercept of the equation. Therefore, the equation (10) will be used in correlation coefficient analysis to define the relationship between the geomagnetic component and SFE. If the ratio of equation (8) is greater than zero, it implies that i is greater than i_o, ΔH_{SFE} is greater than ΔH_o and i_E is greater than i_{oE}. This suggests that solar flare enhance the geomagnetic field.

3. Result and discussion

The enhancement of H component at LKW station during the intense solar flare (X1.7) is shown in Figure 2. The major X-class solar flare occurred on 25th October 2013 at 08:01UTC that generated a 10cm radio burst and coronal mass ejection (CME). This phenomenon caused interference for sensitive receivers including radar and satellite communication. Based on the result, the enhancement of H component seems correlated with occurrence time of solar flare. The amplitude of H component increased from 41497nT to 41510nT from the pre-flare until the maximum peak of a solar flare.

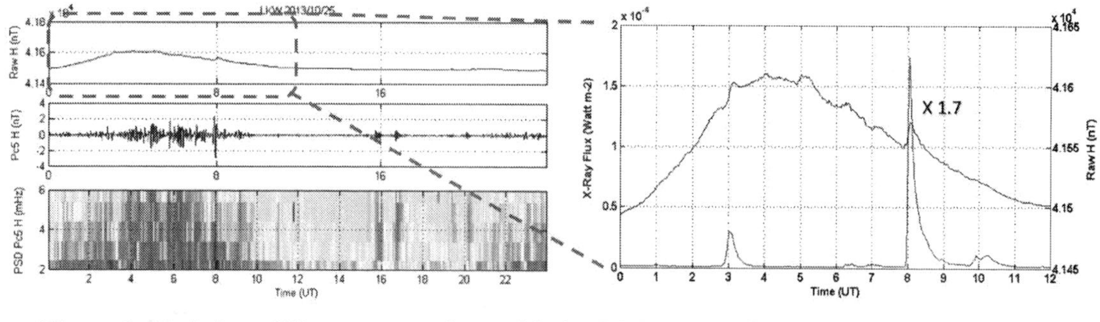

Figure 2. Variation of H component along with the GOES X-ray flux during an intense solar flare on 25th October 2013.

Table 2 shows the details of solar flare events used in this study and the calculated values of ΔH_{SFE} and ΔZ_{SFE} respectively for the different stations. At LKW station having eight solar flares events depicted five solar flares whose ratios of $\Delta H_{SFE}/\Delta H_o$ ranging between 0.27 and 0.65 and two solar flares had ratios 1.35 and 4.93. In the mid-latitude (LZH), the ratios are around 0.02 and 0.47 and only one solar flare event had value more than 1. The ranging ratios at high latitude (YAK) showed ratios between 0.01 and 0.74 for seven solar events, and one solar flare event with ration 2.87. Figure 3 shows a clear correlation between ΔH_{SFE} and ΔH_o with a positive slope at low (LKW) and mid (LZH) latitude stations and negative slope at high (YAK) latitude station. Table 4 also tabulated the intercept deviate significantly from the theoretical prediction of zero. The results strongly show that SFE enhances the H and Z components of the geomagnetic field. The results of $\Delta Z_{SFE}/\Delta Z_o$ were generally larger than the SFE at H component and had similar in the pattern. These are shows the SFE consist positive impulse in H and Z component at low latitude region. The could be attributed to the simultaneous existence of zonal and meridional currents that responsible for Sq currents. A solar flare has affected the geomagnetic field and is not as an augmentation at the pre-flare ionospheric current [2].

Table 2. The calculated value of $\Delta H_{SFE}/\Delta H_o$ and of $\Delta Z_{SFE}/\Delta Z_o$ in all stations.

Date	Flares Class	X-ray flux (Watt m-2)	$\frac{\Delta H_{SFE}}{\Delta H_O}$			$\frac{\Delta Z_{SFE}}{\Delta Z_O}$		
			LKW	LZH	YAK	LKW	LZH	YAK
24-Sep-11	X1.9	1.10E-01	0.4692	0.0630	0.0086	1.0000	0.4035	0.3437
25-Oct-13	X1.7	9.00E-02	0.2653	0.1703	0.0617	0.5807	0.0625	0.0185
15-May-13	X1.2	1.20E-01	1.3547	0.1196	2.8656	2.8136	0.6190	0.7391
19-Oct-14	X1.1	3.90E-01	0.0237	0.4676	0.3688	0.9079	0.2680	0.0000
10-Nov-13	X1.1	3.50E-02	0.4409	3.8689	0.1041	0.5251	0.2752	0.4516
8-Nov-13	X1.1	2.80E-02	0.4909	0.2226	0.7413	0.8160	0.0802	0.4242
19-Nov-13	X1.0	6.60E-02	4.9307	0.0587	0.0345	13.5783	0.3750	0.0270
28-Oct-13	X1.0	8.40E-02	0.6565	0.4015	0.6445	1.3571	0.5231	0.7857

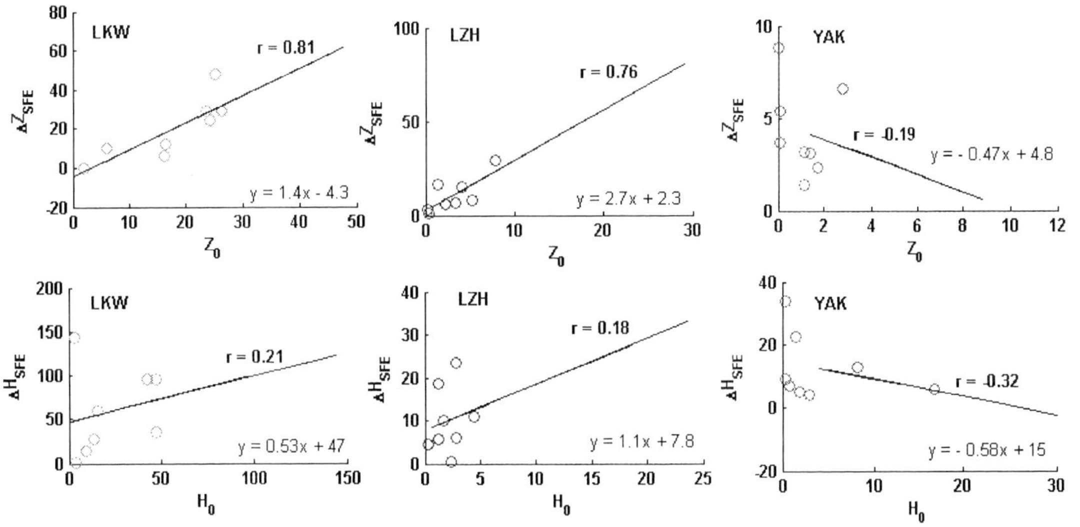

Figure 3. H and Z components enhancement against pre-flare in LKW, LZH and YAK stations.

Table 3. Regression analysis of H and Z component due to SFE

H-component			
Station	Correlation Coe.,r	Intercept, ϕ	Slope, S
LKW	0.21	47	0.53
LZH	0.18	7.8	1.1
YAK	-0.32	15	-0.58
Z-component			
LKW	0.81	-4.3	1.4
LZH	0.76	2.3	2.7
YAK	-0.19	4.8	-0.47

4.Conclusion

This study revealed the solar flare effect on H and Z component of the geomagnetic field. The results of the study show the SFE enhance the geomagnetic field across all latitudes which consistent with the results by Figure 1until Figure 4. This proven the SFE is not a simple augmentation at the pre-flare ionospheric current. The correlation also existed between the intense solar flare (X-class) to the SFE as shown Figure 4. All the results are statistically significant to suggest the SFE enhance the H and Z components of the geomagnetic field at all latitudes.

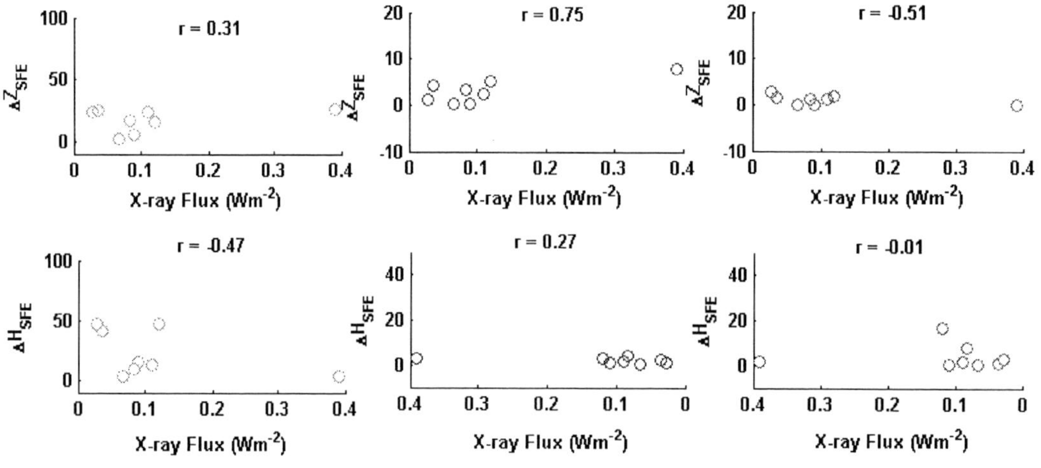

Figure 4. The correlation coefficient of H and Z components against X-ray Flux at LKW, LZH and YAK stations.

Acknowledgment

We acknowledge National Space Agency Malaysia (ANGKASA) and INTERMAGNET for providing the data of H and Z components of the geomagnetic field. Also, we thank National Centers for Environmental Information (NOAA), Boulder, USA and World Data Center for Geomagnetism, Kyoto, Japan for making their data available to access.

References

[1] Chapman S 1961 *Regular motions in the ionosphere: Electrical and magnetic relationships* (America-Bulletin of the American Society) pp 85-100

[2] Rastogi RG 1996 *Effect of Solar Disturbance on the Geomagnetic H, Y and Z fields in American Equatorial Electrojet Stations* (J. Ind. Geophys. Union) pp 45-53

IOP Conf. Series: Journal of Physics: Conf. Series **1152** (2019) 012031 doi:10.1088/1742-6596/1152/1/012031

[3] Ohshio M, Fukushima and Nagata T 1967 *Solar Flare Effects on Geomagnetic Field* (Rep. Ionos. Space Res.Japan) pp77-114

[4] Nagata T *Solar Flare Effect on The Geomagnetic Field* 1966 (J.Geomag.Geoelectr.)pp 197-219

[5] Okeke FN, Ichoja A 2011 *Impact of Severe Solar Flare Effects and Sudden Storm Commencement on Geomagnetic Fields, Across Latitudes* (J.Emerg.Trends Eng.Appl.Sci) pp 219-225 5

[6] Okeke FN, Okpala KC 2005 *Geomagnetic Solar Flare Effect in H, Y, and Z fields in the Northern Hemisphere.*(NJSR) pp 39:99 6

[7] Ugonabo OJ, Okeke FN and Ugwu 2013 *Preliminary analyses of solar flare effects on Geomagnetic H Component at Equatorial and Low Latitudes* (Int.J.Recent Trends Sci.Technol.) pp 235-237 7

[8] Volland H, Taubenheim J (1958) On The Ionospheric Current System of The Geomagnetic Solar Flare Effect. (Journal Atmos.Terr.Phys.) pp 258-260 8

IOP Conf. Series: Journal of Physics: Conf. Series **1152** (2019) 012032 doi:10.1088/1742-6596/1152/1/012032

Drift frequency of kilometric radio burst type III with solar wind speed and density

W.Z.A. Wan Mokhtar[1,2], Z. S. Hamidi[1], Zamri Zainal Abidin[3], Zainol Abidin Ibrahim[3]

[1]Institute of Science, Universiti Teknologi MARA, 40450 Shah Alam, Selangor, Malaysia

[2] Department of Physics, Faculty of Science and Mathematics, Universiti Pendidikan Sultan Idris, 35900 Tanjung Malim, Perak, Malaysia

[3]Physics Department, Faculty of Science, University of Malaya, 50603 Kuala Lumpur, Malaysia

E-mail: adliupsi@gmail.com

Abstract. Solar radio burst is a signature of solar flare generated by plasma emission mechanism. Radio burst is a diagnostic tool to study electron beam evolution and plasma background properties in corona and interplanetary medium. Sudden release magnetic energy up to 1032 ergs due to magnetic connection accelerates electron located at accelerated site to propagate near relativistic. This fast group of electron propagating through plasma background induced high level of Langmuir wave due to bump in tail instability. Non linear wave-wave causing radio burst near the local plasma frequency. In this paper, frequency drift of radio burst is calculated only from 1 MHz to 400 kHz and the data is obtained from Coordinated Data Analysis Web (CDAweb). The data is analysed by identifying time during maximum intensity of burst at the particular frequency range. Only 45 burst observed in September 2017 is considered in this paper. Meanwhile the data of solar wind speed and density is collected from Advanced Composition Explorer (ACE) satellite. Drift frequency of burst then compared to average of solar wind and density. From the analysis, drift frequency of burst decreasing against solar wind speed and solar wind density. But both relationship shows very weak correlation between drift frequency and solar wind properties.

1. Introduction

Solar radio bursts have been studied more than 60 years and many properties of the radio burst is explored. Solar radio burst type III (SRBT III) is the most frequent radio burst occurred due to solar activity and it always associated with solar flare [1], [2]. Radio burst generated from the corona to interplanetary medium is due to propagation of near relativistic electron beam outward the Sun along open magnetic line through plasma emission [2]–[4]. This shows radio burst type III is a signature of non-thermal energetic electron beam travelling from acceleration site via plasma environment outward to the Earth, so the generated SRBT III is also depend on the solar wind and local plasma frequency [5].

One of the properties which are often observed is drift frequency of radio burst [6]. The sign of drift frequency indicated the motion of electron either moving away or moving toward the Sun[6], [7] .

Most of the drifts are started from high to low frequency. The drift frequency is important tool in understanding the origin of burst source[8]. The drift frequency also can be used to estimate the speed of energetic electron beam along the open magnetic field [9].

In this paper, 45 SRBT III events occurred at kilometric wavelength by considering maximum intensity of the burst. Each drift frequency is calculated by tracing maximum intensity of radio burst that change the frequency against time. The average solar wind speed, solar wind density and average intensity of burst also calculated and compared to the drift frequency to determine their correlation.

2. Observations

Kilometric SRBT III consider in this paper only occurred in September 2017. Only 45 SRBT III events are considered started from 2 September 2017 to 19 September 2017. All the data is obtained from CDAWeb hosted by National Aeronautics and Space Administration (NASA). In this paper, data observed by radio receiver band 1 (RAD1) on the WIND spacecraft only considered. RAD1 sensor measured radio emission from 20 to 1040 kHz consist of 256 channels [10]. Time resolution of RAD1 data provided by CDAweb is 1 minute.

SRBT III can have starting frequency from MHz down to kHz. In this paper, the drift frequency is calculated started from 1 MHz to 400 kHz even the starting frequency started at higher frequency. The duration of burst is obtained by determine the start and end time of burst type III at 1 MHz and 400 kHz as shown by red arrow in Figure 1. The obtained duration is used to calculate the average solar wind speed and density during the burst.

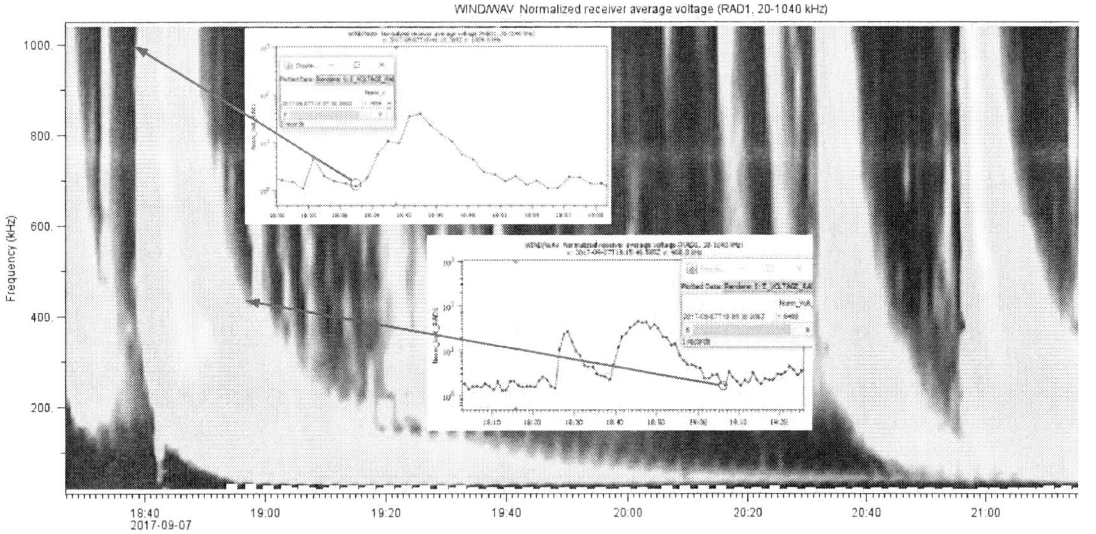

Figure 1. SRBT III on 7 September 2017 observed by RAD1 on board WIND spacecraft. Inset shows the start and end time of burst at 1 MHz and 400 kHz respectively.

The maximum intensity of the selected burst is traced then for each time, t_i we find the frequency f_i of maximum intensity. This work done by using MATLAB software. Then drift frequency is defined as

$$\frac{df}{dt} = \frac{f_f - f_i}{t_f - t_i}$$

where f_f is final frequency, f_i is initial frequency, t_f is final time and t_i is initial time. The frequency f_i is usually does not lie on smooth line and concentrate on particular time as shown in Figure 2. In this paper, the graph of frequency against time of particular burst is plotted and linear fitting is applied to obtain the drift frequency.

The solar wind speed and density is obtained from the Solar Wind Electron, Proton, and Alpha Monitor (SWEPAM) on board the Advanced Composition Explorer (ACE) satellite. Solar wind level 2 data obtained from CDAWeb has 64 second resolution detect the ion with energy between 0.26 and 36 keV. The average of solar wind speed and density during SRBT III occurred from 1 MHz to 400 kHz is calculated. This data then compared to drift frequency to obtain the correlation.

3. Data and Analysis

During September 2017 2 –1 9, 45 single SRBT III is obtained which all have negative drift. All the burst have different duration start and end frequencies, frequency bandwidth, intensity and drift rates. The drift rate is analysed base on solar wind speed and density. The present study only considered 45 bursts where the shortest duration of burst is 480 s while the longest duration of burst is 3540 s. All of the bursts discuss in this paper has negative drift frequency.

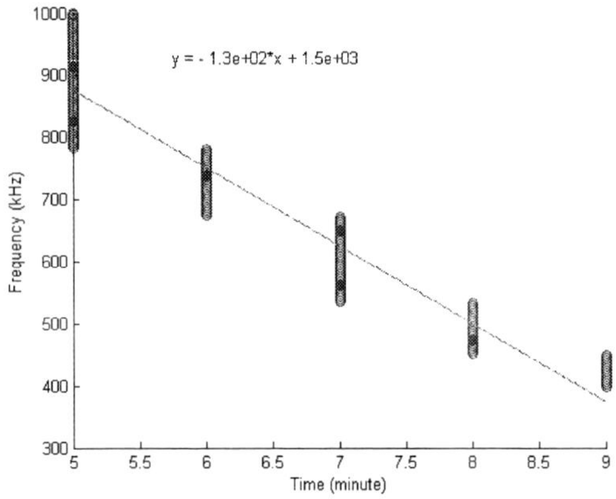

Figure 2. Frequency against time of maximum intensity of radio burst is plotted. The frequency f_i is concentrated at particular time. The linear regression fitting is applied to obtain frequency drift.

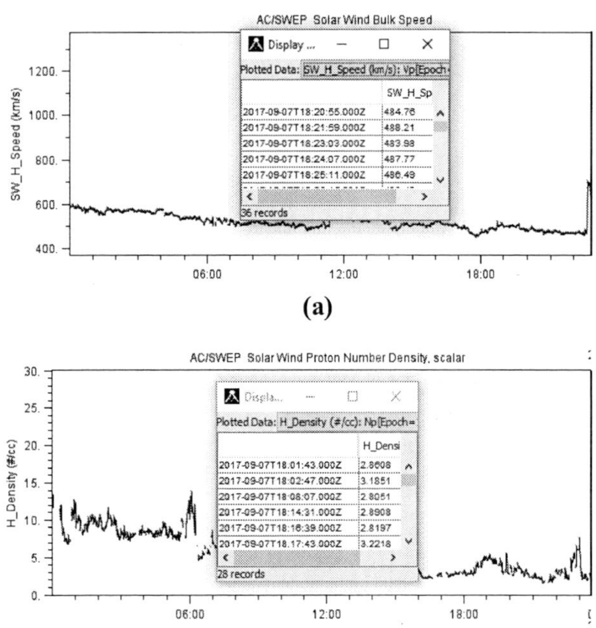

(a)

(b)

Figure 3. Example of speed and density of solar wind data. The inset shows numerical data for specific time. (a) Solar wind speed measured by ACE satellite on 7 September 2017. (b) Solar wind proton density measured by ACE satellite on 7 September 2017

3.1. Drift frequency and solar wind speed

There are 45 bursts detected in September 2017 occurred at different solar wind speed and density. From 45 bursts, 21 bursts occurred below 500 km/s solar wind speed, 8 bursts between 500 km/s and 599 km/s, 12 bursts between 600 km/s and 699 km/s and 4 bursts for above 700 km/s. The distribution of drift frequency in the solar wind speed scale shows no strong correlation. But it still shows the tendency of decreasing drift frequency as increasing the solar wind speed (Figure 4).

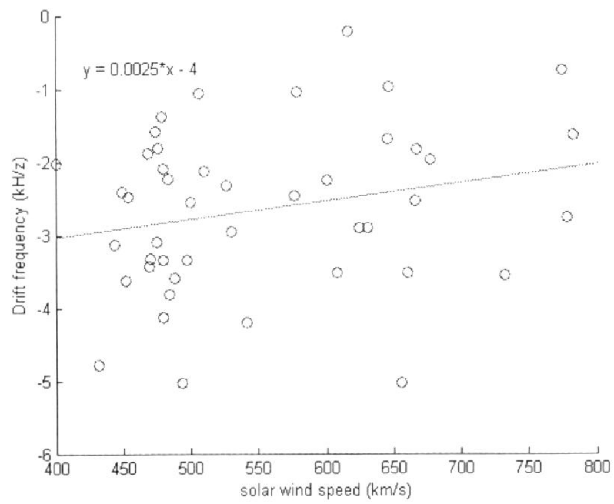

Figure 4. Scatter plot of drift frequency against average solar wind speed for all data shows weak correlation.

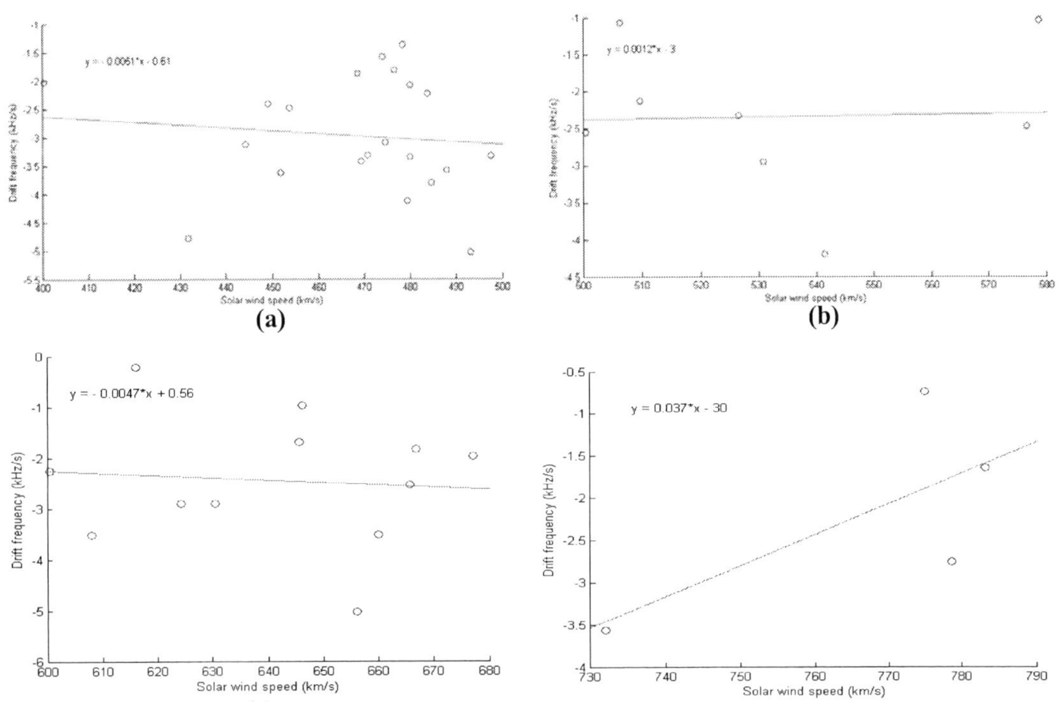

Figure 5. Scatter plot of drift frequency against solar wind speed for different range of solar wind. **(a)** For the range of wind speed between 400 km/s to 499 km/s. **(b)** For the range of wind speed between 500 km/s to 599 km/s. **(c)** For the range of wind speed between 600 km/s to 699 km/s. **(d)** For the range of wind speed between 400 km/s to 499 km/s.

3.2. Drift frequency and solar wind density

Figure 6 represent scatter plot of drift frequency against average solar wind density. To be considered in the average density, the density during of the bursts event is from 1 MHz to 400 kHz. Most of the burst concentrated in the range of 2 to 4 number of proton per cm^3. The linear fit is applied to find out the correlation between drift frequency and solar wind density. The gradient shows that the drift frequency of radio burst decreasing while solar wind density increasing. From the scatter plot, the drift frequency shows quite large distribution against solar wind density meaning that the obtained correlation is very weak.

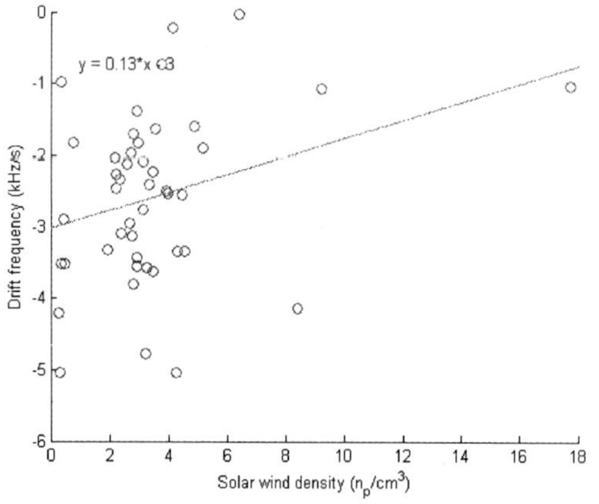

Figure 6 Scatter plot drift frequency against average solar wind density for all 45 bursts. Linear regression fitting is applied and the results shows higher solar wind density causes lower drift frequency.

4. Discussion

In this paper, we have selected 45 kilometric radio burst and performed regression fit analysis to ascertain whether drift frequency of burst depend on solar wind properties or not. For all cases, we found that all drift frequencies is negative. This shows that energetic electron beam travelling outward from the Sun. We also found the drift frequency decreasing as speed and density of solar wind increasing but the correlation is very poor. After further investigation, for speed of solar wind between 400 km/s to 499 km/s and 600 km/s to 699 km/s, drift frequency increasing as the speed increase but not for the range of solar wind speed between 500 km/s to 599 km/s and 700 km/s to 799 km/s. The source of the radio can be originate directly from the solar corona that travelling along open magnetic field or originate from other sources faces many possibilities while travelling through interplanetary space. This suggests that drift frequency radio burst doesn't depend solely to the speed solar wind and more data is needed to confirm their relationship.

The data of frequency f_i concentrate on certain time also affect the accuracy of getting the drift frequency. The low time resolution of RAD1 contributes to this inefficiency. The data also shows that the drift frequency in the range between 0.5 kHz/s to 5.5 kHz/s. This means the drift frequency of burst at kilometric range is slower compared at the corona. The energetic electron beam which is responsible in emitting radio burst due to plasma emission mechanism also depend on open magnetic field line and velocity dispersion. The location of radio source also plays important in generating radio burst. Plasma density also plays important role in emitting radio emission through plasma emission mechanism. Low plasma density causes low local plasma frequency and radio emission easily emitted [11]. This is one of the factor leads to the lower drift frequency at kilometric range.

5. Acknowledgement

We are grateful to NASA for supplying solar wind and radio emission data through CDAweb. We also thanks to D. J. McComas at Southwest Research Institute (SWRI) and M. L. Kalear at (NASA's Goddard Space Flight Center) NASA/GSFC for making their data available for use.

References

[1] S. Krucker *et al.*, "Coronal Hard X-Ray Emission Associated with Radio Type III Bursts," *Astrophys. J.*, vol. 681, no. 1, pp. 644–649, 2008.

[2] I. H. Cairns, V. V Lobzin, A. Donea, S. J. Tingay, P. I. Mccauley, and D. Oberoi, "Low Altitude Solar Magnetic Reconnection , Type III Solar Radio Bursts , and X-ray Emissions," *Sci. Rep.*, vol. 8, no. 1676, pp. 1–12, 2018.

[3] V. Krupar *et al.*, "Statistical Survey of Type III Radio Bursts at Long Wavelengths Observed by the Solar Terrestrial Relations Observatory (STEREO)/Waves Instruments: Radio Flux Density Variations with Frequency," *Sol. Phys.*, vol. 289, no. 8, pp. 3121–3135, 2014.

[4] H. A. S. Reid and N. Vilmer, "Coronal type III radio bursts and their X-ray flare and interplanetary type III counterparts," *Astron. Astrophys. Vol. 597, id.A77, 11 pp.*, vol. 597, p. A77 (1-11), 2017.

[5] H. A. S. Reid, N. Vilmer, and E. P. Kontar, "The Low-High-Low Trend of Type III Radio Burst Starting Frequencies and Solar Flare Hard X-rays," *Astron. Astrophys.*, vol. 85, no. February 2002, pp. 1–15, 2014.

[6] V. N. Melnik *et al.*, "Decameter Type III bursts with changing frequency drift-rate signs," *Sol. Phys.*, vol. 290, no. 1, pp. 193–203, 2015.

[7] H. A. S. Reid and H. Ratcliffe, "A review of solar type III radio bursts," *Res. Astron. Astrophys.*, vol. 14, no. 7, pp. 773–804, 2014.

[8] A. Stanislavsky, A. Konovalenko, and Y. Volvach, "Frequency drift rate of solar decameter „ drift pair " bursts," *Res. Astron. Astrophys.*, vol. 17, no. 97–107, 2017.

[9] H. Ratcliffe, E. P. Kontar, and H. A. S. Reid, "Large-scale simulations of solar type III radio bursts: flux density, drift rate, duration, and bandwidth," *Astron. Astrophys.*, vol. 572, p. A111, 2014.

[10] J. L. Bougeret *et al.*, "WAVES: The radio and plasma wave investigation on the wind spacecraft," *Space Sci. Rev.*, vol. 71, pp. 231–263, 1995.

[11] N. Ramli, Z. S. Hamidi, Z. Z. Abidin, and S. N. Shahar, "The Relation between Solar Radio Burst Types II , III and IV Due to Solar Activities," in *The Relation Between Solar Radio Burst Types II, III and iV Due to Solar Activities*, 2015, pp. 123–127.

IOP Conf. Series: Journal of Physics: Conf. Series **1152** (2019) 012033 doi:10.1088/1742-6596/1152/1/012033

On effective ULF frequency ranges for geomagnetic earthquake precursor

KA Yusof[1,2], M Abdullah[1,3], NSA Hamid[4*] and S Ahadi[5]

[1]Space Science Center, Institute of Climate Change, Universiti Kebangsaan Malaysia, Malaysia
[2]Department of Physics, Faculty of Science, Universiti Putra Malaysia, Malaysia
[3]Centre of Advanced Electronic and Communication Engineering, Faculty of Engineering and Built Environment, Universiti Kebangsaan Malaysia, Malaysia
[4]Physics Program, Faculty of Science and Technology, Universiti Kebangsaan Malaysia, Malaysia
[5]Department of Geophysics, Indonesian Agency for Meteorology, Climatology and Geophysics, Jakarta, Indonesia

Correspondence to: shazana.ukm@gmail.com

Abstract. This paper reviews several effective ULF frequency ranges of geomagnetic field used by prior studies to detect earthquake precursor. Since the chosen frequency ranges are rather arbitrary across different studies, it is important to determine whether any particular range is more effective than others to predict earthquakes having different characteristics, i.e. magnitude, depth and epicentral distance. It is found that there is no significant correlation between any earthquake parameter with effective ULF frequency ranges. It is also concluded that frequency ranges of 0.02 – 0.04 Hz and 0.06 Hz are more optimal for earthquake prediction purpose.

1. Introduction

Physical destructions and fatalities caused by earthquakes has been urging researchers around the world to study the possibility of predicting the disaster before it happens. Earthquake prediction, according to Hayakawa [1], are classified into three types based on its time scale namely long-term, medium-term and short-term prediction. The last one whose time scale is between a few weeks to a month is considered as the most important and useful to mitigate the disaster. Hayakawa [2] also maintains that non-seismological approach to accomplish reliable earthquake prediction is the most feasible way; one of the promising methods being via geomagnetic field perturbation observation.

Earth geomagnetic field is continuously varied by external and internal sources. External sources include solar-terrestrial and magnetospheric effects while internal sources include seismological lithospheric processes [3]. One of the challenges encountered in achieving practical geomagnetic earthquake prediction is discrimination of weak seismogenic emission which is possibly originated from earthquake epicenter from external sources [4]. Several methods have been proposed and applied by prior studies which can be generally classified into ultra-low frequency (ULF) analysis and non-ULF analysis. ULF analysis will be the center element in this paper.

ULF emission, as emphasized by Hayakawa et al. [5] in one of the earliest studies on geomagnetic earthquake prediction, is defined to be between 0.01 – 0.1 Hz. Despite the widely accepted definition of ULF range, the decision of using narrower frequency ranges within the aforementioned range is typically practiced by previous studies. To add the arbitrariness, different frequency ranges were used by different studies, each has successfully predicted past earthquake events. However, the decisions to choose particular frequency range are often made without unambiguous justifications.

In this paper, we review several ULF ranges adopted by previous studies which appeared to be effective in predicting earthquake occurrences. The relationship between each used frequency range with its respective earthquake parameters will be investigated. Ultimately, the aim is to determine either it is case-by-case basis or there is an optimum frequency range. The included past studies will be listed in later section.

2. ULF for earthquake prediction

To achieve reliable earthquake prediction, it is important to identify the characteristics of electromagnetic emission produced during earthquake preparation phase. ULF is proven to be the best candidate due to its properties of having high skin depth and gets less attenuated. This allows for greater penetration into the Earth crust thus increasing the possibility of eventually being detected on the ground [6]. Moreover, ULF is less likely to be contaminated thus clear signal can be identified as probable earthquake precursor. The "official" frequency range of ULF is rather wider, that is anything below 3 Hz [7]. However, frequency below 0.01 Hz is identified to be quasi-DC while measuring instruments exhibit less sensitivity for frequency above 0.1 Hz [5]. Hence, 0.01 – 0.1 Hz is agreed by most previous studies to be the most practical frequency range of ULF for earthquake prediction purpose [6–14].

Prattes et al. (2008 & 2011) and Li et al. (2011) [12, 14, 15] suggest the use of skin depth effect relationship to predetermine the effective ULF frequency range. The effect is described by the formula as follows:

$$\delta = \sqrt{\frac{2\rho}{\mu\omega}} \tag{1}$$

where δ is the hypocentral depth, μ is the magnetic permeability, ρ is the resistivity and ω is the angular frequency, where $\omega = 2\pi f$. From the relationship, frequency, f is inversely proportional to the square of depth, δ. Therefore, lower frequency ULF emission is expected to have higher chance of being detected from deeper lithospheric region. Moreover, this study hypothesizes that effective ULF frequency range used for earthquake prediction has inverse relationship with earthquake depth. This hypothesis will be empirically tested using results obtained from previous studies as presented in subsequent sections.

3. Studied ULF frequency ranges

This study describes the chosen effective ULF frequency ranges by prior studies and their findings in this section. 18 studies are reviewed which cumulatively present 26 individual earthquake events. To the best of our knowledge, all past studies which adopted similar signal processing methods are included in this paper to avoid biased findings. Table 1 lists included prior studies with several relevant parameters of studied earthquakes. Note that for studies with multiple earthquakes, each earthquake parameter is arranged in orderly manner. Besides that, it is also important to mention that some studies used single-value frequency while others used a range of frequency.

Table 1. List of prior studies on earthquake prediction with relevant earthquake parameters. Dash (-) symbol indicates the detail is not stated in the study.

Prior study	Location (Year)	Magnitude(s)	Depth(s) (km)	Epicentral distance(s) (km)	Frequency range (Hz)
Armansyah & Ahadi (2017) [16]	Papua Island, Indonesia, (2013 – 2014)	4.1, 4.5, 4.0	10, 10, 12	26, 13, 16	0.022
Ahadi et al. (2014) [17]	Sumatra, Indonesia (2012)	5.7, 6.1	-	233, 442	0.022
Febriani et al. (2014) [18]	West Java, Indonesia (2009)	7.5	57	135	0.007 – 0.013
Kanata et al. (2014) [19]	Tohoku, Japan (2011)	9.0	30	301	0.001 – 0.1
Takla et al. (2013) [20]	Tohoku, Japan (2011)	9.0	24	80	0.022 – 0.1
Takla et al. (2012) [21]	Pisco, Peru (2007)	8.0	39	180	0.022 – 0.1
Hirano & Hattori (2011) [22]	Iwate-Miyagi Nairiku, Japan (2008)	7.2	8	47	0.096
Li et al. (2011) [14]	Kashi & Wenan, China (2003 & 2006)	5.9, 5.9, 5.1	15, 15, 15	120, 137, 53	0.05 – 0.15
Prattes et al. (2011) [12]	L'Aquila, Italy (2009)	6.3	10	5	0.01 – 0.015
Dudkin et al. (2010) [23]	Koyna-Warna, India (2006)	4.7, 4.2	3.9, 5.1	25, 40	0.01 – 0.03
Masci et al. (2009) [24]	L'Aquila, Italy (2007)	4.0, 3.9	16, 8	23, 29	0.007 – 0.013
Saroso et al. (2009) [25]	Sumatra, Indonesia (2004 & 2005)	9.0, 8.7	30, 30	620, 439	0.03
Yumoto et al. (2009) [26]	Kushiro, Japan (1999)	6.4	104	61	0.022 – 0.1
Ida et al. (2008) [27]	Kashi, China (2003)	6.1	10	116	0.01
Prattes et al. (2008) [15]	Bovec, Slovenia (2004)	5.5	6	153	0.02 – 0.1
Molchanov et al. (2003) [28]	Kamchatka, Russia (2001)	6.2	-	-	0.01–0.03
Akinaga et al. (2001) [29]	Chi-Chi, Taiwan (1999)	7.6	11	120	0.007 – 0.013
Hayakawa et al. (1996) [5]	Guam Island, USA (1993)	7.1	60	65	0.01 – 0.05

4. Relationships between frequency ranges and earthquake parameters

In this section, the relationships, or lack thereof between effective ULF frequency ranges and earthquake parameters are presented. Figure 1 – 3 illustrate the effective frequency ranges used by previous studies based on the earthquake magnitudes, depths and epicentral distances respectively.

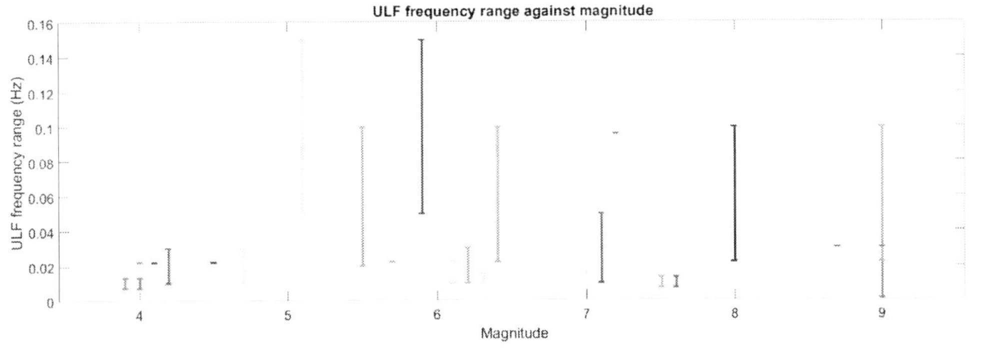

Figure 1. Effective ULF frequency range against earthquake magnitude, plotted using distinct colors to indicate different studies.

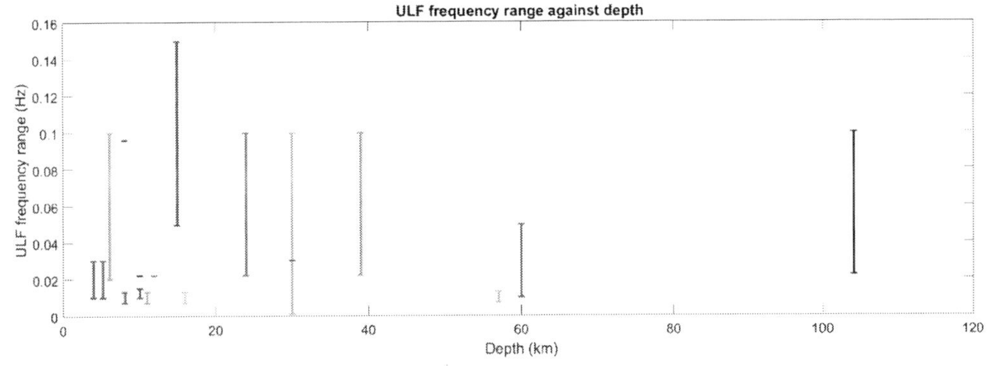

Figure 2. Effective ULF frequency range against depth, plotted using distinct colors to indicate different studies.

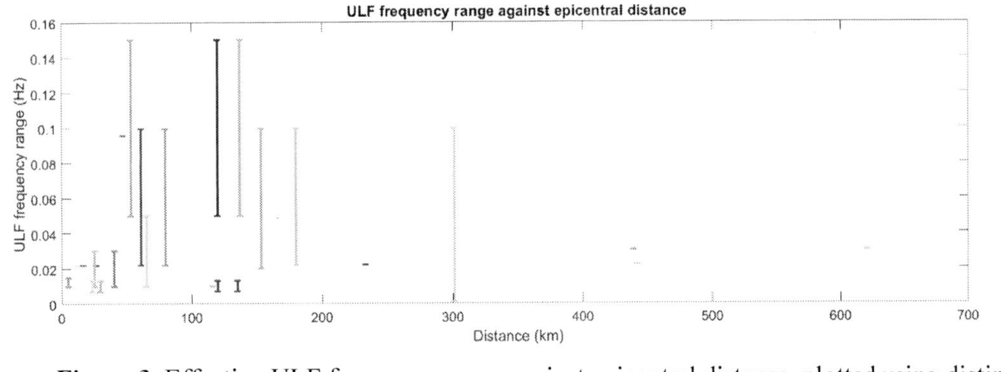

Figure 3. Effective ULF frequency range against epicentral distance, plotted using distinct colors to indicate different studies.

From figure 1, it is shown that earthquakes with varying magnitudes have been predicted with almost similar rate. This is due to the fact that larger earthquakes which occur less frequently have higher chance to be predicted. Meanwhile, weaker earthquakes which happen more often have lower chance to be predicted [10]. These attributes contribute to the near similar rate of prediction across magnitudes. From figure 2 and 3, plots are skewed more towards the left side, which correspond to shallower depths and shorter epicentral distances respectively. Shallower earthquakes have higher possibility of being predicted due to shorter distance needed for seismogenic emission to travel to the ground. Shorter distance causes less attenuation on the emission thus producing stronger anomalous signal for more effective prediction. Similarly, earthquakes with shorter epicentral distances, i.e. distance between earthquake epicenter and magnetometer station, have higher chance to be predicted [10].

Through eyeball observation, there is no apparent correlation between effective ULF frequency ranges with magnitudes, depths and epicentral distances in figure 1, 2 and 3 respectively. To study the correlation statistically, this study performed linear regression analysis and found negligibly small correlations between effective ULF frequency ranges with all three parameters. R^2 value for each parameter is shown in table 2.

Table 2. R^2 value for earthquake parameters

Parameter	**R^2 value**
Magnitude	0.0380
Depth	0.0048
Epicentral distance	0.0007

Hence, it is implied that there is no clear correlation between chosen ULF frequency ranges by prior studies with any of earthquake parameter namely magnitude, depth and epicentral distance.

5. Optimum frequency range

In order to determine whether there exists an optimal effective ULF frequency range for predicting earthquake, the percentage of success for each frequency interval, Δf with 0.01 Hz increment, from 0.01 Hz to 0.15 Hz is calculated. For single-value frequencies, the original value is rounded off to the nearest 0.01 and is taken it as it is. For example, 0.022 Hz is categorized into 0.02 Hz interval. Meanwhile, for ranging frequencies, both upper and lower limits are rounded off and all values between the two extremes, for every 0.01 Hz increment are taken. For example, for frequency range of 0.022 – 0.5 Hz, the values included are 0.02, 0.03, 0.04, 0.05 and 0.06 Hz. Next, the percentage of success is calculated by the following formula:

$$\text{Percentage of success (\%)} = \frac{\text{Number of successes}}{\text{Total successes}} \times 100\% \quad (2)$$

The results are shown in the bar plot in figure 4. $\Delta f = 0.02$ Hz (representing 0.02 – 0.029 Hz range) has the highest percentage of success (39.3%), followed by $\Delta f = 0.03$ and $\Delta f = 0.06$ Hz (representing 0.03 – 0.039 and 0.06 – 0.069 Hz ranges respectively) with almost equal percentage of ~23%. Other frequency ranges especially those having higher values ($\Delta f > 0.06$ Hz) exhibit far lower percentage of successes. Hence, this study suggests that lower frequency ranges, i.e. around 0.02 – 0.03 Hz and 0.06 Hz are optimum ULF frequency ranges for the purpose of earthquake prediction.

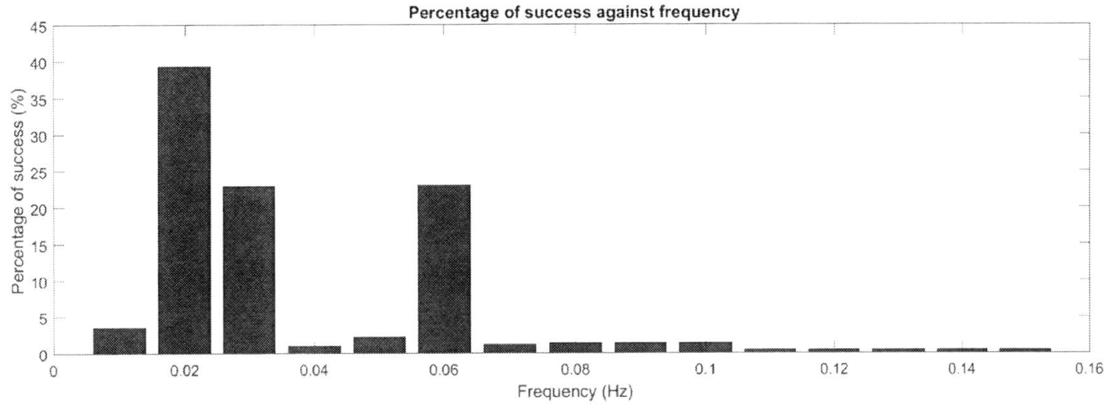

Figure 4. Percentage of success against ULF frequency range.

6. Conclusion and recommendation

In this paper, we reviewed several frequency ranges used by prior studies in predicting upcoming earthquakes. The regression analysis and relationship plots reveal no significant correlation between any earthquake parameter with effective ULF frequency ranges. This is contradictory to the hypothesis we stated before where effective ULF frequency has inverse relationship with hypocentral depth. It is determined that frequencies around 0.02 – 0.03 Hz and 0.06 Hz to be optimal for detecting earthquakes, deduced from the percentage of success calculated. To arrive at more definitive conclusion, a study which examines outputs from using multiple frequency ranges in various earthquake events is recommended. The predictions should be evaluated not only in term of successfulness or the presence of precursor, but also the reliability, indicated by the anomaly strength or amplitude to better determine the correlation.

7. Acknowledgement

The work of NSA Hamid is supported by Universiti Kebangsaan Malaysia under research grant GUP-2016-016.

References

[1] Hayakawa M 2015 *Earthquake prediction with radio techniques* (Singapore: John Wiley & Sons Pte Ltd)

[2] Hayakawa M 2016 Earthquake prediction with electromagnetic phenomena *The Irago Conference 2015 Degree Outlook on Critical Scientific and Technological Challenges for a Sustainable Society: Aichi, Japan, 22-23 October 2015 Progress in Applied Mathematics in Science and Engineering Proceedings (Bali, Indonesia, 29 September–1 October 2015) (Conference collection* volume 1709*)* ed A Sandhu and H Okada (Melville, New York: AIP Publishing) 020002(1)–020002(12)

[3] Takla E M, Yumoto K, Liu J Y, Kakinami Y, Uozumi T, Abe S and Ikeda A 2011 Anomalous Geomagnetic Variations Possibly Linked with the Taiwan Earthquake (Mw=6.4) on 19 December 2009 *International Journal of Geophysics* **2011** 1–10

[4] Currie J L and Waters C L 2014 On the use of geomagnetic indices and ULF waves for earthquake precursor signatures *J. Geophys. Res. Space Physics* **119** 992–1003

[5] Hayakawa M, Kawate R, Molchanov O A and Yumoto K 1996 Results of ultra-low-frequency magnetic field measurements during the Guam Earthquake of 8 August 1993 *Geophys. Res. Lett.* **23** 241–4

[6] Chauhan V, Singh O P, Pandey U, Singh B, Arrora B R, Rawat G, Pathan B M, Sinha A K, Sharma A K and Patil A V 2012 A search for precursor of earthquakes from multi-station ULF observation and TEC measurements in India *Indian J. Radio Space* **41** 543–56

[7] Ohta K, Izutsu J, Schekotov A and Hayakawa M 2013 The ULF/ELF electromagnetic radiation before the 11 March 2011 Japanese earthquake *Radio Sci.* **48** 589–96

[8] Chavez O, Pérez-Enríquez R, Cruz-Abeyro J A, Millan-Almaraz J R, Kotsarenko A and Rojas E 2011 Detection of electromagnetic anomalies of three earthquakes in Mexico with an improved statistical method *Nat. Hazards Earth Syst. Sci.* **11** 2021–7

[9] Scoville J, Heraud J and Freund F 2014 Pre-earthquake Magnetic Pulses *arXiv*

[10] Hayakawa M, Hattori K and Ohta K 2007 Monitoring of ULF (Ultra-Low-Frequency) Geomagnetic Variations Associated with Earthquakes *Sensors* **7** 1108–22

[11] Cohen M B and Marshall R A 2012 ELF/VLF recordings during the 11 March 2011 Japanese Tohoku earthquake *Geophys. Res. Lett.* **39** n/a-n/a

[12] Prattes G, Schwingenschuh K, Eichelberger H U, Magnes W, Boudjada M, Stachel M, Vellante M, Villante U, Wesztergom V and Nenovski P 2011 Ultra Low Frequency (ULF) European multi station magnetic field analysis before and during the 2009 earthquake at L'Aquila regarding regional geotechnical information *Nat. Hazards Earth Syst. Sci.* **11** 1959–68

[13] Xu G, Han P, Huang Q, Hattori K, Febriani F and Yamaguchi H 2013 Anomalous behaviors of geomagnetic diurnal variations prior to the 2011 off the Pacific coast of Tohoku earthquake (Mw9.0) *J. Asian Earth Sci.* **77** 59–65

[14] Li Q, Zhu P, Mamatemin A and Xu X 2011 Detection of ULF electromagnetic emissions as a precursor to two earthquakes in China *Earthq Sci* **24** 601–7

[15] Prattes G, Schwingenschuh K, Eichelberger H U, Magnes W, Boudjada M, Stachel M, Vellante M, Wesztergom V and Nenovski P 2008 Multi-point ground-based ULF magnetic field observations in Europe during seismic active periods in 2004 and 2005 *Nat. Hazards Earth Syst. Sci.* **8** 501–7

[16] Armansyah and Ahadi S 2017 Anomalous ULF Signals and Their Possibility to Etimate the Earthquake Magnitude *International Symposium on Earth Hazard and Disaster Mitigation 2016 The 6th Annual Symposium on Earthquake and Related Geohazard Research for Disaster Risk Reduction (Bandung, Indonesia, 11–12 October 2016) (AIP Conference Proceedings* volume number 1857*)* ed I Meilano and Z Zulhan (Melville, N.Y.: AIP Publishing) pp 1–7

[17] Ahadi S, Puspito N T, Ibrahim G and Saroso S 2014 Determination of the onset time in polarization power ratio Z/H for precursor of Sumatra earthquake *Proceeding of the 3rd International Conference on Theoretical and Applied Physics (ICTAP 2013): Malang, East Java, Indonesia, 10-11 October 2013 3rd ICTAP 2013 (Malang, Indonesia, 10-13 October 2013) (AIP proceedings* volume 1617*)* ed N Mufti *et al* (Meville New York: AIP Publishing) pp 75–8

[18] Febriani F, Han P, Yoshino C, Hattori K, Nurdiyanto B, Effendi N, Maulana I and Gaffar E 2014 Ultra low frequency (ULF) electromagnetic anomalies associated with large earthquakes in Java Island, Indonesia by using wavelet transform and detrended fluctuation analysis *Nat. Hazards Earth Syst. Sci.* **14** 789–98

[19] Kanata B, Zubaidah T, Ramadhani C and Irmawati B 2014 Changes of the Geomagnetic Signals Linked to Tohoku Earthquake on March 11th 2011 *IJTech* **5** 251

[20] Takla E M, Yumoto K, Okano S, Uozumi T and Abe S 2013 The signature of the 2011 Tohoku mega earthquake on the geomagnetic field measurements in Japan *NRIAG Journal of Astronomy and Geophysics* **2** 185–95

[21] Takla E M, Yumoto K, Ishitsuka J, Rosales D, Dutra S, Uozumi T and Abe S 2012 Geomagnetic variations possibly associated with the Pisco earthquake on 15 August 2007, Peru *Tectonophysics* **524-525** 29–36

IOP Conf. Series: Journal of Physics: Conf. Series **1152** (2019) 012033 doi:10.1088/1742-6596/1152/1/012033

[22] Hirano T and Hattori K 2011 ULF geomagnetic changes possibly associated with the 2008 Iwate–Miyagi Nairiku earthquake *J. Asian Earth Sci.* **41** 442–9

[23] Dudkin F, Rawat G, Arora B R, Korepanov V, Leontyeva O and Sharma A K 2010 Application of polarization ellipse technique for analysis of ULF magnetic fields from two distant stations in Koyna-Warna seismoactive region, West India *Nat. Hazards Earth Syst. Sci.* **10** 1513–22

[24] Masci F, Palangio P and Di Persio M 2009 Magnetic anomalies possibly linked to local low seismicity *Nat. Hazards Earth Syst. Sci.* **9** 1567–72

[25] Saroso S, Hattori K, Ishikawa H, Ida Y, Shirogane R, Hayakawa M, Yumoto K, Shiokawa K and Nishihashi M 2009 ULF geomagnetic anomalous changes possibly associated with 2004–2005 Sumatra earthquakes *Phys. Chem. Earth* **34** 343–9

[26] Yumoto K *et al* 2009 A new ULF wave analysis for Seismo-Electromagnetics using CPMN/MAGDAS data *Phys. Chem. Earth* **34** 360–6

[27] Ida Y, Yang D, Li Q, Sun H and Hayakawa M 2008 Detection of ULF electromagnetic emissions as a precursor to an earthquake in China with an improved polarization analysis *Nat. Hazards Earth Syst. Sci.* **8** 775–7

[28] Molchanov O, Schekotov A, Fedorov E, Belyaev G and Gordeev E 2003 Preseismic ULF electromagnetic effect from observation at Kamchatka *Nat. Hazards Earth Syst. Sci.* **3** 203–9

[29] Akinaga Y, Hayakawa M, Liu J Y, Yumoto K and Hattori K 2001 A precursory ULF signature for the Chi-Chi earthquake in Taiwan *Nat. Hazards Earth Syst. Sci.* **1** 33–6

Solar quiet (Sq) and conductivity-depth structure of Asia sub-region using solar quiet field variation

M Abbas[1, 2, 3], M H Jusoh[1] and Z Z Abidin[2]

[1]Applied Electromagnetic Research Group (AERG), Faculty of Electrical Engineering, University of Teknologi MARA, MALAYSIA
[2]Radio Cosmology Laboratory, Faculty of Science, University of Malaya, 50603, MALAYSIA
[3]Department of Physics, Kebbi State University of science and Technology, Aliero, Nigeria

E-mail: huzaimy_uitm@yahoo.com

Abstract. The study used some selected magnetic field records to establish the equivalent external and internal ionospheric current system for the moderate solar quiet year 2012. The month-by-month behaviour of the equivalent current for both external and internal current appears stronger in local summer months (May-August) and weaker in local winter months (November-February). The current vortex appears earlier in local summer than in local winter months and disappeared in December. In each of the month, the latitudinal position of the internal Sq current vortex appeared at a considerable higher latitude than the external. A regression line fitted to the estimated conductivity values gives a profile that starts at 0.02 S/m at a depth of about 30 km and rose gradually to about 0.03 S/m seen at ~120 km depth. Consequently the conductivity profile increases significantly to about 0.07 S/m at depth of about 400 km followed by a sharp steep increase through the transition zone and reached its peak value ~0.3 S/m at depth of about 770 km. An evidence of discontinuity was observed at depth of about 370-440 km and 603-671 km. The sharp steep increase in conductivity profile at depth of about 400 km may be due to phase change of the mantle minerals from olivine to spinel structure.

1. Introduction

The daily geomagnetic field variations at the Earth surface has quite a number of sources which are of external and internal origin to the Earth surface. The field produce by these sources are generally time dependent and changes slowly with latitudes depending on the magnetic activity on the outer space. According to the classical dynamo theory, the main source of the daily geomagnetic field is the electric current flowing at an altitude (90-150 km) in the E-region of the ionosphere generated by the motion of conducting particles across the Earth's main magnetic field, [1]. When magnetic disturbances are absent, the current generated is known as solar quiet (Sq) current and the field produce is known as solar quiet (Sq) field. The sensitivity of the Sq current amplitude to solar activity is different from one season to another such that its strength has an annual variation with higher intensity in summer at mid-latitude region and a semi-annual variation in which it is strongest in equinox at equatorial low latitude, e.g [2] The variability of the Sq current system has been attributed to variation in the ionospheric conductivity, [3]. In a different study, [4] attributed the variability of the Sq current to variation of the tidal and prevailing winds in the upper atmosphere. The equivalent Sq current system during the sunlight hours has two current vortices (one on each hemisphere) centred on the local noon. The current flow in the northern hemisphere is counter-clockwise and clockwise in the southern hemisphere, e.g [5]. The fluctuating current cause a corresponding currents to flow into the conducting Earth structure. The magnitude, direction and depth at which the induced component of the Sq current penetrate is determine by the effective wavelength of the source current and the distribution of electrically conducting materials within the Earth's interior, [6]. The primary aim of this study is to investigate the month-by-month behaviour of the equivalent Sq current system and to estimate the general conductivity profile of the upper mantle beneath Asia sub-region

2.0 Sources of Data

In the data selection, the first criteria adopted to ensure good and quality data are used was to select days in which the geomagnetic disturbance index Kp values are not greater than 3 as earlier suggested by [7]. These groups of quiet days were obtained as published by Geoscience Australia Catalogue at [8]. These days have least solar-terrestrial magnetic disturbances, thus represent the Sq fields generated by dynamo currents in the E region of the ionosphere. The data set used in this study are the International Real-time Magnetic Observatory Network (INTERMAGNET) data for the year 2012. The observatories had their sample records available in minute average values and the original recordings of the field across all the stations were in Universal time (UT) and are all converted to local time. The intermagnet data are recorded as orthogonal north, east and vertically downward components field as X, Y and Z and the X and Y are transformed to their respective northward (H) and eastward (D) coordinate system using the relation given in [9]. The magnetic stations used span from the magnetic equator to high latitude region and table 1.0 shows the list of the stations used in the analysis.

Table 1.0: List of INTERMAGNET data stations and their coordinates system used in the study

		Geographic		Geomagnetic
Name	Code	Lat (°N)	Lon (°E)	Lat (°N)
Dalat	DLT	11.94	108.48	2.60
Phu thuy	PHU	21.03	105.97	10.78
Zhaoqing	GZH	22.00	112.45	12.74
Kanoya	KNY	31.42	130.88	20.50
Kakioka	KAK	36.63	140.19	27.37
Beijing Ming Tom	BMT	40.30	116.20	30.04
Changchum	CNH	44.08	124.86	31.00
Memenbetsu	MMB	43.91	144.15	35.35
Irkutsk	IRT	52.27	104.45	41.84
Parantuka	PKT	52.97	158.25	46.10
Magadan	MGD	60.05	144.19	53.49

3.0 Method of Data Analysis

The analysis started by assuming that as the Earth rotates around the sun and the magnetic observatories across the latitudes are riding with the Earth under a fixed subsolar ionospheric current system. These magnetic observatories sampled the behaviour of the field source current and the current induced to flow in the conducting Earth beneath the observatories. The Sq field variations at any of the latitude in 24 hours represented the measured field through 360° of a sphere. [6] Called this technique "cloning" a sphere from a longitudinal "slice" sample.

Next, a hemispheric mirror modelling technique was applied. This model utilizes the field values of the hemisphere that is of primary interest in the study to create a secondary hemisphere in the opposite hemisphere. This was achieved by simply assuming that the H-field variation in the secondary hemisphere is similar to the primary hemisphere but its D and Z component are caused to be oppositely directed by simply changing the appropriate sign of the two (D and Z) components so as to conform to the reversed source-current vortex as extensively discussed by [6]. In the mirror hemisphere, the variations of the Sq field formed for the three orthogonal components from the primary one were shifted by 6 months such that the summer hemisphere of the primary stations corresponds to the winter hemisphere of the secondary hemisphere. In a similar way, for each day the harmonic coefficients for an analysis sphere was created with field values at all 2.5° latitude increments, pole to pole as input parameter to the SHA analysis. At this stage, a complete sphere has been successfully created consisting of the primary and secondary hemisphere and the outcome of the analysis was strictly apportion to the primary region.

[10], and [9] estimated the spectral contributions to the Sq field and found that only the 24-, 12-, 8- and 6-hr harmonics components are of larger size to contribute significantly to the Fourier analysis. The higher magnitudes is due to the effect of Earth's rotation and not on the characteristics of the signal at any given day. On the basis of this that we determine the 24-, 12-, 8-, and 6-hr monthly Fourier sine and cosine coefficients for each station and component from the detrended Sq field records. The annual and semi-annual Fourier components of 12 (months) daily coefficients in each year were also determined at each 2.5° geomagnetic latitude increments from the geomagnetic equator to northward and southward poles as earlier suggested by [10] The Fourier coefficients obtained were stored in a file and these values represent the year's Sq field variations for each component. For any geomagnetic component W (H, D and Z), the Sq field could be constructed from:

$$\Delta W = \sum_{m=1}^{4}\{C_m\, cos(15mt) + S_m \sin(15mt)\} \quad (1)$$

where ΔW represent the changes for the H, D or Z field component from the daily mean value of the surface field records. The t is the local time (LT) in hours at the stations, while C_m and S_m are the cosine and sine amplitude coefficients as given in the equation below;

$$C_m = A_m^0 + \sum_{Q=1}^{2}\{A_m^{AQ} cos(30MQ) + B_m^{AQ} sin(30MQ)\} \quad (2)$$

$$B_m = B_m^0 + \sum_{Q=1}^{2}\{A_m^{BQ} cos(30MQ) + B_m^{BQ} sin(30MQ)\} \quad (3)$$

where m is the order from 1-4 for the 24,-, 12-, 8-, and 6-hours spectral components. Q is 1 for annual and 2 for semi-annual, while M is the decimal month (represented as 0.00 for January 1 and 12.00 for December 31). The As and Bs are the Fourier amplitude coefficients and the mainline letters are the annual and semi-annual coefficients. Any linear trend on the recorded Sq field variation that will introduce some errors in the sinusoidal amplitude coefficients was removed before the Fourier analysis was performed. The file containing each of the 40 Fourier coefficients representing each of the three orthogonal field directions were carefully arranged according to geomagnetic latitudes of the stations used in the analysis for smoothing. The global observations of Sq fields are more symmetric with respect to geomagnetic axis than the geographic and this cause the SHA coefficients series for the Sq to converge more easily and rapidly with geomagnetic expansion. Therefore, the analysis in this study will proceed with the geomagnetic coordinate system.

When V is determined from the magnetic field record about the Earth, analyses shows that most of the contributions comes from the internal part of the magnetic potential function expansion. The magnetic potential of the quiet field obtained from the daily mean values recorded in universal time (UT) from the external source current and the internally induced current is expressed as;

$$V_{ext} = R_E \sum_{n=1}^{\infty}\left(\frac{r}{R_E}\right)^{n} \sum_{m=0}^{n}[a_n^{me} \cos(m\emptyset) + b_n^{me} \sin(m\emptyset)]\, P_n^m(cos\theta) \quad (4)$$

$$V_{int} = R_E \sum_{n=1}^{\infty}\left(\frac{R_E}{r}\right)^{n+1} \sum_{m=0}^{n}\left[a_n^{mi} \cos(m\emptyset) + b_n^{mi} \sin(m\emptyset)\right] P_n^m(cos\theta) \quad (5)$$

Where C, θ, a, r, ϕ represent constant of integration, geomagnetic colatitude, geocentric distance, Earth's radius and local time of the station. It is obvious from equation 9 and 10 that two magnetic potential components are attributed to external and internal sources. The magnetic potential (V_{ext}) increases with increasing r-indication of increase field strength as the height increases from the surface. On the other hand, the V_{int} increases with decreasing r-indication of increase field strength with depth from the ground. The a_n^{me}, b_n^{me} and a_n^{mi}, b_n^{mi}correspond to the Gauss coefficient with e and i indicating external and internal field sources. It is apparent in equation 4 and 5 that the longitudinal dependence of both magnetic field potential are given by Fourier series, while its latitudinal dependence is represented by a series of Schmidt quasi-normalized associated Legendre functions $P_n^m(cos\theta)$. This function can be estimated using the following recursive formula as provided in [9]

$R_n^m = \sqrt{n^2 - m^2}$
(6a)
$P_0^0 = \cos(\theta) = 1;\ P_1^0\,(cos\theta) = (cos\theta);\ P_1^1\,(cos\theta) = \sin(\theta)$
(6b)
$P_n^m(cos\theta) = \sqrt{\frac{2m-1}{2m}}\sin(\theta)P_{m-1}^{m-1}\ ;\ (cos\theta)\ P_{m-1}^{m-1}\ ;\ for\ m>1,n=m$
(6c)
$P_n^m(cos\theta) = \frac{(2n-1)\cos\theta\ P_{n-1}^m - R_{n-1}^m - P_{n-2}^m}{R_n^m}\ for\ n > m$
(6d)
$\frac{dP_n^m}{d\theta} = \frac{n\cos\theta P_n^m\,(cos\theta) - R_n^m - P_{n-1}^m(cos\theta)}{\sin\theta}\ (except\ for\ \theta = 0\ or\ 180°)$
(6e)
Equation (6e) is undefined at the poles ($for\ \theta = 0\ or\ 180°$).
To determine the spherical harmonic Gauss coefficients, the following values were first obtained;
$a_n^m = \frac{2n+1}{4n(n+1)}\left(\sum_0^{180} X_c^m \frac{dP_n^m}{d\theta}\sin\theta + Y_s^m\, mP_n^m\right)\Delta\theta$
(7a)
$b_n^m = \frac{2n+1}{4n(n+1)}\left(\sum_0^{180} X_s^m \frac{dP_n^m}{d\theta}\sin\theta + Y_c^m\, mP_n^m\right)\Delta\theta$
(7b)
$C_n^m = \frac{1}{4}(2n+1)\sum_0^{180} Z_c^m P_n^m \sin\theta\,\Delta\theta$
(7c)
$d_n^m = \frac{1}{4}(2n+1)\sum_0^{180} Z_s^m P_n^m \sin\theta\,\Delta\theta$
(7d)
Where X_c^m and X_s^m represent the cosine and sine coefficients of the Fourier component m of the northward (ΔH) field change at colatitude (θ). The $\Delta\theta$ in these equations is the step increment which is equivalent to 2.5^0 ($\Delta\theta = 2.5^{\circ}$) used in the analysis. θ is the geomagnetic colatitude.
The spherical harmonic Gauss coefficient for the external and internal is then computed using the equations below;
For the external part

$a_n^{me} = \frac{[(n+1)a_n^m + C_n^m]}{2n+1}$
(8a)

$b_n^{me} = \frac{[(n+1)b_n^m + d_n^m]}{2n+1}$
(8b)
The internal part is given as;

$a_n^{mi} = \frac{[na_n^m - C_n^m]}{2n+1}$
(9a)

$b_n^{mi} = \frac{[nb_n^m - d_n^m]}{2n+1}$
(9b)
The potential function was separated into external and internal source contributions. With reference to Gauss coefficients, the external fields in the orthogonal directions (with X pointing geomagnetic northward, Y eastward and Z vertically downward) may be reconstructed from the external spherical harmonic analysis as series of Fourier cosine and sine of the terms;

$X_c^m(\theta) = \sum_{n=m}^{12} a_n^{me}\frac{dP_n^m}{d\theta}$
(10a)

$X_s^m(\theta) = \sum_{n=m}^{12} b_n^{me}\frac{dP_n^m}{d\theta}$
(10b)

$$Y_c^m(\theta) = -\frac{m}{\sin\theta}\sum_{n=m}^{12} b_n^{me}\frac{dP_n^m}{d\theta}$$
(10c)
$$Y_s^m(\theta) = -\frac{m}{\sin\theta}\sum_{n=m}^{12} a_n^{me}\frac{dP_n^m}{d\theta}$$
(10d)
$$Z_c^m = \sum_{n=m}^{12} na_n^{me} P_n^m$$
(10e)
$$Z_s^m = \sum_{n=m}^{12} nb_n^{me} P_n^m$$
(10f)

where X_c^m, Y_c^m and Z_c^m are the Cosine coefficients of the external orthogonal field components and X_s^m, Y_s^m and Z_s^m are the Sine coefficients. For the internal horizontal component the e may be replaced by I and its vertical internal component is expressed as;

$$Z_c^m(\theta) = -\sum_{n=m}^{12}(n+1)a_n^{mi} P_n^m$$
(11a)
$$Z_s^m(\theta) = -\sum_{n=m}^{12}(n+1)b_n^{mi} P_n^m$$
(11b)

The sphere in which the spherical harmonic analysis is carried out was spherically created to accommodate the boundary conditions of the Gauss method.

The equivalent current function $J(\phi)$ in Amperes (A) for any hour of the day ${}^{\phi}/_{15}$ is expressed as;

$$J(\phi) = \sum_{m=1}^{4}\sum_{n=m}^{12}\{U_n^m \cos(m\phi) + V_n^m \sin(m\phi)\} P_n^m$$
(12)

The external current is expressed by;

$$U_n^m = -K\left[\frac{2n+1}{n+1}\right] a_n^{mi}\left(\frac{a}{R}\right)^n$$
(13a)
$$V_n^m = -K\left[\frac{2n+1}{n+1}\right] b_n^{mi}\left(\frac{R}{a}\right)^{n+1}$$
(13b)

Where $K = \frac{5R}{2\pi}$ and R is the Earth radius given in kilometres, a, is the radius of the sphere of the equivalent current location. Throughout the study, a, is assumed to be approximately equal to R so that the ration $\frac{a}{R}$ and $\frac{R}{a}$ becomes 1 and the factor was omitted.

4.0 Conductivity of the Earth's upper mantle

[11] Provided set of equations for profiling the conductivity with a transfer function equivalent to a substitute conductor. This method required external and internal spherical harmonic coefficients at a given location as input parameters. [12] Modify Schmucker's transfer function. The generalized form of Schmucker's transfer function C_n^m as modify by Campbell and Anderssen is expressed as;

$$C_n^m = Z - iP$$
(14)

The transfer function for a substitute conductor given in equation (1) has a complex number in which the real (Z) and the imaginary (-P) part expressed as;

$$Z = \frac{R}{n(n+1)}\left\{\frac{A_n^m[na_n^{me}-(n+1)a_n^{mi}]+B_n^m[nb_n^{me}-(n+1)b_n^{mi}]}{(A_n^m)^2+(B_n^m)^2}\right\}$$
(15a)
$$P = \frac{R}{n(n+1)}\left\{\frac{A_n^m[nb_n^{me}-(n+1)b_n^{mi}]-B_n^m[na_n^{me}-(n+1)a_n^{mi}]}{(A_n^m)^2+(B_n^m)^2}\right\}$$
(15b)

The R is the radius of the Earth expressed in kilometres; hence the Z and P are expressed in kilometres. The coefficient A_n^m and B_n^m are obtained using the spherical harmonic coefficients for the external and internal field given below;

$$A_n^m = a_n^{me} + a_n^{mi}$$
(16a)

$$B_n^m = b_n^{me} + b_n^{mi}$$
(16b)

For each (m,n) set of spherical harmonic coefficients, the depth (in km) to the uniform substitute was estimated using the relation

$$d_n^m = Z - P$$
(17a)

And has a substitute –layer conductivity (S/m) given by;

$$\sigma_n^m = \frac{5.4 \times 10^4}{m(\pi P)^2}$$
(17b)

The ratio of the internal to the external components of the quiet field variation S_n^m is given as;

$$S_n^m = u + iv$$
(18a)

Where

$$u = \left[\frac{(a_n^{me})(a_n^{mi}) + (b_n^{me})(b_n^{mi})}{[(a_n^{me})]^2 + [(b_n^{me})]^2}\right]$$
(18b)

$$v = \left[\frac{(b_n^{me})(a_n^{mi}) - (a_n^{me})(b_n^{mi})}{[(a_n^{me})]^2 + [(b_n^{me})]^2}\right]$$
(18c)

The validity of equation (a) and (b) is bounded by three conditions: the first condition is given by;

$$0° = arg(C_n^m) \geq -45°;$$
(19a)

The second condition is that

$$80° = arg(S_n^m) \geq 10.5°;$$
(19b)

The third is that the SHA amplitudes (equation 12) not be small because the relative error in the SHA coefficients increases with decrease in amplitudes of the coefficients. Hence the amplitude is set at;

$$[(A_n^m)^2 + (B_n^m)^2]^{0.5} \geq G_m$$
(19c)

5.0 Results and Discussion

Figures 1and 2 shows the month-by-month equivalent external and internal Sq current system. It is obvious from these figures that the external and internal equivalent currents are characterized by a single daytime current vortex at mid-latitude that appears stronger in northern summer months (May-August) and weaker in northern winter (November-February). The maximum intensity of the external current reached 12.6×10^4A in June and its minimum 4.4×10^4A is seen in November. The internal Sq current has its maximum amplitude value -6.2×10^4A in June and its minimum -2.2×10^4A appeared in November. The structure and strength of the Sq current system are primarily controlled by the neutral tidal winds and the conductivity in the dynamo region of the ionosphere. Modelling the Sq current system at different solar epoch, [13] found that changes in ionospheric conductivity cause a corresponding changes in the Sq current intensity. This account for the maximum intensity of the Sq current system observed in northern summer month (June) in the present study. Apart from the intensity of the external Sq current, the latitudinal position of the current vortex shows that it varies on monthly basis.

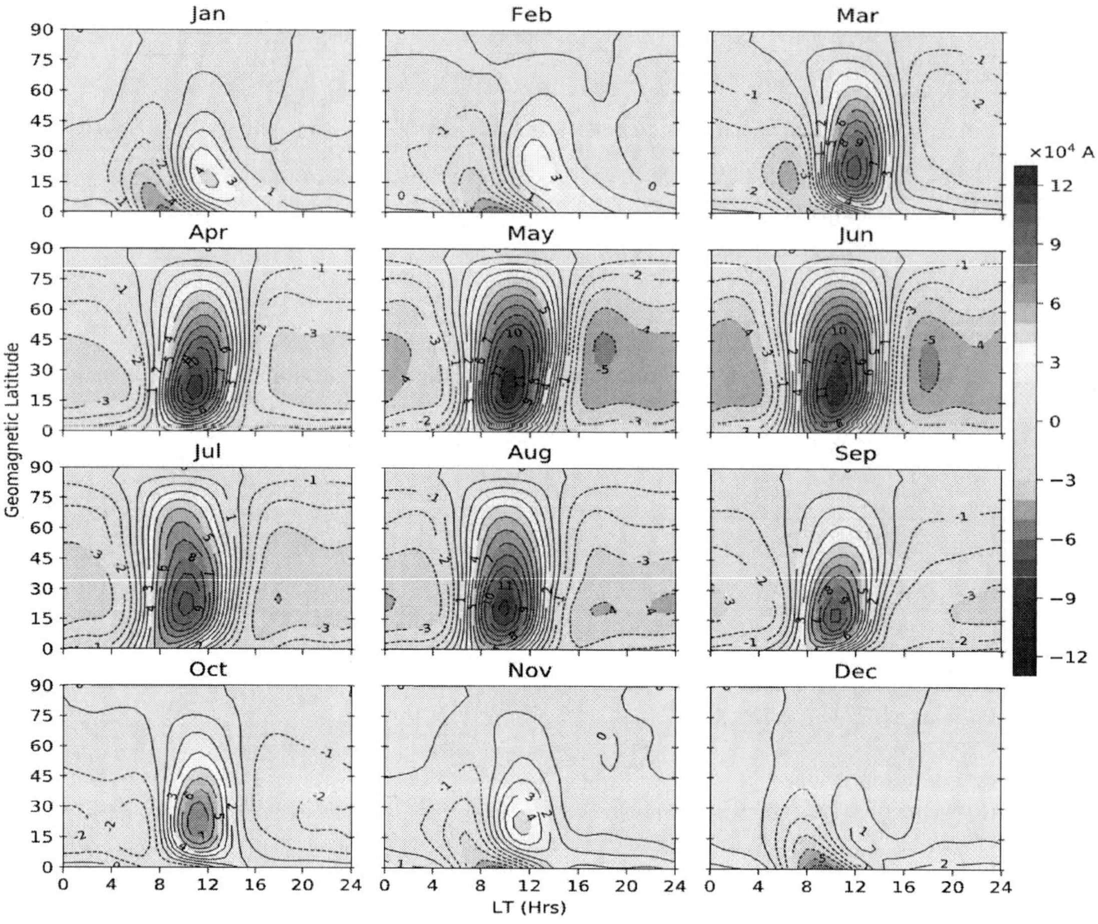

Figure 1.0 Equivalent external Sq current system in 2012

For example, the latitudinal position of the external Sq current vortex ranged between 23° and 25° seen between April to August as depicted in figure 1. During these periods, the internal Sq current vortex moves poleward relative to the external vortex. It ranged between 25° and 28° seen between April to September as shown in figure 2. In each of the month, the latitudinal position of the internal Sq current vortex appeared at a considerable higher latitude than the external. Figures 1 and 2 also show that the current vortex appear earlier in local summer months (May-August) than in local winter months (November-February). This variation in the position of the Sq foci has been largely attributed to interhemispheric field-aligned current (IHFACs). [14] earlier suggested that the field-aligned current of the current vortex produce an eastward geomagnetic field that advances the equivalent current vortex in the summer hemisphere and delays the vortex in the winter hemisphere. The consequences is that the intensity of the Sq current vortex appears at earlier local time in summer than in winter as obvious in the present study. This could be the main reason for the Sq current whorl appearing at later hour of the day in northern winter months. If in deed the noontime southward flow of FACs is responsible for the northern eastward displacement of the Sq current whorl, then it has a shorter time considering the months of later occurrence of Sq current whorl, while the northward flow has longer period of about 6-months. This difference may likely reflect local effect of the upper mantle conductivity upon the induced currents in the Earth's interior.

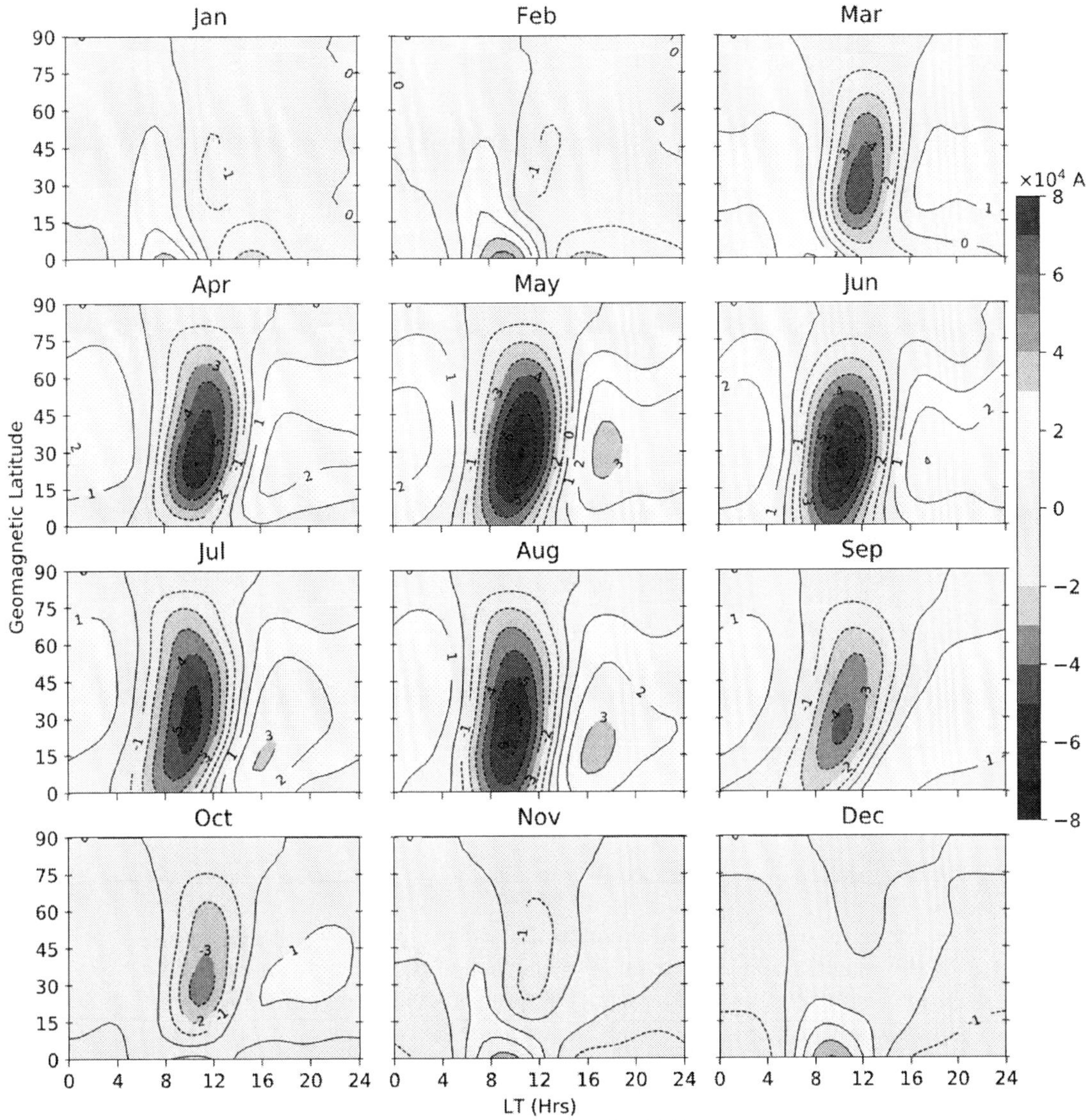

Figure 2.0 Equivalent internal Sq current system in 2012

During times when the external Sq current was minimum in northern winter, the daytime current in November appears very weak with its vortex at a considerable lower latitude (25°) and disappeared in December. The internal Sq current vortex was also observed to disappear in December. This implies that the disappearance of Sq vortex is a consequences of weaker daytime Sq current. This result is in sharp contrast with the early findings of [2] that did not observed any disappearance of the Sq vortex focus in their winter months. However, the present result is consistent with the India-Siberia region reported by [16]. These authors observed disappearance of Sq vortex focus in their winter months. In a similar study by [17], they found for the Central Asia Sq focus vortex magnitude to reach about 16.8 x 10^4A in June and July and virtually disappeared in December and January. The disappearance of sq current vortex particularly in winter months indicates the existence of significant changes in the summer-winter variation pattern of the Sq current system.

Figures 3 show the mantle electrical conductivity-depth profile of Asia sub-region during moderate and low solar activity year 2012. The blue dotted points are the computed conductivity values. The locally weighted regression fitting to the conductivity values is given by the solid line computed using the best Lowess technique. A small group of about 3 conductivity values at depth range between 200 and 400 Km are observed in figure 3. It is not certain whether these are real or outliers or perhaps there are times when the varying source Sq field is inducing a response from a small scale, high conductivity anomaly at these locations. Similar effect was earlier observed in the India-Siberia region by [18]. Apart from these small groups of conductivity values, generally the scattered points are much concentrated within the first few kilometres down to a depth of about 450 km and with increasing depth, the density of the scattered points decreases to a depth of about 770 km.

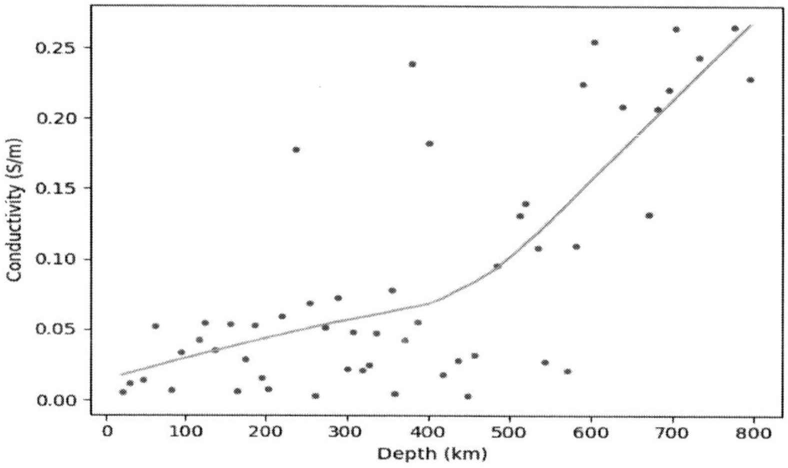

Figure 3. Mantle electrical conductivity-depth profile computed from the monthly spherical harmonic analyses of the Sq fields applied to the equivalent substitute conductor technique. The points that is indicated by • the computed values while the solid line through the points is a locally weighted regression fitting of the computed conductivity values.

These points resulted from series of factors among them includes; error incurred from the field measurements, error from spherical harmonic analysis (SHA) fitting, various sources of currents. The noise in the original data and the effect of lateral inhomogeneity also constitute an error that is very difficult to evaluate accurately, hence what could actually be termed an error is in the measurement of the regression fitting for the scattered conductivity values. It is obvious from both figure 4 that the profile shows downward increase in conductivity with depth from the crustal surface deep down into the lower mantle in consistent with earlier result of [6] that observed conductivity profile to the lower mantle depth beneath Australian region. This downward increase in the conductivity profile is no surprise since for a given earth's interior the temperature increases with depth, it is expected for the electrical conductivity to increase accordingly. The gradual increase at depth range between 30 and 120 km indicate the existence of weak conductivity zone. This observation is in Sharp contrast with the South American sector that showed highest conductivity profile at this depth range suggesting lateral inhomogeneity between the continents. However, the present result is Consistent with East Asia and African observation by [19] and also Malaysian region by [20]. Since the electrical conductivity within few kilometres of the crust depends on the amount of interstitial water that is present, it is likely the weak conductivity may be due to dehydration of the crust through metamorphic process as earlier identified by [21]. A sudden increase in conductivity from 0.03 to 0.04 S/m at depth range between

120 and 270 km correspond to the asthenosphere with global seismic low velocity and high electrical conductivity, e.g [22]. A sharp steep increase in conductivity was observed at depth of about 400 km with corresponding magnitude of 0.07 S/m and increases steeply to about 0.15 S/m at depth of 600 km and continued until it reached about 0.3 S/m at depth of 770 km with no indication of levelling off. The sharp steep increase at depth of 400 km may be due to phase change of the mantle materials from olivine to spinel structure as earlier as earlier discussed by research workers e.g, [18]. The high electrical conductivity profile at depth between 400 and 600 km correspond to the mantle transition zone identified with high electrical conductivity, e.g [20]. The mantle transition zone has been identified as the most likely important water reservoir on Earth due to its ability to store water content of more than 2 wt%, [23]. The enhanced conductivity profile in the transition may likely be influence by the presence of water. The discontinuity profile observed at 370–440 km is likely associated to the transformation of α-phase of olivine to β at high temperature and pressure, [24] while the dissociation of γ is ascribed to the seismic discontinuity at 603–671 km, [25]. The γ-phase split into perovskite and magnessiowustite at the transition zone and even beyond, [26].

6.0 Conclusion

The monthly behaviour of the external and internal Sq current system appears stronger in local summer months (May-August) and weaker in local winter months (November-February). The current vortex appears earlier in local summer than in local winter months. In each of the month, the latitudinal position of the internal Sq current vortex appeared at a considerable higher latitude than the external. An evidence of discontinuity was observed at depth of about 370-440 km and 603-671 km. The sharp steep increase in conductivity profile at depth of about 400 km may be due to phase change of the mantle minerals from olivine to spinel structure.

Acknowledgements

The results presented in this paper rely on data collected at magnetic observatories. The authors are grateful to the national institutes that support them and INTERMGNET for promoting high standard magnetic observatories.

References

[1] Richmond A. D. 1979. *Ionospheric wind dynamo theory: a review*. J. Geomagn. Geoelectr. **31** (3):287-310

[2] Matsushita S Maeda H 1965. On the geomagnetic solar quiet daily variation field during the IGY. J. Geophys. Res. **70**, 2535–2558.

[3] Kirchhoff, V.W.J.H., Carpenter, L.A., 1976. *The day-to-day variability in ionospheric electric fields and currents*. J. Geophys. Res. **81**, 2737– 2742.

[4] Torta, J.M., Curto, J.J., Bencze, P., 1997. *Behavior of the quiet dayionospheric current system in the European region*. J. Geophys. Res.**102**, 2483–2494.

.[5] A.D. Richmond A D 1989. *Modeling the ionosphere wind dynamo: A review quiet daily geomagnetic fields*. Pure Appl. Geophys. **131**, 413–435.

[6] Campbell, W.H., Barton, C.E., Chamalaun, C., Welsh, W., 1998. *Quiet-day ionospheric currents and their application to upper mantle conductivity in Australia*. Earth Planets Space **50**, 347–360.

[7] Campbell, W.H 1997. *Introduction to Geomagnetic fields.* Cambridge University Press, New York.

[8] http://ww.gav.gov.au/oracle/geomag/iqd_form.jsp.

[9] Campbell, W.H 2003. *Introduction to Geomagnetic fields.* Second ed. Cambridge University Press, New York.

]10] Campbell, W.H., Schiffmacher, E.R., 1985. *Quiet ionospheric currents of the Northern Hemisphere derived from geomagnetic field records.* J. Geophys. Res. Space Phys. **90**, 6475–6486.

[11] Schmucker, U., 1970. *An introduction to induction anomalies.* J. Geomagn. Geoelectr. **22**, 9–33.

[12] Campbell, W.H., Anderssen, R.S., 1983. *Conductivity of the subcontinental upper mantle: an analysis using quiet-day geomagnetic records of North America.* J. Geomagn. Geoelectr. **35**, 367–382.

[13] Takeda, M., Yamada, Y., and Araki T., 1986. *Simulation of ionospheric currents and geomagnetic field variations of Sq for different solar activity*, J. Atmos. Terr. Phys., v.48, pp: 277–287.

[14] Fukushima, N., 1979. *Electric potential difference between conjugate points in middle latitudes caused in the ionosphere.* J. Geom. Geo-electr. 31, 401–409.

[15] Campbell, W.H., Arora, B.R., Schiffmacher, E.R., 1993. *External Sq currents in the India–Siberia region.* J. Geophys. Res. 98, 3741–3752.

[16] Campbell, W.H., Schiffmacher, E.R., 1985. *Quiet ionospheric currents of the northern hemisphere derived from Geomagnetic field records.* J. Geophys. Res. 90, 6475–6486

[17] Arora, B.R., Campbell, W.H., Schiffmacher, E.R., 1995. *Upper mantle electrical conductivity in the Himalayan region.* J. Geomagn. Geoelectr. **47**, 653–665.

[18] Campbell, W.H., Schiffmacher, E.R., 1988. *Upper mantle electrical conductivity for seven subcontinental regions of the Earth.* J. Geomagn. Geoelectr. **40**, 1387–1406.

[19] Abidin, Z.Z., M.H. Jusoh, M. Abbas and Yoshikawa A. 2017 *Application of solar quiet (Sq) current in determining mantle conductivity-depth structure in Malaysia,* J. Atm. Sol.Terr. Phys.

[20] Hyndman, R., Hyndman, D., 1968. *Water saturation and high electrical conductivity in the lower continental crust.* Earth Planet Sci. Lett, **4**, 427-432.

[21] Dziewonski, A.M., Anderson, D.L., 1981. *Preliminary reference Earth model.* Phys. Earth planet **25**, 297–356.

[22] Ohtani, E., Sakai, T., 2008. *Recent advances in the study of mantle phase transitions.* Phys. Earth Planet. In. **170**, 240–247.

[23] Katsura, T., Ito, E., 1989. *The system Mg_2SiO_4-Fe_2SiO_4 at high pressures and temperatures: precise determination of stabilities of olivine, modified spinel and spinel.* J. Geophys. Res. **94**, 15663–15670.

[24] Ito, E., Takahashi, E., 1989. Post-spinel transformations in the system Mg_2SiO_4-Fe_2SiO_4 and some geophysical implications. J. Geophys. Res. **94**, 10637–10646.

[25] Chandrasekhar, E., 2011. *Regional electromagnetic induction studies using long period geomagnetic variations.* In: Petrovsky, E., Herrero-Bervera, E., Harinarayana, T., Ivers, D. (Eds.), The Earth's Magnetic Interior, Dordrecht, Heidelberg, London, New York, vol. 1. Springer, pp. 31–42

IOP Conf. Series: Journal of Physics: Conf. Series **1152** (2019) 012035 doi:10.1088/1742-6596/1152/1/012035

Electromagnetic induction responses to geomagnetic disturbances at low-and-mid-latitudes

Mio Nakahara[1], Akimasa Yoshikawa[2], Teiji Uozumi[3] and Akiko Fujimoto[4]

[1] Department of Earth and Planetary Sciences, Graduate School of Sciences Kyushu University, Fukuoka 819-0395, Japan
[2] Department of Earth and Planetary Sciences, Faculty of Sciences, Kyushu University, Fukuoka 819-0395, Japan
[3] International Center for Space Weather Science and Education, Fukuoka 819-0395, Kyushu University, Japan
[4] Kyushu Institute of Technology, Fukuoka 804-8550, Japan

E-mail: nakahara.mio.383@s.kyushu-u.ac.jp

Abstract. The geomagnetically induced current (GIC) is one of the most widely recognized phenomena caused by geomagnetic disturbances. Realistic predictions of magnetic field fluctuations may be used to evaluate the induction of electric fields to ground surfaces, and thus to estimate the occurrence of GICs. Although many GICs occur at high latitudes, they are now being studied at low and mid-latitudes as well. The purpose of this research was to understand the dynamics, observation, and prediction in Japan for GICs occurring at the low and mid-latitudes. In this study, the influence of geomagnetic field variations on Earth's electric field was examined. The magnetic field and the electric field components of 3 observation points for 1 year in 2015 are visually examined, and the characteristics of the fluctuations of the magnetic field and the surface electrical field were also analysed.

1. Introduction

Due to space weather phenomena such as magnetic storms and substorms, Earth's magnetic field changes daily. These phenomena are known to have many influences on our lives, including one such phenomenon, the geomagnetically induced current (GIC). GIC is the current induced in the ground by changes in the geomagnetically induced electric field (GIE) due to geomagnetic disturbance. If GIC flows into a transmission line, the supply of electricity will be hindered; if it flows into an underground pipeline, it will advance the corrosion of the pipe.

Japan is considered to have a magnetic latitude that is lower than the geographical latitude, and it is difficult for the country to be affected by GIC [Watari et al. 2015]. However, in recent years research to observe and predict GICs has nonetheless advanced in Japan.

This study aimed to establish a predictive method of GIE, which is the electromotive force of GIC, at low and mid-latitudes, especially in Japan. As an initial stage of this research, it is emphasized here that the clarification of the relationship between the geomagnetic field and Earth's electric field, especially observational values.

2. Data

T Earth's geomagnetic and electric fields are observed by the Japan Meteorological Agency in Masumitsu (MMB), Kakioka (KAK), and Kanoya (KNY), Japan. These observational values are made available on the web [http://www.kakioka-jma.go.jp/obsdata/metadata/ja]. This research uses that data from January 1, 2015 to December 31, 2015; and uses northward+ (H), eastward+ (D), downward+ (Z) of the magnetic field, and northward+ (Ex), eastward+ (Ey) of the electric field. The difference between the median and the raw data values is denoted as *d*.

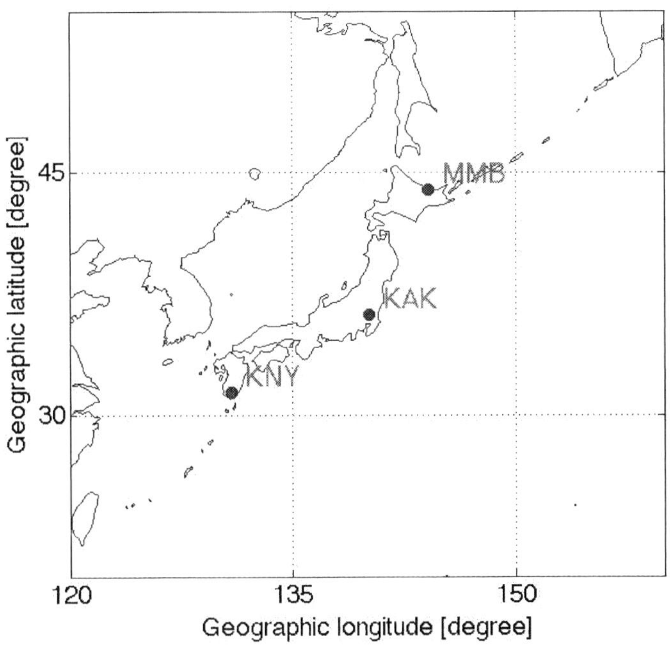

Figure 1. The observation points, Memanbetsu(MMB), Kakioka(KAK), and Kanoya(KNY). All of these locations observe Earth's geomagnetic field and electric field.

3. Analyses

3.1. Development of GIE during aurora substorms onset in MMB

To analyze the origination directions of substorms in MMB, this research used 1-sec geomagnetic field data, 1-sec electric field data, and the auroral electrojet (AE) index (WDC).

First, all daily data from 2015 were examined, and counted the substorm events using the AE index. Among them, those events were limited conditions at MMB to 18-06 LT (09-21UT) because substorms are a phenomenon of the night sky wherein clear changes in Earth's geomagnetic and electric fields are observed. There are 139 events were listed up.

Secondly, because the current path with the substorm has a local time dependence, events were classified as 09-12UT, 12-15UT, 15-18UT, or 18-21 UT. Focusing on the direction of electric field origination for any given substorm, they were classified as northeast (NE), northwest (NW), southeast (SE), and southwest (SW).

Table 1 shows the number of events classified by the origination direction and the time of occurrence. From this table, it can be seen that the NE electric field easily develops in or near to evening time; on the contrary, the SW electric field mainly develops in the times in or near the morning.

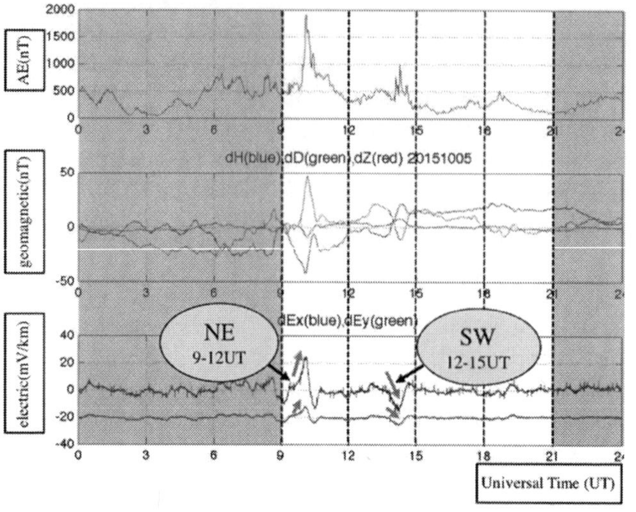

Figure 2. Example of the classification by time and component of events as originating in the northeast (NE), presented as a peak in the difference between the median and raw northward+ and eastward+ component of the electric field, or southwest (SW), presented as a trough in the difference between the median and raw northward-(southward+) and eastward-(westward+) component of the electric field.

	NE	NW	SE	SW
9-12UT (18-21LT)	48	0	0	2
12-15UT (21-24LT)	21	0	0	21
15-18UT (24-3LT)	1	0	1	27
18-21UT (3-6LT)	2	0	0	16

Table 1. Number of events classified by start time and component type. Black shading indicates places where more than 10 events could be confirmed.

3.2. Development of GIE during the magnetic storm

Figure 3 shows the observations of the magnetic and the electric fields for a magnetic storm (St. Patrick's), which occurred in March of 2015. During this magnetic storm, the correlation between Earth's magnetic field and electric field was investigated.

The correlation coefficient was first calculated for each of the three magnetic field components, as well as for the two electric field components (Raw). Second, in order to remove waves of a relatively long period, such as the daily fluctuations of the magnetic field, the correlation coefficient was also calculated using a high-pass filter with a period of 30 minutes (High-pass).

Table 2. Raw and high-pass correlation coefficient of each component on Mar 17, 2015 at MMB, KAK, and KNY. Larger absolute values are shown in gray.

	MMB Raw	MMB High-pass	KAK Raw	KAK High-pass	KNY Raw	KNY High-pass
dH – dEx	-0.30	-0.57	-0.29	-0.62	-0.19	-0.27
dH – dEy	-0.36	-0.72	-0.43	-0.68	-0.40	-0.71
dD – dEx	0.37	0.78	0.03	0.03	0.52	0.21
dD – dEy	0.30	0.70	-0.16	-0.18	-0.19	-0.24
dZ – dEx	-0.28	-0.87	-0.62	-0.62	0.13	-0.17
dZ - dEy	-0.21	-0.70	-0.58	-0.64	-0.54	-0.69

(R: Correlation coefficient)

In Table 2, the correlation is better in most cases when magnetic field fluctuations on the 30-minute scale are excluded. This finding supports the idea that the electric field responds to the short-term magnetic field fluctuations more so than the long-term fluctuations.

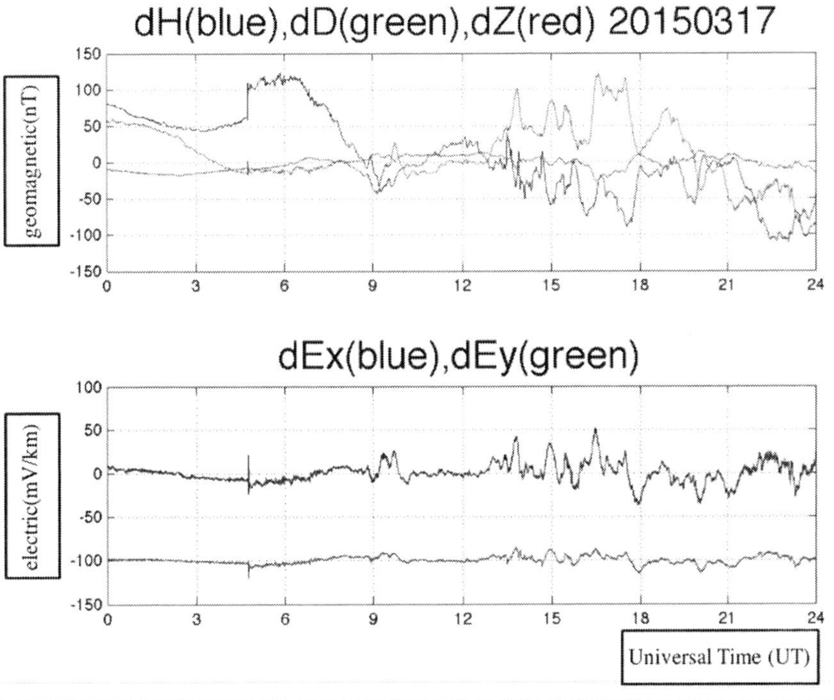

Figure 3. The difference between the median and raw values of the northward+ (dH), eastward+ (dD), and downward+ (dZ) components of the geomagnetic field (upper row), and of the northward+ (dEx) and eastward+ (dEy) components of the electric field (bottom row) at Memanbetsu (MMB). There is a tendency for the electric field to develop more easily from magnetic field fluctuations on the scale of several minutes after 13 UT, rather than from fluctuations of the magnetic field on the scale of several hours near 5 UT.

3.3. Development of GIE by geomagnetic pulsation in MMB.

Geomagnetic pulsation is a known magnetic disturbance phenomenon during the daytime. The daytime geomagnetic pulsation is a periodic fluctuation of 0.2 to 100 seconds.

The pulsation in the electric field during the daytime can be seen in Figure 4. Since this pulsation occurs at the same time as the pulsation of the magnetic field, this suggests that the fluctuation of the magnetic field may cause the electric field pulsation. Currently, proposed GIE generation models are mostly focused on ionospheric currents or equatorial electrojets (EEJs) originating from general GIE. However, due to fluctuation of geomagnetic field itself, the GIE may not depend on large current systems, which also suggests that the electric field may develop.

IOP Conf. Series: Journal of Physics: Conf. Series **1152** (2019) 012035 doi:10.1088/1742-6596/1152/1/012035

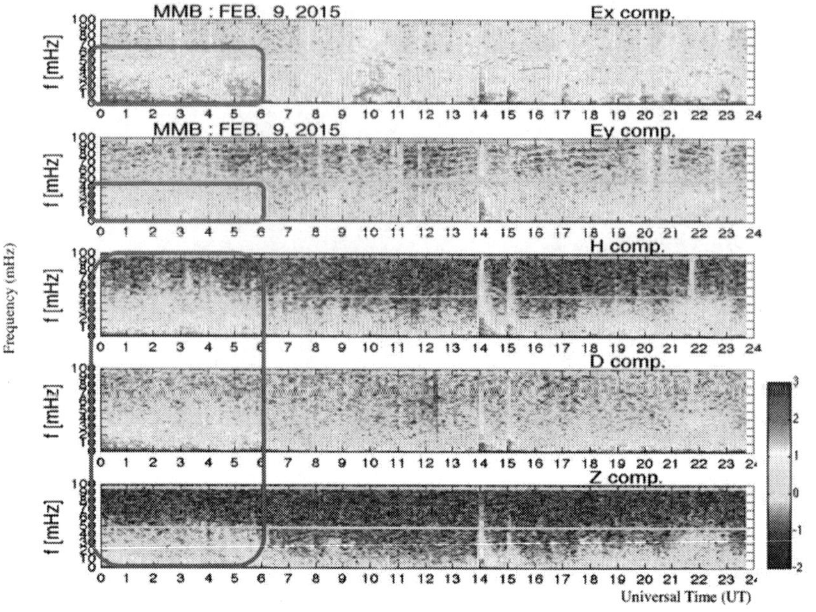

Figure 4. The spectrograms of each component on Feb 9, 2015 at MMB. Each graph is labeled by the kind of observational value (i.e., northward+ of the electric field (Ex), eastward+ of the electric field (Ey), northward+ of the magnetic field (H), eastward+ of the magnetic field (D), and downward+ of the magnetic field (Z). Frequency bands with high wavelengths are shown in red frames.

4. Discussion

The time dependency of the electric field origination direction during substorms is caused by the current system in the sky (Table 1). In the skies of Japan, at the time of substorm occurrence, the field-aligned current (FAC) is mostly toward the south in the evening, and many occur in the northward direction in the morning. Even from the limited number of electric field observations, it is clear that there are many forces that oppose the FAC. However, despite the fact that the FAC is seen from 12-15UT due to the evening darkness, origination of the same number of events from the northeast and southwest, may indicate that the direction of electric field origination is determined by multiple factors, not only FAC.

That the electric field is more likely to develop with shorter variations than the long-time fluctuations, such as the daily fluctuation considered in section 3.2, can be explained by Faraday's Law of electromagnetic induction (Eq.1).

$$\frac{\partial \mathbf{B}}{\partial t} = -\nabla \times \mathbf{E} \tag{1}$$

Equation (1) is Faraday's Law. This equation explains that the electric field develops as the time change of the magnetic field becomes shorter. That is, if the magnetic field rapidly develops over a short time, the electric field also develops.

The electric field responds to direct magnetic field fluctuations (section 3.3). In sections 3.1 and 3.2, a large current system generated in the sky was investigated, and both the magnetic field and the electric field were developed from the sudden current system. The details of the daytime geomagnetic pulsation have not been elucidated yet, but the geomagnetic field is characterized by direct vibration. Our findings presented in sections 3.1 and 3.2 suggest that it is not only the variation of the geomagnetic field, but also the equivalent current system in the sky that may contribute to the development of the electric field. However, this is not necessarily true for the geomagnetic pulsation. It is necessary to investigate each disturbance phenomenon carefully, and also to investigate how much influence the terrain and other components may exert against the electric field with respect to geomagnetism and electric currents.

5. Conclusion

In order to predict the occurrence of GICs in Japan, it was hypothesized that it is necessary to consider the characteristics of GIEs. Therefore, the relationship between the observational values of Earth's geomagnetic and electric fields were examined.

At the time of a substorm, there is a time dependence with respect to the direction of origination of the electric field, and this can be considered to depend on the direction of the current system in the sky (e.g., FAC). There is space for continued discussion as to whether this electric field is reacts directly to currents in the sky, is driven by the magnetic field fluctuations caused by currents in the sky, or both.

This study considers the correlation between Earth's magnetic and electric fields at the time of magnetic storm onset. Evaluation of the correlation between the raw data and the high-pass (30-minute) data, the correlation tended to be higher in the high-pass data. This indicates that the electric field responds to sudden magnetic field fluctuations more so than magnetic field fluctuations with their relatively long fluctuation cycles (e.g., daily fluctuations). This may be intuitively tied to Faraday's Law and may be the key to predicting electric fields in the future.

Regarding daytime geomagnetic pulsations, these were also observed in the electric field at the same time zone as the geomagnetic pulsation. Even in the absence of a large current system in the sky, the electric field fluctuates due to changes in the magnetic field. In the future, it will be necessary to analyse more detailed frequency data to further clarify the relationships presented here.

6. Acknowledgements

Thank the Japan Meteorological Agency for providing with geomagnetic and electrical field data, and the World Data Center for Geomagnetism (WDC), University of Kyoto for providing the auroral electrojet (AE) indices.

References

[1] Berge, Jonathan E. 2011 Impact of Geomagnetically Induced Currents on Power Transformers *Electronic Thesis and Dissertation Repository*. 132.

[2] José Ramírez-Ni˜no, Carlos Haro-Hernández, Joaquín Héctor Rodriguez-Rodriguez, Rito Mijarez, 2016 Core station effects of geomagnetic induced currents in power transformers *Jornal of Applid Research and Technology* **14** 87-92

[3] D. H. Boteler 2001 Space Weather Effects on Power Systems *Space Weather Geophysical Monograph 125*. 347

[4] Watari, S., et al. 2009 Measurements of geomagnetically induced current in a power grid in Hokkaido, Japan *Space Weather* **7** S03002, doi:10.1029/2008SW000417.

[5] Shinichi Watari 2015 Effect of space weather on power grids and measurement of geomagnetically induced current (GIC) in Hokkaido *Conductivity Anomaly Research meeting 2015* pp.1-14

[6] Jeffrey J Love and Andrei Swidinsky 2015 Observatory geoelectric fields induced in a two-layer lithosphere during magnetic storms *Earth, Planets and Space* 67:58 doi 10.1186/s40623-015-0213-3

[7] Yusuke Ebihara, et al. 2016 Geospace Disturbances and Geomagentically Induced Current *Proceedings of the Space Sciences and Technology Conference* **60** 3p 2016-09-06

[8] Krzysztof Lewandowski 2016 Protection of the Smart City against CME *Transportation Research Procedia* **16** 298 – 312

Institute of Physics
Dirac House, Temple Back
Bristol BS1 6BE UK

ISSN: 1742-6588
ISBN 978-1-5108-8274-4

CURRAN ASSOCIATES INC.
proceedings.com